I0476140

Handbook of Mineral Names

By Rob Kanen BSc Hons (Ed.)

ISBN 1451543344

© Copyright, 2004-2010, Compiled by R.A. Kanen (Editor)

All rights reserved. Except for the
purposes of private study, research or
review, as listed under the copyright law,
no part of this publication may be reproduced,
stored in a retrieval system, or transmitted
in any form or by any means without prior
permission of the copyright owner.

Table of Contents

Mineral Data by Mineral Name

Alphabetical listing of mineral name, formula and crystal system listed by mineral name.

Mineral Name	Formula	Crystal System
ABELSONITE	Ni++[C32H36N4]	Triclinic
ABENAKIITE-(Ce)	Na26(Ce,Nd,La,Pr,Th,Sm)6(SiO3)6(PO4)6(CO3)6(S++++O2)O	Trigonal
ABERNATHYITE	K(UO2)(AsO4).4H2O	Tetragonal
ABHURITE	Sn3O(OH)2Cl2	Trigonal
ABSWURMBACHITE	(Cu++,Mn++)Mn+++6(SiO4)O8	Tetragonal
ACANTHITE	Ag2S	Monoclinic
ACETAMIDE	CO(CH3)(NH2)	Trigonal
ACHAVALITE	FeSe	Trigonal
ACTINOLITE	Ca2(Mg,Fe++)5Si8O22(OH)2	Monoclinic
ACUMINITE	SrAlF4(OH).H2O	Monoclinic
ADAMITE	Zn2(AsO4)(OH)	Orthorhombic/Monoclinic
ADELITE	CaMg(AsO4)(OH)	Orthorhombic
ADMONTITE	MgB6O10.7H2O	Monoclinic
AEGIRINE (ACMITE)	NaFe+++Si2O6	Monoclinic
AENIGMATITE	Na2Fe++5TiSi6O20	Triclinic
AERINITE	Ca4(Al,Fe+++,Mg,Fe++)10Si12O36(CO3).12H2O	Monoclinic
AERUGITE	Ni17As6O32	Trigonal
AESCHYNITE-	(Y,Ca,Fe,Th)(Ti,Nb)2(O,OH)6	Orthorhombic
AESCHYNITE-(Ce)	(Ce,Ca,Fe,Th)(Ti,Nb)2(O,OH)6	Orthorhombic
AESCHYNITE-(Nd)	(Nd,Ce,Ca)(Ti,Nb)2(O,OH)6	Orthorhombic
AFGHANITE	(Na,Ca,K)8(Si,Al)12O24(SO4,Cl,CO3)3.H2O	Hexagonal
AFWILLITE	Ca3Si2O4(OH)6	Monoclinic
AGARDITE-	(Y,Ca)Cu6(AsO4)3(OH)6.3H2O	Hexagonal
AGARDITE-(Ca)	CaCu6(AsO4)3(OH)6.3H2O	Hexagonal
AGARDITE-(Ce)	(Ce,Ca)Cu6(AsO4)3(OH)2.3H2O	Hexagonal
AGARDITE-(Dy)	(Dy,La,Ca)Cu6(AsO4)3(OH)6.3H2O	Hexagonal
AGARDITE-(La)	(La,Ca)Cu6(AsO4)3(OH)6.3H2O	Hexagonal
AGARDITE-(Nd)	(Nd,La,Ce,Ca)Cu6(AsO4)3(OH)6.3H2O	Hexagonal
AGRELLITE	NaCa2Si4O10F	Triclinic
AGRINIERITE	(K2,Ca,Sr)U3O10.4H2O	Orthorhombic
AGUILARITE	Ag4SeS	Orthorhombic
AHEYLITE	(Fe++,Zn)Al6(PO4)4(OH)8.4H2O	Triclinic
AHLFELDITE	(Ni,Co)SeO3.2H2O	Monoclinic
AIKINITE	PbCuBiS3	Orthorhombic
AJOITE	(K,Na)Cu7AlSi9O24(OH)6.3H2O	Triclinic
AKAGANEITE	Fe+++(O,OH,Cl)	Monoclinic ps Tetragonal
AKATOREITE	(Mn++,Fe++)Al2Si8O24(OH)8	Triclinic

Mineral Data

Mineral Name	Formula	Crystal System
AKDALAITE	4Al2O3.H2O	Hexagonal
AKERMANITE	Ca2MgSi2O7	Tetragonal
AKHTENSKITE	Mn++++O2	Hexagonal
AKROCHORDITE	Mn4Mg(AsO4)2(OH)4.4H2O	Monoclinic
AKSAITE	MgB6O7(OH)6.2H2O	Orthorhombic
AKTASHITE	Cu6Hg3As4S12	Triclinic
ALABANDITE (ALABANDINE)	MnS	Cubic
ALACRANITE	As8S9	Monoclinic
ALAMOSITE	PbSiO3	Monoclinic
ALARSITE	AlAsO4	Trigonal
ALBITE	NaAlSi3O8	Triclinic
ALBRECHTSCHRAUFITE	Ca4Mg(UO2)2(CO3)6F2.17H2O	Triclinic
ALDERMANITE	Mg5Al12(PO4)8(OH)22.32H2O	Orthorhombic
ALDZHANITE	CaMgB2O4Cl.7H2O	Orthorhombic
ALEKSITE	PbBi2Te2S2	Trigonal
ALFORSITE	Ba5(PO4)3Cl	Hexagonal
ALGODONITE	Cu1-xAsx	Hexagonal
ALIETTITE	Ca0,2Mg6(Si,Al)8O20(OH)4.4H2O	
ALLACTITE	Mn7(AsO4)2(OH)8	Monoclinic
ALLANITE-(YTTROORTHITE)	(Y,Ce,Ca)2(Al,Fe+++)3(SiO4)3(OH)	Monoclinic
ALLANITE-(Ce) (ORTHITE)	(Ce,Ca,Y)2(Al,Fe+++)3(SiO4)3(OH)	Monoclinic
ALLANITE-(La)	Ca(La,Ce)(Fe++,Mn++)(Al,Fe+++)2(SiO4)(Si2O7)O(OH)	Monoclinic
ALLARGENTUM	Ag1-xSbx	Hexagonal
ALLEGHANYITE	Mn5(SiO4)2(OH)2	Monoclinic
ALLOCLASITE	(Co,Fe)AsS	Monoclinic
ALLOPHANE	Al2O3.(SiO2)1.3-2.(H2O)2.5-3	Amorphous
ALLUAIVITE	Na19(Ca,Mn++)6(Ti,Nb)3(Si3O9)2(Si10O28)2Cl.2H2O	Trigonal
ALLUAUDITE	NaCaFe++(Mn,Fe++,Fe+++,Mg)2(PO4)3	Monoclinic
ALMANDINE (ALMANDITE)	Fe++3Al2(SiO4)3	Isometric
ALMBOSITE	Fe++5Fe+++4V+++++4Si3O27	
ALSTONITE	BaCa(CO3)2	Triclinic ps Orthorhombic
ALTAITE	PbTe	Isometric
ALTHAUSITE	Mg2(PO4)(OH,F,O)	Orthorhombic
ALTHUPITE	ThAl(UO2)7(PO4)4(OH)5.15H2O	Triclinic
ALTISITE	(K,Na)6Na3Ti2Al2Si8O26Cl3	Monoclinic
ALUMINITE	Al2(SO4)(OH)4.7H2O	Monoclinic ps Orthorhombic
ALUMINIUM (ALUMINUM)	Al	Isometric
ALUMINOCOPIAPITE	Al2/3Fe+++4(SO4)6O(OH)2.20H2O	Triclinic
ALUMINOKATOPHORITE	Na2Ca(Fe++,Mg)4Al(Si7Al)O22(OH)2	Monoclinic
ALUMOHYDROCALCITE	CaAl2(CO3)2(OH)4.3H2O	Triclinic
ALUMOKLYUCHEVSKITE	K3Cu3(Al,Fe+++)O2(SO4)4	Monoclinic
ALUMOTANTITE	AlTaO4	Orthorhombic

Mineral Data

Mineral Name	Formula	Crystal System
ALUMOTUNGSTITE	(W,Al)(O,OH)3	Cubic ps Cubic
ALUNITE	KAl3(SO4)2(OH)6	Trigonal
ALUNOGEN	Al2(SO4)3.17H2O	Triclinic
ALVANITE	(Zn,Ni)Al4(VO3)2(OH)12.2H2O	Monoclinic
AMAKINITE	(Fe++,Mg)(OH)2	Trigonal
AMARANTITE	Fe+++(SO4)(OH).3H2O	Triclinic
AMARILLITE	NaFe++(SO4)2.6H2O	Monoclinic
AMBER (SUCCINITE,BERNSTEIN)	[C,H,O]	Amorphous
AMBLYGONITE	(Li,Na)Al(PO4)(F,OH)	Triclinic
AMEGHINITE	NaB3O3(OH)4	Monoclinic
AMESITE	Mg2Al(SiAl)O5(OH)4	Triclinic
AMICITE	K2Na2Al4Si4O16.5H2O	Monoclinic
AMINOFFITE	Ca2(Be,Al)Si2O7(OH).H2O	Tetragonal
AMMONIOALUNITE	(NH4)Al3(SO4)2(OH)6	Trigonal
AMMONIOBORITE	(NH4)2B10O16.5H2O	Monoclinic
AMMONIOJAROSITE	(NH4)Fe+++3(SO4)2(OH)6	Trigonal
AMMONIOLEUCITE	(NH4,K)AlSi2O6	Tetragonal
AMSTALLITE	CaAl(Si,Al)4O8(OH)4.(H2O,Cl)	Monoclinic
ANALCIME (ANALCITE)	NaAlSi2O6.H2O	Isometric
ANANDITE	(Ba,K)(Fe++,Mg)3(Si,Al,Fe)4O10(O,OH)2	Monoclinic
ANAPAITE	Ca2Fe++(PO4)2.4H2O	Triclinic
ANATASE	TiO2	Tetragonal
ANCYLITE-(Ce)	SrCe(CO3)2(OH).H2O	Orthorhombic
ANDALUSITE	Al2SiO5=Al[6]Al[5]OSiO4	Orthorhombic
ANDERSONITE	Na2Ca(UO2)(CO3)3.6H2O	Trigonal
ANDESINE	(Na,Ca)(Si,Al)4O8	Triclinic
ANDORITE	PbAgSb3S6	Orthorhombic
ANDRADITE	Ca3Fe+++2(SiO4)3	Isometric
ANDREMEYERITE	BaFe(Fe++,Mn,Mg)Si2O7	Monoclinic
ANDROSITE-(La)	(Mn++,Ca)(La,Ce,Nd)Mn++(Al,Mn+++)(SiO4)(Si2O7)O(OH)	Monoclinic
ANDUOITE	(Ru,Os)As2	Orthorhombic
ANGELELLITE	Fe+++4(AsO4)2O3	Triclinic
ANGLESITE	PbSO4	Orthorhombic
ANHYDRITE	CaSO4	Orthorhombic
ANILITE	Cu7S4	Orthorhombic
ANKANGITE	Ba(Ti,V+++,Cr+++)8O16	Tetragonal
ANKERITE	Ca(Fe++,Mg,Mn)(CO3)2	Trigonal
ANNABERGITE	Ni3(AsO4)2.8H2O	Monoclinic
ANNITE	KFe++3AlSi3O10(OH,F)2	Monoclinic
ANORTHITE	Ca2Al2Si2O8	Triclinic
ANORTHOCLASE	(Na,K)AlSi3O8	Triclinic
ANTARCTICITE	CaCl2.6H2O	Trigonal
ANTHOINITE	WAl(O,OH)3	Triclinic
ANTHONYITE	Cu(OH,Cl)2.3H2O	Monoclinic
ANTHOPHYLLITE	(Mg,Fe++)7Si8O22(OH)2	Orthorhombic

Mineral Data

Mineral Name	Formula	Crystal System
ANTIGORITE	(Mg,Fe++)3Si2O5(OH)4	Monoclinic
ANTIMONPEARCEITE	(Ag,Cu)16(Sb,As)2S11	Monoclinic
ANTIMONSELITE	Sb2Se3	Orthorhombic
ANTIMONY	Sb	Trigonal
ANTLERITE	Cu3(SO4)(OH)4	Orthorhombic
ANYUIITE	Au(Pb,Sb)2	Tetragonal
APACHITE	Cu9Si10O29.11H2O[Ino.]	Monoclinic
APHTHITALITE (GLASERITE)	(K,Na)3Na(SO4)2	Trigonal
APJOHNITE	MnAl2(SO4)4.22H2O	Monoclinic
APLOWITE	(Co,Mn,Ni)SO4.4H2O	Monoclinic
APUANITE	Fe++Fe+++4Sb+++4O12S	Tetragonal
ARAGONITE	CaCO3	Orthorhombic
ARAMAYOITE	Ag(Sb,Bi)S2	Triclinic
ARAVAIPAITE	Pb3AlF9.H2O	Triclinic
ARCANITE	K2SO4	Orthorhombic
ARCANITE (TAYLORITE)	K2SO4	Orthorhombic
ARCHERITE	(K,NH4,)H2PO4	Tetragonal
ARCTITE	Na2Ca4(PO4)3F	Trigonal
ARCUBISITE	Ag6CuBiS4	
ARDAITE	Pb19Sb13S35Cl7	Monoclinic
ARDEALITE	Ca2(SO4)(HPO4).4H2O	Monoclinic
ARDENNITE	(Mn,Ca,Mg)4(Al,Mn,Fe,Mg)6(As,V,P,Si)(O,OH)4(SiO4)2Si3O10(OH,O)6	Orthorhombic
ARFVEDSONITE	Na3(Fe++,Mg)4Fe+++Si8O22(OH)2	Monoclinic
ARGENTOJAROSITE	AgFe+++3(SO4)2(OH)6	Trigonal
ARGENTOPENTLANDITE	Ag(Fe,Ni)8S8	Isometric
ARGENTOPYRITE	AgFe2S3	Orthorhombic
ARGENTOTENNANTITE	(Ag,Cu)10(Zn,Fe)2(As,Sb)4S13	Isometric
ARGUTITE	GeO2	Tetragonal
ARGYRODITE	Ag8GeS6	Orthorhombic ps Cubic
ARHBARITE	Cu2(AsO4)(OH).6H2O	Monoclinic
ARISTARAINITE	Na2MgB12O20.8H2O	Monoclinic
ARMALCOLITE	(Mg,Fe++)Ti2O5	Orthorhombic
ARMANGITE	Mn++26As+++18O50(OH)4(CO3)=Mn26(AsO2(OH))4(AsO3)14	Trigonal
ARMENITE	BaCa2Al6Si9O30.2H2O	Orthorhombic
ARMSTRONGITE	CaZrSi6O15.2.5H2O	Monoclinic
ARROJADITE	KNa4CaMn++4Fe+++10Al(PO4)12(OH,F)2	Monoclinic
ARSENBRACKEBUSCHITE	Pb2(Fe++,Zn)(AsO4)2.H2O	Monoclinic
ARSENDESCLOIZITE	PbZn(AsO4)(OH)	Orthorhombic
ARSENIC	As	Trigonal
ARSENIOPLEITE	NaCaMn(Mn,Mg)2(AsO4)3	Monoclinic
ARSENIOSIDERITE	Ca2Fe+++3(AsO4)3O2.3H2O	Monoclinic
ARSENOBISMITE	Bi2(AsO4)(OH)3	
ARSENOCLASITE	Mn5(AsO4)2(OH)4	Orthorhombic
ARSENOCRANDALLITE	(Ca,Sr)Al3[(As,P)O4]2(OH)5.H2O	Trigonal

Mineral Data

Mineral Name	Formula	Crystal System
ARSENOFLORENCITE-(Ce)	(Ce,La)Al3(AsO4)2(OH)6	Trigonal
ARSENOFLORENCITE-(La)	(La,Sr)Al3(AsO4,SO4,PO4)2(OH)6	Trigonal
ARSENOFLORENCITE-(Nd)	(Nd,La,Ce,Ba)(Al,Fe+++)3(AsO4,PO4)2(OH)6	Trigonal
ARSENOGORCEIXITE	BaAl3AsO3(OH)(AsO4,PO4)(OH,F)6	Trigonal
ARSENOGOYAZITE (WEILERITE)	(Sr,Ca,Ba)Al3(AsO4,PO4)(OH)5F.H2O	Trigonal
ARSENOHAUCHECORNITE	Ni18Bi3AsS16	Tetragonal
ARSENOLAMPRITE	As	Orthorhombic
ARSENOLITE	As2O3	Isometric
ARSENOPALLADINITE	Pd8(As,Sb)3	Triclinic
ARSENOPYRITE (MISPICKEL)	FeAsS	Monoclinic ps Orthorhombic
ARSENOSULVANITE	Cu3(As,V)S4	Isometric
ARSENPOLYBASITE	(Ag,Cu)16(As,Sb)2S11	Monoclinic
ARSENTSUMEBITE	Pb2Cu(AsO4)(SO4)(OH)	Monoclinic
ARSENURANOSPATHITE	HAl(UO2)4(AsO4)4.40H2O	Tetragonal
ARSENURANYLITE	Ca(UO2)4(AsO4)2(OH)4.6H2O	Orthorhombic
ARTHURITE	CuFe+++2(AsO4,PO4,SO4)2(O,OH)2.4H2O	Monoclinic
ARTINITE	Mg2(CO3)(OH)2.3H2O	Monoclinic
ARTROEITE	PbAlF3(OH)2	Triclinic
ARUPITE	(Ni,Fe++)3(PO4)2.8H2O	Monoclinic
ARZAKITE	Hg3S2(Br,Cl)2	Monoclinic/Triclinic
ARZRUNITE	Cu4Pb2SO4O2Cl6.4H2O	Orthorhombic
ASBECASITE	Ca3(Ti,Fe,Sn)(Be,B,Al)2(As+++,Sb+++)3O6(SiO4)2	Trigonal
ASBOLAN	(Co,Ni)1-y(Mn++++O2)2-x(OH)2-2y+2x.nH2O	Hexagonal
ASCHAMALMITE	Pb6Bi2S9	Monoclinic
ASHANITE	(Nb,Ta,U,Fe,Mn)4O8	Orthorhombic
ASHBURTONITE	HPb4Cu++4Si4O12(HCO3)4(OH)4Cl	Tetragonal
ASHCROFTINE-	K5Na5(Y,Ca)12Si28O70(OH)2(CO3)8.8H2O	Tetragonal
ASHOVERITE	Zn(OH)2	Tetragonal
ASISITE	Pb7SiO8Cl2	Tetragonal
ASSELBORNITE	(Pb,Ba)(UO2)6(BiO)4(AsO4)2(OH)12.3H2O	Isometric
ASTROCYANITE-(Ce)	Cu+2(Ce,Nd,La,Pr,Sm,Ca,Y)2(UO2)(CO3)5(OH)2.3H2O	Hexagonal
ASTROPHYLLITE	(K,Na)3(Fe++,Mn)7Ti2Si8O24(O,OH)7	Triclinic
ATACAMITE	Cu2Cl(OH)3	Orthorhombic
ATELESTITE	Bi2(AsO4)O(OH)	Monoclinic
ATHABASCAITE	Cu5Se4	Orthorhombic
ATHENEITE	(Pd,Hg)3As	Hexagonal
ATLASOVITE	Cu6Fe+++Bi+++O4(SO4)5.KCl	Tetragonal
ATOKITE	(Pd,Pt)3Sn	Isometric
ATTAKOLITE (ATTACOLITE)	(Ca,Sr)Mn++(Al,Fe+++)4[(Si,P)O4]H(PO4)3(OH)4	Monoclinic
AUBERTITE	CuAl(SO4)2Cl.14H2O	Triclinic
AUGELITE	Al2(PO4)(OH)3	Monoclinic
AUGITE	(Ca,Na)(Mg,Fe,Al,Ti)(Si,Al)2O6	Monoclinic
AURICHALCITE	(Zn,Cu)5(CO3)2(OH)6	Monoclinic

Mineral Name	Formula	Crystal System
AURICUPRIDE	Cu_3Au	Orthorhombic
AUROANTIMONATE	$AuSbO_3$	
AURORITE	$(Mn,Ag,Ca)Mn^{++++}_3O_7.3H_2O$	Triclinic
AUROSTIBITE	$AuSb_2$	Isometric
AUSTINITE	$CaZn(AsO_4)(OH)$	Orthorhombic
AUTUNITE	$Ca(UO_2)_2(PO_4)_2.10-12H_2O$	Tetragonal
AVICENNITE	Tl_2O_3	Isometric
AVOGADRITE	$(K,Cs)BF_4$	Orthorhombic
AWARUITE	$Ni_2FetoNi_3Fe$	Hexagonal
AZOPROITE	$(Mg,Fe^{++})_2(Fe^{+++},Ti,Mg)BO_5$	Orthorhombic
AZURITE	$Cu_3(CO_3)_2(OH)_2$	Monoclinic
BABEFPHITE	$BaBe(PO_4)(F,O)$	Tetragonal
BABINGTONITE	$Ca_2(Fe^{++},Mn)Fe^{+++}Si_5O_{14}(OH)$	Triclinic
BADDELEYITE	ZrO_2	Monoclinic
BAFERTISITE	$Ba(Fe^{++},Mn)_2TiSi_2O_7(O,OH)_2$	Monoclinic
BAGHDADITE	$Ca_3(Zr,Ti)Si_2O_9$	Monoclinic
BAHIANITE	$Al_5Sb^{+++++}_3O_{14}(OH)_2$	Monoclinic
BAILEYCHLORE	$(Zn,Fe^{++},Al,Mg)_6(Si,Al)_4O_{10}(OH)_8$	Triclinic
BAIYUNEBOITE-(Ce)	$BaNaCe_2(CO_3)_4F$	Hexagonal
BAKERITE	$Ca_4B_4(BO_4)(SiO_4)_3(OH)_3.H_2O$	Monoclinic
BALANGEROITE	$(Mg,Fe^{+++},Fe^{++},Mn^{++})_{42}Si_{16}O_{54}(OH)_{40}$	Monoclinic
BALAVINSKITE	$Sr_2B_6O_{11}.4H_2O$	
BALIPHOLITE	$BaMg_2LiAl_3Si_4O_{12}(OH,F)_8$	Orthorhombic
BALKANITE	$Cu_9Ag_5HgS_8$	Orthorhombic
BALYAKINITE	$CuTeO_3$	Orthorhombic
BAMBOLLAITE	$Cu(Se,Te)_2$	Tetragonal
BANALSITE	$BaNa_2Al_4Si_4O_{16}$	Orthorhombic
BANDYLITE	$CuB(OH)_4Cl$	Tetragonal
BANNERMANITE	$(Na,K)_xV^{++++}_xV^{+++++}_{6-x}O_{15}$	Monoclinic
BANNISTERITE	$KCa(Mn,Fe^{++},Zn)_{21}(Si,Al)_{32}O_{76}(OH)_{16}.12H_2O$	Monoclinic
BAOTITE	$Ba_4(Ti,Nb)_8Si_4O_{28}Cl$	Tetragonal
BARARITE	$(NH_4)_2SiF_6$	Hexagonal
BARATOVITE	$KCa_7Li_3(Ti,Zr)_2(Si_6O_{18})_2(OH,F)_2$	Monoclinic ps Hexagonal
BARBERIITE	$(NH_4)BF_4$	Orthorhombic
BARBERTONITE	$Mg_6Cr_2(CO_3)(OH)_{16}.4H_2O$	Hexagonal
BARBOSALITE	$Fe^{++}Fe^{+++}_2(PO_4)_2(OH)_2$	Monoclinic
BARENTSITE	$Na_7AlH_2(CO_3)_4F_4$	Triclinic
BARIANDITE	$V_2O_4.4V_2O_5.12H_2O$	Monoclinic
BARICITE	$(Mg,Fe^{++})_3(PO_4)_2.8H_2O$	Monoclinic
BARIOMICROLITE (RIJKEBOERITE)	$Ba(Ta,Nb)_2(O,OH)_7$	Isometric
BARIOORTHOJOAQUINITE	$(Ba,Sr)_4Fe^{++}_2Ti_2Si_8O_{26}.H_2O$	Orthorhombic
BARIOPYROCHLORE (PANDAITE)	$(Ba,Sr)_2(Nb,Ti)_2(O,OH)_7$	Isometric
BARITE (BARYTE,BARYTINE)	$BaSO_4$	Orthorhombic
BARIUMALUMOPHARMACO SIDERITE	$Ba(Al,Fe^{+++})_4(AsO_4)_3(OH)_5.5H_2O$	Tetragonal ps Cubic

Mineral Data

Mineral Name	Formula	Crystal System
BARIUMBANNISTERITE	(K,H3O)(Ba,Ca)(Mn++,Fe++,Mg)21(Si,Al)32O80(O,OH)16.4-12H2O	Monoclinic
BARIUMBREWSTERITE	(Ba,Sr)Al2Si6O16.5H2O	Monoclinic
BARIUMPHARMACOSIDERITE	BaFe+++4(AsO4)3(OH)5.5H2O	Tetragonal ps Cubic
BARNESITE	Na2V6O16.3H2O	Monoclinic
BARRERITE	(Na,K,Ca)2Al2Si7O18.7H2O	Orthorhombic
BARRINGERITE	(Fe,Ni)2P	Hexagonal
BARRINGTONITE	MgCO3.2H2O	Triclinic
BARROISITE	NaCa(Mg,Fe++)3Al2(Si7Al)O22(OH)2	Monoclinic
BARSTOWITE	3PbCl2.PbCO3.H2OorPb4Cl6CO3.H2O	Monoclinic
BARTELKEITE	PbFe++Ge3O8	Monoclinic
BARTONITE	K3Fe10S14	Tetragonal
BARYLITE	BaBe2Si2O7	Orthorhombic ps Hexagonal
BARYSILITE	Pb8Mn(Si2O7)3	Trigonal
BARYTOCALCITE	BaCa(CO3)2	Monoclinic
BARYTOLAMPROPHYLLITE	(Na,K)2(Ba,Ca,Sr)2(Ti,Fe)3(SiO4)4(O,OH)2	Monoclinic
BASALUMINITE	Al4(SO4)(OH)10.5H2O	Hexagonal
BASSANITE	2CaSO4.H2O	Monoclinic ps Hexagonal
BASSETITE	Fe++(UO2)2(PO4)2.8H2O	Monoclinic
BASTNASITE-(BASTNAESITE-)	(Y,Ce)(CO3)F	Hexagonal
BASTNASITE-(Ce) (BASTNAESITE-(Ce))	(Ce,La)(CO3)F	Hexagonal
BASTNASITE-(La) (BASTNAESITE-(La))	(La,Ce)(CO3)F	Hexagonal
BATISITE	(Ba,K,Na)3Ti2Si4O14	Orthorhombic
BAUMHAUERITE	Pb3As4S9	Triclinic
BAUMHAUERITE-2a	Pb11Ag0,7As17.2Sb0,4S36or(Pb,Ag)3As4S9	Monoclinic
BAURANOITE	BaU2O7.4-5H2O	
BAVENITE	Ca4Be2Al2Si9O26(OH)2	Orthorhombic
BAYANKHANITE	Cu6HgS4	Hexagonal
BAYERITE	Al(OH)3	Monoclinic
BAYLDONITE	(Cu,Zn)3Pb(AsO4)2(OH)2	Monoclinic
BAYLEYITE	Mg2(UO2)(CO3)3.18H2O	Monoclinic
BAYLISSITE	K2Mg(CO3)2.4H2O	Monoclinic
BAZHENOVITE	CaS5.CaS2O3.6Ca(OH)2.20H2O	Monoclinic
BAZIRITE	BaZrSi3O9	Hexagonal
BAZZITE	Be3(Sc,Al)2Si6O18	Hexagonal
BEARSITE	Be2(AsO4)(OH).4H2O	Monoclinic
BEARTHITE	Ca2Al(PO4)2(OH)	Monoclinic
BEAVERITE	PbCu++(Fe+++,Al)2(SO4)2(OH)6	Trigonal
BECHERERITE	(Zn,Cu)3Zn(S,Si)(O,OH)4(OH)13	Trigonal
BECQUERELITE	Ca(UO2)6O4(OH)6.8H2O	Orthorhombic
BEHIERITE	(Ta,Nb)BO4	Tetragonal
BEHOITE	Be(OH)2	Orthorhombic
BEIDELLITE	(Na,Ca0,5)0,3Al2(Si,Al)4O10(OH)2.nH2O	Monoclinic

Mineral Data

Mineral Name	Formula	Crystal System
BELENDORFFITE	Cu7Hg6	Trigonal ps Cubic
BELKOVITE	Ba3(Nb,Ti)6(Si2O7)2O12	Hexagonal
BELLBERGITE	(K,Ba,Sr)(Ca,Sr,Na)4Al9Si9O36.15H2O	Hexagonal
BELLIDOITE	Cu2Se	Tetragonal
BELLINGERITE	Cu++3(IO3)6.2H2O	Triclinic
BELOVITE	(Sr,Ce,Na,Ca)5(PO4)3(OH)	Hexagonal
BELYANKINITE	Ca1-2(Ti,Zr,Nb)5O12.9H2O	Amorphous
BEMENTITE	Mn8Si6O15(OH)10	Monoclinic
BENAVIDESITE	Pb4(Mn,Fe)Sb6S14	Monoclinic
BENITOITE	BaTiSi3O9	Hexagonal
BENJAMINITE	(Ag,Cu)3(Bi,Pb)7S12	Monoclinic
BENLEONARDITE	Ag8(Sb,As)Te2S3	Tetragonal
BENSTONITE	(Ba,Sr)6(Ca,Mn)6Mg(CO3)13	Trigonal
BENTORITE	Ca6(Cr,Al)2(SO4)3(OH)12.26H2O	Hexagonal
BENYACARITE	(K,H2O)(Mn++,Fe++)2(Fe+++,Ti,Al)2Ti(PO4)4(O,F)2.14H2O	Orthorhombic
BERAUNITE	Fe++Fe+++5(PO4)4(OH)5.4H2O	Monoclinic
BERBORITE-1T	Be2(BO3)(OH,F).H2O	Trigonal
BERBORITE-2T	Be2(BO3)(OH,F).H2O	Trigonal
BERBORITE-3H	Be2(BO3)(OH,F).H2O	Hexagonal
BERDESINSKIITE	V+++2TiO5	Monoclinic
BERGENITE	(Ba,Ca)2(UO2)3(PO4)2(OH)4,5.5H2O	Monoclinic
BERGSLAGITE	CaBe(AsO4)(OH)	Monoclinic
BERLINITE	AlPO4	Trigonal
BERMANITE	Mn++Mn+++2(PO4)2(OH)2.4H2O	Monoclinic
BERNALITE	Fe(OH)3	Orthorhombic ps Cubic
BERNARDITE	Tl(As1-xSbx)5S8orTl(As1-x)5S9	Monoclinic
BERNDTITE-C27	SnS2	Hexagonal
BERNDTITE-C6	SnS2	Trigonal
BERRYITE	Pb3(Ag,Cu)5Bi7S16	Monoclinic
BERTHIERINE	(Fe++,Fe+++,Mg)2-3(Si,Al)2O5(OH)4	Monoclinic
BERTHIERITE	FeSb2S4	Orthorhombic
BERTOSSAITE	(Li,Na)2CaAl4(PO4)4(OH,F)4	Orthorhombic
BERTRANDITE	Be4Si2O7(OH)2	Orthorhombic
BERYL	Be3Al2Si6O18	Hexagonal
BERYLLITE	Be3SiO4(OH)2.H2O	Orthorhombic
BERYLLONITE	NaBePO4	Monoclinic
BERZELIANITE	Cu2Se	Isometric
BERZELIITE	(Ca,Na)3(Mg,Mn)2(AsO4)3	Isometric
BETAFITE	(Ca,Na,U)2(Ti,Nb,Ta)2O6(OH)	Isometric
BETEKHTINITE	Cu10(Fe,Pb)S6	Orthorhombic
BETPAKDALITE	(H,K)6Ca4Fe+++6As+++++4Mo++++++16O74.24-40H2O	Monoclinic
BEUDANTITE	PbFe+++3(AsO4)(SO4)(OH)6	Trigonal
BEUSITE	(Mn++,Fe++,Ca,Mg)3(PO4)2	Monoclinic
BEYERITE	(Ca,Pb)Bi2(CO3)2O2	Tetragonal
BEZSMERTNOVITE	Au4Cu(Te,Pb)	Orthorhombic

Mineral Data

Mineral Name	Formula	Crystal System
BIANCHITE	(Zn,Fe++)(SO4).6H2O	Monoclinic
BICCHULITE	Ca2Al2SiO6(OH)2	Isometric
BIDEAUXITE	Pb2AgCl3(F,OH)2	Isometric
BIEBERITE	CoSO4.7H2O	Monoclinic
BIJVOETITE-	(Y,Dy)2(UO2)4(CO3)4(OH)6.11H2O	Orthorhombic
BIKITAITE	LiAlSi2O6.H2O	Monoclinic
BILIBINSKITE	Au3Cu2PbTe2	ps Cubic
BILINITE	Fe++Fe+++2(SO4)4.22H2O	Monoclinic
BILLIETITE	Ba(UO2)6O4(OH)6.8H2O	Orthorhombic
BILLINGSLEYITE	Ag7AsS6	Isometric
BINDHEIMITE	Pb2Sb2O6(O,OH)	Isometric
BIOTITE	K(Mg,Fe++)3(Al,Fe+++)Si3O10(OH,F)2	Monoclinic
BIPHOSPHAMMITE	(NH4,K)H2PO4	Tetragonal
BIRINGUCCITE	Na2B5O8(OH).H2O	Monoclinic
BIRNESSITE	Na4Mn14O27.9H2O	Orthorhombic
BISCHOFITE	MgCl2.6H2O	Monoclinic
BISMITE	Bi2O3	Monoclinic
BISMOCLITE	BiOCl	Tetragonal
BISMUTH	Bi	Trigonal
BISMUTHINITE	Bi2S3	Orthorhombic
BISMUTITE	Bi2(CO3)O2	Tetragonal
BISMUTOCOLUMBITE	Bi(Nb,Ta)O4	Tetragonal
BISMUTOFERRITE	BiFe+++2(SiO4)2(OH)	Monoclinic
BISMUTOHAUCHECORNITE	Ni9Bi2S8	Tetragonal
BISMUTOMICROLITE (WESTGRENITE)	(Bi,Ca)(Ta,Nb)2O6(OH)	Isometric
BISMUTOSTIBICONITE	Bi(Sb+++++,Fe+++)2O7	Isometric
BISMUTOTANTALITE	Bi(Ta,Nb)O4	Orthorhombic
BITYITE	CaLiAl2(AlBeSi2)O10(OH)2	Monoclinic
BIXBYITE	(Mn+++,Fe+++)2O3	Isometric
BJAREBYITE	(Ba,Sr)(Mn++,Fe++,Mg)2Al2(PO4)3(OH)3	Monoclinic
BLAKEITE	Aferrictellurite	
BLATTERITE	(Mn++,Mg)2(Mn+++,Sb+++,Fe+++)(BO3)O2	Orthorhombic
BLIXITE	Pb2Cl(O,OH)2	Orthorhombic
BLODITE (BLOEDITE)	Na2Mg(SO4)2.4H2O	Monoclinic
BLOSSITE	Cu2V+++++2O7	Orthorhombic
BOBFERGUSONITE	Na2Mn++5Fe+++Al(PO4)6	Monoclinic
BOBIERRITE	Mg3(PO4)2.8H2O	Monoclinic
BOGDANOVITE	(Au,Te,Pb)3(Cu,Fe)	Isometric
BOGGILDITE (BOEGGILDITE)	Sr2Na2Al2(PO4)F9	Monoclinic
BOGGSITE	NaCa2(Al5Si19O48).17H2O	Orthorhombic
BOGVADITE	Na2SrBa2Al4F20	Orthorhombic
BOHDANOWICZITE	AgBiSe2	Hexagonal
BOHMITE (BOEHMITE)	AlO(OH)	Orthorhombic
BOKITE	KAl3Fe+++6V++++6V+++++20O76.30H2O	
BOLDYREVITE	NaCaMgAl3F14.4H2O	Amorphous

Mineral Name	Formula	Crystal System
BOLEITE	Pb26Ag10Cu24Cl62(OH)48.3H2O	Isometric
BOLIVARITE	Al2(PO4)(OH)3.4-5H2O	Amorphous
BOLTWOODITE	HK(UO2)(SiO4).1,5H2O	Monoclinic
BONACCORDITE	Ni2Fe+++BO5	Orthorhombic
BONATTITE	CuSO4.3H2O	Monoclinic
BONSHTEDTITE	Na3Fe++(PO4)(CO3)	Monoclinic ps Orthorhombic
BOOTHITE	CuSO4.7H2O	Monoclinic
BORACITE	Mg3B7O13Cl	Orthorhombic ps Cubic
BORAX	Na2B4O5(OH)4.8H2O	Monoclinic
BORCARITE	Ca4MgB4O6(OH)6(CO3)2	Triclinic
BORISHANSKIITE	Pd1+x(As,Pb)2,x=0-0,2	Orthorhombic
BORNEMANITE	BaNa4Ti2NbSi4O17(F,OH).Na3PO4	Orthorhombic
BORNHARDTITE	Co++Co+++2Se4	Isometric
BORNITE	Cu5FeS4	Orthorhombic ps Cubic
BORODAEVITE	Ag5(Bi,Sb)9S16	Monoclinic
BOROMUSCOVITE	KAl2(Si3B)O10(OH,F)2	Monoclinic
BOROVSKITE	Pd3SbTe4	Isometric
BOSTWICKITE	CaMn+++6Si3O16.7H2O	Orthorhombic
BOTALLACKITE	Cu2Cl(OH)3	Monoclinic
BOTRYOGEN	MgFe+++(SO4)2(OH).7H2O	Monoclinic
BOTTINOITE	NiSb+++++2(OH)12.6H2O	Hexagonal
BOULANGERITE	Pb5Sb4S11	Monoclinic
BOURNONITE	PbCuSbS3	Orthorhombic
BOUSSINGAULTITE	(NH4)2Mg(SO4)2.6H2O	Monoclinic
BOWIEITE	(Rh,Ir,Pt)1,77S3	Orthorhombic
BOYLEITE	(Zn,Mg)SO4.4H2O	Monoclinic
BRABANTITE	CaTh(PO4)2	Monoclinic
BRACEWELLITE	Cr+++O(OH)	Orthorhombic
BRACKEBUSCHITE	Pb2(Mn,Fe++)(VO4)2.H2O	Monoclinic
BRADLEYITE	Na3Mg(PO4)(CO3)	Monoclinic
BRAGGITE	(Pt,Pd,Ni)S	Tetragonal
BRAITSCHITE-(Ce)	(Ca,Na2)7(Ce,La)2B22O43.7H2O	Hexagonal
BRAMMALLITE	(Na,H3O)(Al,Mg,Fe)2(Si,Al)4O10[(OH)2,H2O]	Monoclinic
BRANDTITE	Ca2(Mn,Mg)(AsO4)2.2H2O	Monoclinic
BRANNERITE	(U,Ca,Y,Ce)(Ti,Fe)2O6	Monoclinic
BRANNOCKITE	KSn2Li3Si12O30	Hexagonal
BRASS (LAITON,MESSING)	Cu3Zn2	Cubic
BRASSITE	MgHAsO4.4H2O	Orthorhombic
BRAUNITE	Mn++Mn+++6SiO12	Tetragonal
BRAUNITEII	Ca(Mn+++,Fe+++)14SiO24	Tetragonal
BRAZILIANITE	NaAl3(PO4)2(OH)4	Monoclinic
BREDIGITE	Ca7Mg(SiO4)4	Orthorhombic ps Hexagonal
BREITHAUPTITE	NiSb	Hexagonal
BRENKITE	Ca2(CO3)F2	Orthorhombic
BREWSTERITE	(Sr,Ba,Ca)Al2Si6O16.5H2O	Monoclinic

Mineral Name	Formula	Crystal System
BREZINAITE	Cr3S4	Monoclinic
BRIANITE	Na2CaMg(PO4)2	Monoclinic
BRIANYOUNGITE	Zn3(CO3,SO4)(OH)4	Monoclinic ps Orthorhombic
BRIARTITE	Cu2(Zn,Fe)GeS4	Tetragonal
BRINDLEYITE (NIMESITE)	(Ni,Mg,Fe++)2Al(SiAl)O5(OH)4	Monoclinic
BRITHOLITE- (ABUKUMALITE)	(Y,Ca)5(SiO4,PO4)3(OH,F)	Hexagonal
BRITHOLITE-(Ce)	(Ce,Ca)5(SiO4,PO4)3(OH,F)	Hexagonal
BRIZZIITE	NaSb+++++O3	Trigonal
BROCHANTITE	Cu4(SO4)(OH)6	Monoclinic
BROCKITE	(Ca,Th,Ce)(PO4).H2O	Hexagonal
BROKENHILLITE	(Mn,Fe)32[Si24O60]OH29Cl11	Hexagonal
BROMARGYRITE (BROMYRITE)	AgBr	Isometric
BROMELLITE	BeO	Hexagonal
BROOKITE	TiO2	Orthorhombic
BROWNMILLERITE	Ca2(Al,Fe+++)2O5	Orthorhombic
BRUCITE	Mg(OH)2	Trigonal
BRUGGENITE (BRUEGGENITE)	Ca(IO3)2.H2O	Monoclinic
BRUGNATELLITE	Mg6Fe+++(CO3)(OH)13.4H2O	Hexagonal
BRUNOGEIERITE	(Ge++,Fe++)Fe+++2O4	Isometric
BRUSHITE	CaHPO4.2H2O	Monoclinic
BUCHWALDITE	NaCaPO4	Orthorhombic
BUCKHORNITE	AuPb2BiTe2S3	Orthorhombic
BUDDINGTONITE	(NH4)AlSi3O8.0,5H2O	Monoclinic
BUERGERITE	NaFe+++3Al6(BO3)3Si6O21F	Trigonal
BUKOVITE	Tl2Cu3FeSe4	Tetragonal
BUKOVSKYITE	Fe+++2(AsO4)(SO4)(OH).7H2O	Monoclinic
BULACHITE	Al2(AsO4)(OH)3.3H2O	Orthorhombic
BULTFONTEINITE	Ca2SiO2(OH,F)4	Triclinic
BUNSENITE	NiO	Isometric
BURANGAITE	(Na,Ca)2(Fe++,Mg)2Al10(PO4)8(OH,O)12.4H2O	Monoclinic
BURBANKITE	(Na,Ca)3(Sr,Ba,Ce)3(CO3)5	Hexagonal
BURCKHARDTITE	Pb2(Fe+++,Mn+++)Te++++(AlSi3)O12(OH)2.H2O	Monoclinic ps Hexagonal
BURKEITE	Na6(CO3)(SO4)2	Orthorhombic
BURPALITE	Na2CaZrSi2O7F2	Monoclinic
BURSAITE	Pb5Bi4S11	Monoclinic
BURTITE	CaSn(OH)6	Isometric
BUSTAMITE	(Mn,Ca)3Si3O9	Trigonal
BUTLERITE	Fe+++(SO4)(OH).2H2O	Monoclinic
BUTSCHLIITE (BUETSCHLIITE)	K2Ca(CO3)2	Trigonal
BUTTGENBACHITE	Cu19Cl4(NO3)2(OH)32.2H2O	Hexagonal
BYELORUSSITE-(Ce) (BELORUSSITE-(Ce))	NaBa2(Ce,La)2Mn++Ti2Si8O26(F,OH).H2O	Orthorhombic

Mineral Data

Mineral Name	Formula	Crystal System
BYSTRITE	(Na,K)7Ca(Si6Al6)O24S1,5.H2O	Trigonal
BYSTROMITE (BYSTROEMITE)	MgSb2O6	Tetragonal
BYTOWNITE	(Ca,Na)(Si,Al)4O8	Triclinic
CABRIITE	Pd2SnCu	Orthorhombic
CACOXENITE	(Fe+++,Al)25(PO4)17O6(OH)12.75H2O	Hexagonal
CADMIUM	Cd	Hexagonal
CADMOSELITE	CdSe	Hexagonal
CADWALADERITE	Al(OH)2Cl.4H2O	Amorphous
CAFARSITE	Ca8(Ti,Fe++,Fe+++,Mn)6-7(As+++O3)12.4H2O	Isometric
CAFETITE	Ca(Fe+++,Al)2Ti4O12.4H2O	Orthorhombic
CAHNITE	Ca2B(AsO4)(OH)4	Tetragonal
CALAVERITE	AuTe2	Monoclinic
CALCIBORITE	CaB2O4	Orthorhombic
CALCIOANCYLITE-(Ce)	(Ca,Sr)Ce(CO3)2(OH).H2O	Orthorhombic
CALCIOANCYLITE-(Nd)	Ca(Nd,Ce,Gd,Y)3(CO3)4(OH)3.H2O	Orthorhombic
CALCIOBETAFITE	Ca2(Nb,Ti)2(O,OH)7	Isometric
CALCIOBURBANKITE	Na3(Ca,Ce,Sr,La,Nd)3(CO3)5	Hexagonal
CALCIOCOPIAPITE	CaFe+++4(SO4)6(OH)2.19H2O	Triclinic
CALCIOFERRITE	Ca4Fe++(Fe+++,Al)4(PO4)6(OH)4.12H2O	Monoclinic
CALCIOHILAIRITE	CaZrSi3O9.3H2O	Trigonal
CALCIOTANTITE	CaTa4O11	Hexagonal
CALCIOURANOITE	(Ca,Ba,Pb)U2O7.5H2O	Amorphous
CALCITE	CaCO3	Trigonal
CALCIUMCATAPLEIITE	CaZrSi3O9.2H2O	Hexagonal
CALCJARLITE	Na(Ca,Sr)3Al3(F,OH)16	Monoclinic
CALCLACITE	Ca[Cl2/CH3COO].10H2O	Monoclinic/Triclinic
CALCURMOLITE	Ca(UO2)3(MoO4)3(OH)2.11H2O	Monoclinic
CALCYBEBOROSILITE-	(Y,Ca)2(B,Be)2Si2O8(OH)2	Monoclinic
CALDERITE	(Mn++,Ca)3(Fe+++,Al)2(SiO4)3	Isometric
CALEDONITE	Pb5Cu2(CO3)(SO4)3(OH)6	Orthorhombic
CALKINSITE-(Ce)	(Ce,La)2(CO3)3.4H2O	Orthorhombic
CALLAGHANITE	Cu2Mg2(CO3)(OH)6.2H2O	Monoclinic
CALOMEL	Hg2Cl2	Tetragonal
CALUMETITE	Cu(OH,Cl)2.2H2O	Orthorhombic
CALZIRTITE	CaZr3TiO9	Tetragonal
CAMEROLAITE	Cu4Al2[HSbO4,SO4]OH10(CO3).2H2O	Monoclinic
CAMERONITE	AgCu7Te10	Tetragonal
CAMGASITE	CaMg(AsO4)(OH).5H2O	Monoclinic
CAMINITE	Mg7(SO4)5(OH)4.H2O	Tetragonal
CAMPIGLIAITE	Cu4Mn(SO4)2(OH)6.4H2O	Monoclinic
CANAPHITE	CaNa2P2O7.4H2O	Monoclinic
CANASITE	(Na,K)6Ca5Si12O30(OH,F)4	Monoclinic
CANAVESITE	Mg2(CO3)(HBO3).5H2O	Monoclinic
CANCRINITE	Na6Ca2Al6Si6O24(CO3)2	Hexagonal
CANCRISILITE	Na7Al5Si7O24(C03).3H2O	Hexagonal
CANFIELDITE	Ag8SnS6	Orthorhombic ps Cubic

Mineral Data

Mineral Name	Formula	Crystal System
CANNIZZARITE	Pb4Bi5(S,Se)11,5	Monoclinic
CANNONITE	Bi2O(OH)2S04	Monoclinic
CAPGARONNITE	HgAg(Cl,Br,I)S	Orthorhombic
CAPPELENITE-	Ba(Y,Ce)6Si3B6O24F2	Trigonal
CARACOLITE	Na3Pb2(SO4)3Cl	Monoclinic ps Hexagonal
CARBOBORITE	Ca2Mg(CO3)2B2(OH)8.4H2O	Monoclinic
CARBOCERNAITE	(Ca,Na)(Sr,Ce,Ba)(CO3)2	Orthorhombic
CARBOIRITE	Fe++Al2GeO5(OH)2	Triclinic
CARBONATECYANOTRICHITE	Cu4Al2(CO3,SO4)(OH)12.2H2O	Orthorhombic
CARBONATEFLUORAPATITE	Ca5(PO4,CO3)3F	Hexagonal
CARBONATEHYDROXYLAPATITE	Ca5(PO4,CO3)3(OH)	Hexagonal
CARLETONITE	KNa4Ca4Si8O18(CO3)4(OH,F).H2O	Tetragonal
CARLFRIESITE	CaTe++++2Te++++++O8	Monoclinic
CARLHINTZEITE	Ca2AlF7.H2O	Triclinic ps Monoclinic
CARLINITE	Tl2S	Trigonal
CARLOSRUIZITE	K6(Na,K)4Na6Mg10(Se++++++O4)12(IO3)12.12 H2O	Trigonal
CARLOSTURANITE	(Mg,Fe++,Ti)21(Si,Al)12O28(OH)34	Monoclinic
CARLSBERGITE	CrN	Isometric
CARMINITE	PbFe+++2(AsO4)2(OH)2	Orthorhombic
CARNALLITE	KMgCl3.6H2O	Orthorhombic
CARNOTITE	K2(UO2)2V2O8.3H2O	Monoclinic
CAROBBIITE	KF	Isometric
CARPHOLITE	MnAl2Si2O6(OH)4	Orthorhombic
CARRBOYDITE	(Ni,Cu)14Al9(SO4,CO3)6(OH)43.7H2O	Hexagonal
CARROLLITE	Cu(Co,Ni)2S4	Isometric
CARYINITE	Na(Ca,Pb)(Ca,Mn)(Mn,Mg)2(AsO4)3	Monoclinic
CARYOPILITE	(Mn++,Mg)6Si4O10(OH)8	Monoclinic
CASCANDITE	Ca(Sc,Fe++)Si3O8(OH)	Triclinic
CASSEDANNEITE	Pb5(VO4)2(CrO4)2.H2O	Monoclinic
CASSIDYITE	Ca2(Ni,Mg)(PO4)2.2H2O	Triclinic
CASSITERITE	SnO2	Tetragonal
CASTAINGITE	CuMo2S5	Hexagonal
CASWELLSILVERITE	NaCrS2	Trigonal
CATAPLEIITE	Na2ZrSi3O9.2H2O	Hexagonal
CATTIERITE	CoS2	Isometric
CAVANSITE	Ca(VO)Si4O10.4H2O	Orthorhombic
CAYSICHITE-	Ca3Y4GdSi8O20(CO3)6(OH).2H20	Orthorhombic
CEBAITE-(Ce)	Ba3Ce2(CO3)5F2	Monoclinic
CEBAITE-(Nd)	Ba3(Nd,Ce)2(CO3)5F2	Monoclinic
CEBOLLITE	Ca2(Mg,Fe++,Al)Si2(O,OH)7	Orthorhombic
CECHITE	Pb(Fe++,Mn)(VO4)(OH)	Orthorhombic
CELADONITE	K(Mg,Fe++)(Fe+++,Al)Si4O10(OH)2	Monoclinic
CELESTINE (CELESTITE)	SrSO4	Orthorhombic
CELSIAN	BaAl2Si2O8	Monoclinic

Mineral Data

Mineral Name	Formula	Crystal System
CERIANITE-(Ce)	(Ce++++,Th)O2	Isometric
CERIOPYROCHLORE-(Ce) (MARIGNACITE)	(Ce,Ca,Y)2(Nb,Ta)2O6(OH,F)	Isometric
CERITE-(Ce)	Ce+++9Fe+++(SiO4)6[(SiO3)(OH)](OH)3	Trigonal
CERNYITE	Cu2CdSnS4	Tetragonal
CEROLITE (KEROLITE)	(Mg,Ni)3Si4O10(OH)2.H2O	Monoclinic
CEROTUNGSTITE-(Ce)	CeW2O6(OH)3	Monoclinic
CERULEITE	Cu2Al7(AsO4)4(OH)13.12H2O	Triclinic
CERUSSITE	PbCO3	Orthorhombic
CERVANDONITE-(Ce)	(Ce,Nd,La)(Fe+++,Fe++,Ti++++,Al)3SiAs(Si,As)O13	Monoclinic
CERVANTITE	Sb+++Sb+++++O4	Orthorhombic
CERVELLEITE	Ag4TeS	Isometric
CESANITE	Na3Ca2(SO4)3(OH)	Hexagonal
CESAROLITE	PbH2Mn++++3O8	
CESBRONITE	Cu5(TeO3)2(OH)6.2H2O	Orthorhombic
CESIUMKUPLETSKITE	(Cs,K,Na)3(Mn,Fe++)7(Ti,Nb)2Si8O24(O,OH,F)7	Triclinic
CESPLUMTANTITE	(Cs,Na)2(Pb,Sb+++)3Ta8O24	Tetragonal
CESSTIBTANTITE	(Cs,Na)SbTa4O12	Isometric
CETINEITE	(K,Na)3+x(Sb2O3)3(Sb2S3)(OH)x.(2,8-x)H2O	Hexagonal
CHABAZITE	CaAl2Si4O12.6H2O	Trigonal
CHABOURNEITE	(Tl,Pb)21(Sb,As)91S147	Triclinic
CHAIDAMUNITE	CuFe(SO4)2(OH).4H2O	Triclinic ps Monoclinic
CHALCANTHITE	CuSO4.5H2O	Triclinic
CHALCOALUMITE	CuAl4(SO4)(OH)12.3H2O	Monoclinic
CHALCOCITE	Cu2S	Monoclinic
CHALCOCYANITE	CuSO4	Orthorhombic
CHALCOMENITE	CuSeO3.2H2O	Orthorhombic
CHALCONATRONITE	Na2Cu(CO3)2.3H2O	Monoclinic
CHALCOPHANITE	(Zn,Fe++,Mn++)Mn++++3O7.3H2O	Trigonal
CHALCOPHYLLITE	Cu++18Al2(AsO4)3(SO4)3(OH)27.33H2O	Trigonal
CHALCOPYRITE	CuFeS2	Tetragonal
CHALCOSIDERITE	CuFe+++6(PO4)4(OH)8.4H2O	Triclinic
CHALCOSTIBITE	CuSbS2	Orthorhombic
CHALCOTHALLITE	(Tl,K)2Cu5,5(Fe,Ag)SbS4	Orthorhombic ps Tetragonal
CHAMBERSITE	Mn3B7O13Cl	Orthorhombic
CHAMEANITE	(Cu,Fe)4As(Se,S)4	Isometric
CHAMOSITE	(Fe++,Mg,Fe+++)5Al(Si3Al)O10(OH,O)8	Monoclinic
CHANGBAIITE	PbNb2O6	Trigonal
CHANTALITE	CaAl2SiO4(OH)4	Tetragonal
CHAOITE	C	Hexagonal
CHAPMANITE	Sb+++Fe+++2(SiO4)2(OH)	Monoclinic
CHARLESITE	Ca6(Al,Si)2(SO4)2B(OH)4(OH,O)12.26H2O	Hexagonal
CHAROITE	K(Ca,Na)2Si4O10(OH,F).H2O	Monoclinic
CHATKALITE	Cu6Fe++Sn2S8	Tetragonal
CHAYESITE	K(Mg,Fe++)4Fe+++(Si12O30)	Hexagonal

Mineral Data

Mineral Name	Formula	Crystal System
CHEKHOVICHITE	(Bi,Pb,Fe)2Te4O11	Monoclinic
CHELKARITE	CaMgB2O4Cl2.7H2O	Orthorhombic
CHELYABINSKITE	(Ca,Mg)3Si(OH)6(SO4,CO3)2.9H2O	Orthorhombic
CHENEVIXITE	Cu2Fe+++2(AsO4)2(OH)4.H2O	Monoclinic
CHENGDEITE	Ir3Fe	Isometric
CHENITE	Pb4Cu(SO4)2(OH)6	Triclinic
CHERALITE-(Ce)	(Ce,Ca,Th,U)(P,Si)O4	Monoclinic
CHEREMNYKHITE	Zn3Pb3Te++++O6(VO4,AsO4)2	Orthorhombic
CHEREPANOVITE	RhAs	Orthorhombic
CHERNIKOVITE (HYDROGENAUTUNITE)	(H3O)2(UO2)2(PO4)2.6H2O	Tetragonal
CHERNOVITE-	YAsO4	Tetragonal
CHERNOVITE-(Ce)	(Ce,Y)(AsO4)	Tetragonal
CHERNYKHITE	(Ba,Na)V+++,Al)2(Si,Al)4O10(OH)2	Monoclinic
CHERVETITE	Pb2V2O7	Monoclinic
CHESSEXITE	Na4Ca2(Mg,Zn)3Al8(SiO4)2(SO4)10(OH)10.40H2O	Orthorhombic
CHESTERITE	(Mg,Fe++)17Si20O54(OH)6	Orthorhombic
CHESTERMANITE	Mg2(Fe+++,Mg,Al,Sb+++++)BO3O2	Orthorhombic
CHEVKINITE-(Ce)	(Ce,La)4(Ti,Fe)5Si4O12O10	Monoclinic
CHIAVENNITE	CaMnBe2Si5O13(OH)2.2H2O	Orthorhombic
CHILDRENITE	Fe++Al(PO4)(OH)2.H2O	Monoclinic
CHILUITE (CHILUNITE)	Bi6Te++++++2Mo++++++2O21	Hexagonal
CHIOLITE	Na5Al3F14	Tetragonal
CHKALOVITE	Na2BeSi2O6	Orthorhombic
CHLADNIITE	Na2Ca(Mg,Fe++)7(PO4)6	Trigonal
CHLORALUMINITE	AlCl3.6H2O	Trigonal
CHLORAPATITE	Ca5(PO4)3Cl	Monoclinic
CHLORARGYRITE (CERARGYRITE)	AgCl	Isometric
CHLORELLESTADITE	Ca5(SiO4,PO4,SO4)3(Cl,F)	Orthorhombic
CHLORITOID	(Fe++,Mg,Mn)2Al4Si2O10(OH)4	Monoclinic/Triclinic
CHLORMAGALUMINITE	(Mg,Fe++)4Al2(OH)12(Cl2,CO3).2H2O	Hexagonal
CHLOROCALCITE	KCaCl3	Orthorhombic ps Cubic
CHLOROMAGNESITE	MgCl2	Tetragonal
CHLOROMANGANOKALITE	K4MnCl6	Trigonal
CHLOROPHOENICITE	(Mn,Mg)3Zn2(AsO4)(OH,O)6	Monoclinic
CHLOROTHIONITE	K2Cu(SO4)Cl2	Orthorhombic
CHLOROXIPHITE	Pb3CuCl2(OH)2O2	Monoclinic
CHOLOALITE	PbCu(Te++++O3)2	Isometric
CHONDRODITE	(Mg,Fe++)5(SiO4)2(F,OH)2	Monoclinic
CHRISTITE	TlHgAsS3	Monoclinic
CHROMATITE	CaCrO4	Tetragonal
CHROMDRAVITE	NaMg3(Cr,Fe+++)6(BO3)3Si6O18(OH)4	Trigonal
CHROMFERIDE	Fe3Cr1-x(x=0,6)	Isometric
CHROMITE	Fe++Cr2O4	Isometric
CHROMIUM	Cr	Isometric

Mineral Name	Formula	Crystal System
CHRYSOBERYL	BeAl2O4	Orthorhombic
CHRYSOCOLLA	(Cu,Al)2H2Si2O5(OH)4.nH2O	Monoclinic
CHUBUTITE	Pb7O6Cl2	Orthorhombic ps Tetragonal
CHUDOBAITE	(Mg,Zn)5H2(AsO4)4.10H2O	Triclinic
CHUKHROVITE-	Ca3(Y,Ce)Al2(SO4)F13.10H2O	Isometric
CHUKHROVITE-(Ce)	Ca3(Ce,Y)Al2(SO4)F13.10H2O	Isometric
CHURCHITE-(WEINSCHENKITE)	YPO4.2H2O	Monoclinic
CHURCHITE-(Dy)	(Dy,Sm,Gd,Nd)(PO4).2H2O	Monoclinic
CHURCHITE-(Nd)	Nd(PO4).2H2O	Monoclinic
CHURSINITE	Hg+Hg++(AsO4)	Monoclinic
CHVALETICEITE	(Mn++,Mg)SO4.6H2O	Monoclinic
CHVILEVAITE	Na(Cu,Fe,Zn)2S2	Hexagonal
CIANCIULLIITE	Mn++++(Mg,Mn++)2Zn+2(OH)10.2-4H2O	Monoclinic
CINNABAR (CINABRE,ZINNOBER)	HgS	Trigonal
CLAIRITE	(NH4)2(Fe+++,Mn+++)3(SO4)4(OH)3.3H2O	Triclinic
CLARAITE	(Cu,Zn)3(CO3)(OH)4.4H2O	Tetragonal ps Hexagonal
CLARINGBULLITE	Cu++4(OH)6(Cl,OH)2	Hexagonal
CLARKEITE	(Na,Ca,Pb)2U2(O,OH)7	Orthorhombic
CLAUDETITE	As2O3	Monoclinic
CLAUSTHALITE	PbSe	Isometric
CLIFFORDITE	UTe3O9	Isometric
CLINOATACAMITE	Cu2(OH)3Cl	Monoclinic
CLINOBEHOITE	Be(OH)2	Monoclinic
CLINOBISVANITE	BiVO4	Monoclinic
CLINOCHALCOMENITE	CuSeO3.2H2O	Monoclinic
CLINOCHLORE	(Mg,Fe++)5Al(Si3Al)O10(OH)8	Monoclinic
CLINOCHRYSOTILE	Mg3Si2O5(OH)4	Monoclinic
CLINOCLASE	Cu3(AsO4)(OH)3	Monoclinic
CLINOENSTATITE	Mg2Si2O6	Monoclinic
CLINOFERROSILITE	(Fe++,Mg)2Si2O6	Monoclinic
CLINOHEDRITE	CaZnSiO4.H2O	Monoclinic
CLINOHOLMQUISTITE	Li2(Mg,Fe++)3Al2Si8O22(OH)2	Monoclinic
CLINOHUMITE	(Mg,Fe++)9(SiO4)4(F,OH)2	Monoclinic
CLINOJIMTHOMPSONITE	(Mg,Fe++)5Si6O16(OH)2	Monoclinic
CLINOKURCHATOVITE	Ca(Mg,Fe++,Mn)B2O5	Monoclinic
CLINOMIMETITE	Pb5(AsO4)3Cl	Monoclinic ps Hexagonal
CLINOPHOSINAITE	Na3CaPSiO7	Monoclinic
CLINOPTILOLITE	(Na,K,Ca)2-3Al3(Al,Si)2Si13O36.12H2O	Monoclinic
CLINOSAFFLORITE	(Co,Fe,Ni)As2	Monoclinic
CLINOTOBERMORITE	Ca5Si5(O,OH)18.5H2O	Monoclinic
CLINOTYROLITE	Ca2Cu9[(As,S)O4]4(O,OH)10.10H2O	Monoclinic
CLINOUNGEMACHITE	K2Na3Fe(SO4)6(OH)3.9H2O	Monoclinic psTrigonal
CLINOZOISITE	Ca2Al3(SiO4)3(OH)=Ca2AlAl2(SiO4)(Si2O7)O(OH)	Monoclinic
CLINTONITE	Ca(Mg,Al)3(Al3Si)O10(OH)2	Monoclinic

Mineral Data

Mineral Name	Formula	Crystal System
COALINGITE	Mg10Fe+++2(CO3)(OH)24.2H2O	Trigonal
COBALTAUSTINITE	CaCo(AsO4)(OH)	Monoclinic
COBALTITE	CoAsS	Isometric
COBALTKORITNIGITE	(Co,Zn)(As+++++O3)(OH).H2O	Triclinic
COBALTOMENITE	CoSeO3.2H2O	Monoclinic
COBALTPENTLANDITE	Co9S8	Isometric
COBALTZIPPEITE	Co++0,5(UO2)2(SO4)(OH)3.H2O	Monoclinic
COCCINITE	Hg++I2	Tetragonal
COCHROMITE	(Co,Ni,Fe++)(Cr,Al)2O4	Isometric
COCONINOITE	Fe+++2Al2(UO2)2(PO4)4(SO4)(OH)2.20H2O	Monoclinic
COERULEOLACTITE	(Ca,Cu)Al6(PO4)4(OH)8.4-5H2O	Triclinic
COESITE	SiO2	Monoclinic
COFFINITE	U(SiO4)1-x(OH)4x	Tetragonal
COHENITE	(Fe,Ni,Co)3C	Orthorhombic
COLEMANITE	Ca2B6O11.5H2O	Monoclinic
COLLINSITE	Ca2(Mg,Fe++)(PO4)2.2H2O	Triclinic
COLORADOITE	HgTe	Isometric
COLQUIRIITE	CaLiAlF6	Trigonal
COLUSITE	Cu12-13V(As,Sb,Sn,Ge)3S16	Isometric
COMANCHEITE	Hg13(Cl,Br)8O9	Orthorhombic
COMBEITE	Na2Ca2Si3O9	Trigonal
COMBLAINITE	(Ni++x,Co+++1-x)(OH)2(CO3)(1-x)/2.yH2O	Trigonal
COMPREIGNACITE	K2(UO2)6O4(OH)6.8H2O	Orthorhombic
CONGOLITE	(Fe++,Mg,Mn)3B7O13Cl	Trigonal
CONICHALCITE	CaCu(AsO4)(OH)	Orthorhombic
CONNELLITE	Cu19Cl4(SO4)(OH)32.3H2O	Hexagonal
COOKEITE	LiAl4(Si3Al)O10(OH)8	Monoclinic
COOMBSITE	K(Mn++,Fe++,Mg)13(Si,Al)18O42(OH)14	Trigonal
COOPERITE	(Pt,Pd,Ni)S	Tetragonal
COPIAPITE (FERROCOPIAPITE)	Fe++Fe+++4(SO4)6(OH)2.20H2O	Triclinic
COPPER (CUIVRE,KUPFER)	Cu	Isometric
COQUANDITE	Sb6O8(SO4).H2O	Triclinic
COQUIMBITE	Fe+++2(SO4)3.9H2O	Trigonal
CORDEROITE	Hg3S2Cl2	Isometric
CORDIERITE	Mg2Al4Si5O18	Orthorhombic
CORDYLITE-(Ce)	Ba(Ce,La)2(CO3)3F2	Hexagonal
CORKITE	PbFe+++3(PO4)(SO4)(OH)6	Trigonal
CORNETITE	Cu3(PO4)(OH)3	Orthorhombic
CORNUBITE	Cu5(AsO4)2(OH)4	Triclinic
CORNWALLITE	Cu5(AsO4)2(OH)4	Monoclinic
CORONADITE	Pb(Mn++++,Mn++)8O16	Monoclinic ps Tetragonal
CORRENSITE	(Mg,Fe,Al)9(Si,Al)8O20(OH)10.nH2O	Orthorhombic
CORUNDUM	Al2O3	Trigonal
CORVUSITE	V++++2V+++++12O34.nH2O	Orthorhombic
COSALITE	Pb2Bi2S5	Orthorhombic
COSTIBITE	CoSbS	Orthorhombic

Mineral Name	Formula	Crystal System
COTUNNITE	$PbCl_2$	Orthorhombic
COULSONITE	$Fe^{++}V^{+++}_2O_4$	Isometric
COUSINITE	$MgU_2Mo_2O_{13}.6H_2O$	
COVELLITE	CuS	Hexagonal
COWLESITE	$CaAl_2Si_3O_{10}.5-6H_2O$	Orthorhombic
COYOTEITE	$NaFe_3S_5.2H_2O$	Triclinic
CRANDALLITE	$CaAl_3(PO_4)_2(OH)_5.H_2O$	Trigonal
CRAWFORDITE	$Na_3Sr(PO_4)(CO_3)$	Monoclinic ps Orthorhombic
CREASEYITE	$Pb_2Cu_2Fe^{+++}_2Si_5O_{17}.6H_2O$	Orthorhombic
CREDNERITE	$CuMnO_2$	Monoclinic
CREEDITE	$Ca_3Al_2(SO_4)(F,OH)_{10}.2H_2O$	Monoclinic
CRERARITE	$(Pt,Pb)Bi_3(S,Se)_{4-x}(x=0,7)$	Isometric
CRICHTONITE	$(Sr,La,Ce,Y)(Ti,Fe^{+++},Mn)_{21}O_{38}$	Trigonal
CRIDDLEITE	$TlAg_2Au_3Sb_{10}S_{10}$	Monoclinic ps Tetragonal
CRISTOBALITE	SiO_2	Tetragonal
CROCOITE	$PbCrO_4$	Monoclinic
CRONSTEDTITE	$Fe^{++}_2Fe^{+++}(SiFe^{+++})O_5(OH)_4$	Monoclinic
CROOKESITE	$Cu_7(Tl,Ag)Se_4$	Tetragonal
CROSSITE	$Na_2(Mg,Fe^{++})_3(Al,Fe^{+++})_2Si_8O_{22}(OH)_2$	Monoclinic
CRYOLITE	Na_3AlF_6	Monoclinic
CRYOLITHIONITE	$Na_3Li_3Al_2F_{12}$	Isometric
CRYPTOHALITE	$(NH_4)_2SiF_6$	Isometric
CRYPTOMELANE	$K(Mn^{++++},Mn^{++})_8O_{16}$	Monoclinic ps Tetragonal
CUALSTIBITE	$Cu_6Al_3Sb^{+++++}_3O_{18}.16H_2O$	Trigonal
CUBANITE	$CuFe_2S_3$	Orthorhombic
CUMENGITE (CUMENGEITE)	$Pb_{21}Cu_{20}Cl_{42}(OH)_{40}$	Tetragonal
CUMMINGTONITE	$(Mg,Fe^{++})_7Si_8O_{22}(OH)_2$	Monoclinic
CUPALITE	$(Cu,Zn)Al$	Orthorhombic
CUPRITE	Cu_2O	Isometric
CUPROBISMUTITE	$Cu_{10}Bi_{12}S_{23}$	Monoclinic
CUPROCOPIAPITE	$CuFe^{+++}_4(SO_4)_6(OH)_2.20H_2O$	Triclinic
CUPROIRIDSITE	$CuIr_2S_4$	Isometric
CUPROPAVONITE	$AgPbCu_2Bi_5S_{10}$	Monoclinic
CUPRORHODSITE	$CuRh_2S_4$	Isometric
CUPRORIVAITE	$CaCuSi_4O_{10}$	Tetragonal
CUPROSKLODOWSKITE	$(H_3O)_2Cu(UO_2)_2(SiO_4)_2.2H_2O$	Triclinic
CUPROSPINEL	$(Cu,Mg)Fe^{+++}_2O_4$	Isometric
CUPROSTIBITE	$Cu_2(Sb,Tl)$	Tetragonal
CUPROTUNGSTITE	$Cu^{++}_3(WO_4)_2(OH)_2$	Tetragonal
CURETONITE	$Ba(Al,Ti^{++++})(PO_4)_4(O,OH)F$	Monoclinic
CURIENITE	$Pb(UO_2)_2V_2O_8.5H_2O$	Orthorhombic
CURITE	$Pb_2U_5O_{17}.4H_2O$	Orthorhombic
CUSPIDINE	$Ca_4Si_2O_7(F,OH)_2$	Monoclinic
CUZTICITE	$Fe^{+++}_2Te^{++++++}O_6.3H_2O$	Hexagonal
CYANOCHROITE	$K_2Cu(SO_4).6H_2O$	Monoclinic

Mineral Name	Formula	Crystal System
CYANOPHILLITE	Cu5Al2Sb+++3O12(OH).12H2O	Orthorhombic
CYANOTRICHITE	Cu4Al2(SO4)(OH)12.2H2O	Orthorhombic
CYLINDRITE	Pb3Sn4FeSb2S14	Triclinic
CYMRITE	BaAl2Si2(O,OH)8.H2O	Monoclinic
CYRILOVITE	NaFe+++3(PO4)2(OH)4.2H2O	Tetragonal
DACHIARDITE	(Ca,Na2,K2)5Al10Si38O96.25H2O	Monoclinic
DADSONITE	Pb10+xSb14-xS31-xClx	Triclinic
DALYITE	K2ZrSi6O15	Triclinic
DAMARAITE	Pb4O3Cl2	Orthorhombic
DANALITE	Fe++4Be3(SiO4)3S	Isometric
DANBAITE	CuZn2	Isometric
DANBURITE	CaB2(SiO4)2	Orthorhombic
DANIELSITE	(Cu,Ag)14HgS8	Orthorhombic
DANNEMORITE	Mn2(Fe++,Mg)5Si8O22(OH)2	Monoclinic
DANSITE	Na21Mg(SO4)10Cl3	Isometric
DAOMANITE	CuPtAsS2	Orthorhombic
DAQINGSHANITE-(Ce)	(Sr,Ca,Ba)3(Ce,La)(PO4)(CO3)3-x(OH,F)x	Hexagonal
DARAPIOSITE	KNa2Li(Mn,Zn)2ZrSi12O30	Hexagonal
DARAPSKITE	Na3(SO4)(NO3).H2O	Monoclinic
DATOLITE	CaBSiO4(OH)	Monoclinic
DAUBREEITE	BiO(OH,Cl)	Tetragonal
DAUBREELITE	Fe++Cr2S4	Isometric
DAVANITE	K2TiSi6O15	Triclinic
DAVIDITE-	Y(Ti,Fe)21O38	Trigonal
DAVIDITE-(Ce)	(Ce,La)(Y,U,Fe++)(Ti,Fe+++)20(O,OH)38	Trigonal
DAVIDITE-(La)	(La,Ce)(Y,U,Fe++)(Ti,Fe+++)20(O,OH)38	Trigonal
DAVREUXITE	MnAl6Si4O17(OH)2	Monoclinic
DAVYNE	Na4K2Ca2Si6Al6O24(SO4)Cl2	Hexagonal
DAWSONITE	NaAL(CO3)(OH)2	Orthorhombic
DAYINGITE	CuCoPtS4	Isometric
DEANESMITHITE	Hg+2Hg++3Cr++++++O5S2	Triclinic
DEERITE	(Fe++,Mn)6(Fe+++,Al)3Si6O20(OH)5	Monoclinic ps Orthorhombic
DEFERNITE	Ca6(CO3)2-x(SiO4)x(OH)7(Cl,OH)1-2x(x=0,5)	Orthorhombic
DELAFOSSITE	Cu+Fe+++O2	Trigonal
DELHAYELITE	(Na,K)10Ca5Al6Si32O80(Cl2,F2,SO4)3.18H2O	Orthorhombic
DELINDEITE	(Na,K)3(Ba,Ca)4(Ti,Fe,Al)6Si8O26(OH)4	Monoclinic
DELLAITE	Ca6Si3O11(OH)2	Monoclinic
DELORYITE	Cu++4(UO2)(MoO4)2(OH)6	Monoclinic
DELRIOITE	CaSrV2O6(OH)2.3H2O	Monoclinic
DELVAUXITE	CaFe+++4(PO4,SO4)2(OH)8.4-6H2O	Amorphous
DEMESMAEKERITE	Pb2Cu5(UO2)2(SeO3)6(OH)6.2H2O	Triclinic
DENISOVITE	(K,Na)Ca2Si3O8(F,OH)	Monoclinic
DENNINGITE	(Mn,Zn)Te2O5	Tetragonal
DERBYLITE	Fe+++,Fe++,Ti)7Sb+++O13(OH)	Monoclinic
DERRIKSITE	Cu4(UO2)(SeO3)2(OH)6	Orthorhombic
DERVILLITE	Ag2AsS2	Monoclinic

Mineral Name	Formula	Crystal System
DESAUTELSITE	$Mg_6Mn^{+++}2(CO_3)(OH)16.4H_2O$	Trigonal
DESCLOIZITE	$PbZn(VO_4)(OH)$	Orthorhombic
DESPUJOLSITE	$Ca_3Mn^{++++}(SO_4)2(OH)6.3H_2O$	Hexagonal
DEVILLINE (DEVILLITE)	$CaCu_4(SO_4)2(OH)6.3H_2O$	Monoclinic
DEWINDTITE	$Pb_3[H(UO_2)3O_2(PO_4)2]2.12H_2O$	Orthorhombic
DIABOLEITE	$Pb_2CuCl_2(OH)4$	Tetragonal
DIADOCHITE (DESTINEZITE)	$Fe^{+++}2(PO_4)(SO_4)(OH).5H_2O$	Triclinic
DIAMOND	C	Isometric
DIAOYUDAOITE	$NaAl_{11}O_{17}$	Hexagonal
DIAPHORITE	$Pb_2Ag_3Sb_3S_8$	Monoclinic
DIASPORE	$AlO(OH)$	Orthorhombic
DICKINSONITE	$(K,Ba)(Na,Ca)5(Mn^{++},Fe^{++},Mg)14Al(PO_4)12(OH,F)2$	Monoclinic
DICKITE	$Al_2Si_2O_5(OH)4$	Monoclinic
DIENERITE	Ni_3As	Isometric
DIETRICHITE	$(Zn,Fe^{++},Mn)Al_2(SO_4)4.22H_2O$	Monoclinic
DIETZEITE	$Ca_2(IO_3)2(CrO_4)$	Monoclinic
DIGENITE	Cu_9S_5	Isometric
DIMORPHITE	As_4S_3	Orthorhombic
DINITE	$C_{20}H_{36}$	Orthorhombic
DIOMIGNITE	$Li_2B_4O_7$	Tetragonal
DIOPSIDE	$CaMgSi_2O_6$	Monoclinic
DIOPTASE	$CuSiO_2(OH)2$	Trigonal
DISSAKISITE-(Ce)	$Ca(Ce,La)(Mg,Fe^{++})(Al,Fe^{+++})2Si_3O_{12}(OH)$	Monoclinic
DITTMARITE	$(NH_4)Mg(PO_4).H_2O$	Orthorhombic
DIXENITE	$Cu^+Mn^{++}14Fe^{+++}(As^{+++}O_3)5(SiO_4)2(As^{+++++}O_4)(OH)6$	Trigonal
DJERFISHERITE	$K_6Na(Fe,Cu)24S_{26}Cl$	Isometric
DJURLEITE	$Cu_{31}S_{16}$	Monoclinic
DMISTEINBERGITE	$CaAl_2Si_2O_8$	Hexagonal
DOLEROPHANITE	$Cu_2(SO_4)O$	Monoclinic
DOLLASEITE-(Ce)	$CaCeMg_24AlSi_3O_{11}(OH,F)2$	Monoclinic
DOLOMITE	$CaMg(CO_3)2$	Trigonal
DOLORESITE	$H_8V_6O_{16}$	Monoclinic
DOMEYKITE	Cu_3As	Isometric
DONATHITE	$(Fe^{++},Mg)(Cr,Fe^{+++})2O_4$	Tetragonal
DONHARRISITE	$Ni_8Hg_3S_9$	Monoclinic
DONNAYITE-	$Sr_3NaCaY(CO_3)6.3H_2O$	Triclinic psTrigonal
DONPEACORITE	$(Mn,Mg)MgSi_2O_6$	Orthorhombic
DORALLCHARITE	$(Tl,K)Fe^{+++}3(SO_4)2(OH)6$	Orthorhombic
DORFMANITE	$Na_2HPO_4.2H_2O$	Orthorhombic
DORRITE	$Ca_2(Mg_2,Fe^{+++}4)Al_4Si_2O_{20}$	Triclinic
DOUGLASITE	$K_2Fe^{++}Cl_4.2H_2O$	Monoclinic
DOWNEYITE	SeO_2	Tetragonal
DOYLEITE	$Al(OH)3$	Triclinic
DOZYITE	$Mg_7(Al,Fe^{+++},Cr)2Al_2Si_4O_{15}(OH)12$	Monoclinic

Mineral Data

Mineral Name	Formula	Crystal System
DRAVITE	NaMg3Al6(BO3)3Si6O18(OH)4	Trigonal
DRESSERITE	BaAl2(CO3)2(OH)4.H2O	Orthorhombic
DREYERITE	BiVO4	Tetragonal
DRUGMANITE	Pb2(Fe+++,Al)H(PO4)2(OH)2	Monoclinic
DRYSDALLITE	Mo(Se,S)2	Hexagonal
DUFRENITE	Fe++Fe+++4(PO4)3(OH)5.2H2O	Monoclinic
DUFRENOYSITE	Pb2As2S5	Monoclinic
DUFTITE	PbCu(AsO4)(OH)	Orthorhombic
DUGGANITE	Pb3Zn3Te(As,V,Si)2(O,OH)14	Hexagonal
DUHAMELITE	Pb2Cu4Bi(VO4)4(OH)3.8H2O	Orthorhombic
DUMONTITE	Pb2(UO2)3O2(PO4)2.5H2O	Monoclinic
DUMORTIERITE	Al6,5-7(BO3)(SiO4)3(O,OH)3	Orthorhombic
DUNDASITE	PbAl2(CO3)2(OH)4.H2O	Orthorhombic
DURANGITE	NaAl(AsO4)F	Monoclinic
DURANUSITE	As4S	Orthorhombic
DUSSERTITE	BaFe+++3(AsO4)2(OH)5	Trigonal
DUTTONITE	V++++O(OH)2	Monoclinic
DWORNIKITE	(Ni,Fe++)SO4.H2O	Monoclinic
DYPINGITE	Mg5(CO3)4(OH)2.5H2O	Monoclinic
DYSCRASITE	Ag3Sb	Orthorhombic
DZHALINDITE	In(OH)3	Isometric
DZHARKENITE	FeSe2	Isometric
DZHEZKAZGANITE	Lead or copper rhenium sulfide	Amorphous
EAKERITE	Ca2SnAl2Si6O18(OH)2.2H2O	Monoclinic
EARLANDITE	Ca3[C(OH)(CH2)2(COO)3]2.4H2O	Monoclinic
EARLSHANNONITE	(Mn,Fe++)Fe+++2(PO4)2(OH)2.4H2O	Monoclinic
ECANDREWSITE	(Zn,Fe++,Mn++)TiO3	Trigonal
ECDEMITE	Pb6As+++2O7Cl4	Tetragonal
ECKERMANNITE	Na3(Mg,Fe++)4AlSi8O22(OH)2	Monoclinic
ECLARITE	Pb9(Cu,Fe)Bi12S28	Orthorhombic
EDENHARTERITE	TlPbAs3S6	Orthorhombic
EDENITE	NaCa(Mg,Fe++)5(Si7Al)O22(OH)2	Monoclinic
EDGARBAILEYITE	Hg+6Si2O7	Monoclinic
EDINGTONITE	BaAl2Si3O10.4H2O	Orthorhombic/Tetragonal
EDOYLERITE	Hg++3Cr++++++O4S2	Monoclinic
EFFENBERGERITE	BaCuSi4O10	Tetragonal
EFREMOVITE	(NH4)2Mg2(SO4)3	Isometric
EGGLETONITE	(Na,K,Ca)2(Mn,Fe)8(Si,Al)12O29(OH)7.11H2O	Monoclinic
EGLESTONITE	Hg6Cl3O(OH)	Isometric
EHRLEITE	Ca2ZnBe(PO4)2(PO3OH).4H2O	Triclinic
EIFELITE	KNa3Mg4Si12O30	Hexagonal
EITELITE	Na2Mg(CO3)2	Trigonal
EKANITE	ThCa2Si8O20	Tetragonal
EKATERINITE	Ca2B4O7(Cl,OH)2.2H2O	Hexagonal
EKMANITE	(Fe++,Mg,Mn,Fe+++)3(Si,Al)4O10(OH)2.2H2O	Orthorhombic
ELBAITE	Na(Li,Al)3Al6(BO3)3Si6O18(OH)4	Trigonal
ELLENBERGERITE	Mg6TiAl6Si8O28(OH)10	Hexagonal

Mineral Name	Formula	Crystal System
ELLISITE	Tl3AsS3	Trigonal
ELPASOLITE	K2NaAlF6	Isometric
ELPIDITE	Na2ZrSi6O15.3H2O	Orthorhombic
ELYITE	Pb4Cu(SO4)(OH)8	Monoclinic
EMBREYITE	Pb5(CrO4)2(PO4)2.H2O	Monoclinic
EMELEUSITE	Na4Li2Fe+++2Si12O30	Orthorhombic ps Hexagonal
EMMONSITE	Fe+++2Te++++3O9.2H2O	Triclinic
EMPLECTITE	CuBiS2	Orthorhombic
EMPRESSITE	AgTe	Orthorhombic
ENARGITE	Cu3AsS4	Orthorhombic
ENGLISHITE	K3Na2Ca10Al15(PO4)21(OH)7.26H2O	Orthorhombic
ENSTATITE	Mg2Si2O6	Orthorhombic
EOSPHORITE	MnAl(PO4)(OH)2.H2O	Monoclinic
EPHESITE	NaLiAl2(Al2Si2)O10(OH)2	Monoclinic/Triclinic
EPIDIDYMITE	NaBeSi3O7(OH)	Orthorhombic
EPIDOTE	Ca2(Fe+++,Al)3(SiO4)3(OH)=Ca2(Fe,Al)Al2(SiO4)(Si2O7)O(OH)	Monoclinic
EPISTILBITE	CaAl2Si6O16.5H2O	Monoclinic
EPISTOLITE	Na5Ti+++Nb2Si4O12O5F.5H2O	Triclinic
EPSOMITE	MgSO4.7H2O	Orthorhombic
ERDITE	NaFeS2.2H2O	Monoclinic
ERICAITE	(Fe++,Mg,Mn)3B7O13Cl	Orthorhombic
ERICSSONITE	BaMn2Fe+++OSi2O7(OH)	Monoclinic
ERIOCHALCITE	CuCl2.2H2O	Orthorhombic
ERIONITE	(K2,Ca,Na2)2Al4Si14O36.15H2O	Hexagonal
ERLIANITE	(Fe++,Fe+++,Mg)24(Fe+++V)6Si36O90(OH,O)48	Orthorhombic
ERLICHMANITE	OsS2	Isometric
ERNIENICKELITE	NiMn++++3O7.3H2O	Trigonal
ERNIGGLIITE	Tl2SnAs2S6	Hexagonal
ERNSTITE	(Mn++1-xFe+++x)Al(PO4)(OH)2-xOx	Monoclinic
ERSHOVITE	Na4K3(Fe++,Mn++,Ti)2Si8O20(OH)4.4H2O	Triclinic
ERTIXIITE	Na2Si4O9	Isometric
ERYTHRITE	Co3(AsO4)2.8H2O	Monoclinic
ERYTHROSIDERITE	K2Fe+++Cl5.H2O	Orthorhombic
ESKEBORNITE	CuFeSe2	Tetragonal
ESKIMOITE	Ag7Pb10Bi15S36	Monoclinic
ESKOLAITE	Cr2O3	Trigonal
ESPERITE	PbCa3Zn4(SiO4)4	Monoclinic
ESSENEITE	CaFe+++AlSiO6	Monoclinic
ETTRINGITE	Ca6Al2(SO4)3(OH)12.26H2O	Hexagonal
EUCAIRITE	CuAgSe	Orthorhombic
EUCHLORINE	KNaCu++3(SO4)3O	Monoclinic
EUCHROITE	Cu2(AsO4)(OH).3H2O	Orthorhombic
EUCLASE	BeAlSiO4(OH)	Monoclinic
EUCRYPTITE	LiAlSiO4	Trigonal

Mineral Data

Mineral Name	Formula	Crystal System
EUDIALYTE	Na4(Ca,Ce)2(Fe++,Mn,Y)ZrSi8O22(OH,Cl)2	Trigonal
EUDIDYMITE	NaBeSi3O7(OH)	Monoclinic
EUGENITE	Ag9Hg2	Isometric
EUGSTERITE	Na4Ca(SO4)3.2H2O	Monoclinic
EULYTITE	Bi4(SiO4)3	Isometric
EUXENITE-	(Y,Ca,Ce,U,Th)(Nb,Ta,Ti)2O6	Orthorhombic
EVANSITE	Al3(PO4)(OH)6.6H2O	Amorphous
EVEITE	Mn2(AsO4)(OH)	Orthorhombic
EVENKITE	(CH3)2(CH2)22	Monoclinic
EWALDITE	(Ba,Sr)(Ca,Na,Y,Ce)(CO3)2.10-12H2O	Hexagonal
EYLETTERSITE	(Th,Pb)1-xAl3(PO4,SiO4)2(OH)6	Trigonal
EZCURRITE	Na4B10O17.7H2O	Triclinic
EZTLITE	Pb2Fe+++6(Te++++O3)3(Te++++++O6)(OH)10.8H2O	Monoclinic
FABIANITE	CaB3O5(OH)	Monoclinic
FAHEYITE	(Mn,Mg)Fe+++2Be2(PO4)4.6H2O	Hexagonal
FAHLEITE	Zn5CaFe+++2(AsO4)6.14H2O	Orthorhombic
FAIRBANKITE	PbTe++++O3	Triclinic
FAIRCHILDITE	K2Ca(CO3)2	Hexagonal
FAIRFIELDITE	Ca2(Mn,Fe++)(PO4)2.2H2O	Triclinic
FALCONDOITE	(Ni,Mg)4Si6O15(OH)2.6H2O	Orthorhombic
FALKMANITE	Pb5Sb4S11	Monoclinic
FAMATINITE	Cu3SbS4	Tetragonal
FANGITE	Tl3AsS4	Orthorhombic
FARRINGTONITE	Mg3(PO4)2	Monoclinic
FAUJASITE	(Na2,Ca)Al2Si4O12.8H2O	Isometric
FAUSTITE	(Zn,Cu)Al6(PO4)4(OH)8.4H2O	Triclinic
FAYALITE (OLIVINE)	Fe++2SiO4	Orthorhombic
FEDORITE	(K,Na)5(Ca,Na)14Si32O76(OH,F)4.2H2O	Triclinic
FEDOROVSKITE	Ca2(Mg,Mn)2B4O7(OH)6	Orthorhombic
FEDOTOVITE	K2Cu++3O(SO4)3or(K,Na)2(Cu,Zn,Pb)3S3O13	Monoclinic
FEITKNECHTITE	Mn+++O(OH)	Hexagonal
FELSOBANYAITE (FELSOBANYITE)	Al4(SO4)(OH)10.5H2O	Orthorhombic
FENAKSITE	(K,Na,Ca)4(Fe++,Fe+++,Mn)2Si8O20(OH,F)	Triclinic
FERBERITE	Fe++WO4	Monoclinic
FERCHROMIDE	Cr3Fe1-x(x=0,6)	Isometric
FERDISILICITE	FeSi2	Isometric
FERGUSONITE-	YNbO4	Tetragonal
FERGUSONITE-(Ce)	(Ce,Nd,La)NbO4.0,3H2O	Monoclinic
FERGUSONITE-(Nd)	(Nd,Ce)(Nb,Ti)O4	Monoclinic
FERGUSONITE-BETA-	YNbO4	Monoclinic
FERGUSONITE-BETA-(Ce)	(Ce,La,Nd)NbO4	Monoclinic
FERGUSONITE-BETA-(Nd)	(Nd,Ce)NbO4	Monoclinic
FERMORITE	(Ca,Sr)5(AsO4,PO4)3(OH)	Hexagonal
FEROXYHYTE	Fe+++O(OH)	Hexagonal
FERRARISITE	Ca5H2(AsO4)4.9H2O	Triclinic

Mineral Data

Mineral Name	Formula	Crystal System
FERRAZITE	(Pb,Ba)3(PO4)2.8H2O	Triclinic
FERRIANNITE (FERRI-ANNITE)	K(Fe++,Mg)3(Fe+++,Al)Si3O10(OH)2	Monoclinic
FERRICOPIAPITE	Fe+++2/3Fe+++4(SO4)6(OH)2.20H2O	Triclinic
FERRIERITE	(Na,K)2Mg(Si,Al)18O36(OH).9H2O	Orthorhombic
FERRIHYDRITE	5Fe+++2O3.9H2O	Hexagonal
FERRIKATOPHORITE	Na2Ca(Fe++,Mg)4Fe+++(Si7Al)O22(OH)2	Monoclinic
FERRILOTHARMEYERITE	Ca(Zn,Cu++)(Fe+++,Zn)[(As+++++O3(OH)2](OH)3	Monoclinic
FERRIMOLYBDITE	Fe+++2(MoO4)3.8H2O	Orthorhombic
FERRINATRITE	Na3Fe+++(SO4)3.3H2O	Trigonal
FERRIPYROPHYLLITE	Fe+++2Si4O10(OH)2	Monoclinic
FERRISICKLERITE	Li(Fe+++,Mn++)PO4	Orthorhombic
FERRISTRUNZITE	Fe+++Fe+++2(PO4)2(OH)3.5H2O	Triclinic
FERRISURITE	(Pb,Cu)2-3(CO3)1,5-2(OH,F)0,5-1[(Fe,Al)2Si4O10(OH)2].nH2O	Monoclinic
FERRISYMPLESITE	Fe+++3(AsO4)2(OH)3.5H2O	Amorphous
FERRITUNGSTITE	(K,Ca,Na)(W,Fe+++)2(O,OH)6.H2O	Isometric
FERRIWINCHITE	NaCaMg4Fe+++Si8O22(OH)2	Monoclinic
FERROACTINOLITE	Ca2(Fe++,Mg)5Si8O22(OH)2	Monoclinic
FERROALLUAUDITE	NaCaFe++(Fe++,Mn,Fe+++,Mg)2(PO4)3	Monoclinic
FERROALUMINOTSCHERMAKITE	Ca2Fe++3Al2(Si7Al)O22(OH)2	Monoclinic
FERROANTHOPHYLLITE	(Fe++,Mg)7Si8O22(OH)2	Orthorhombic
FERROAXINITE	Ca2Fe++Al2BO3Si4O12(OH)	Triclinic
FERROBARROISITE	NaCa(Fe++,Mg)3Al2(Si7Al)O22(OH)2	Monoclinic
FERROBUSTAMITE	Ca(Fe++,Ca,Mn)Si2O6	Triclinic
FERROCARPHOLITE	(Fe++,Mg)Al2Si2O6(OH)4	Orthorhombic
FERROCLINOHOLMQUISTITE	Li2(Fe++,Mg)3Al2Si8O22(OH)2	Monoclinic
FERROCOLUMBITE	Fe++Nb2O6	Orthorhombic
FERROECKERMANNITE	Na3(Fe++,Mg)4AlSi8O22(OH)2	Monoclinic
FERROEDENITE	NaCa2(Fe++,Mg)5(Si7Al)O22(OH)2	Monoclinic
FERROFERRITSCHERMAKITE	Ca2(Fe++,Mg)3Fe+++2(Si7Al)O22(OH)2	Monoclinic
FERROGEDRITE	(Fe++,Mg)5Al2(Si6Al2)O22(OH)2	Orthorhombic
FERROGLAUCOPHANE	Na2(Fe++,Mg)3Al2Si8O22(OH)2	Monoclinic
FERROHAGENDORFITE	(Na,Ca)2Fe++(Fe++,Fe+++)2(PO4)3	Monoclinic
FERROHEXAHYDRITE	Fe++SO4.6H2O	Monoclinic
FERROHOLMQUISTITE	Li2(Fe++,Mg)3Al2Si8O22(OH)2	Orthorhombic
FERROHORNBLENDE	Ca2(Fe++,Mg)4Al(Si7Al)O22(OH,F)2	Monoclinic
FERROKAERSUTITE	NaCa2(Fe++,Mg)4Ti(Si6Al2)O22(OH)2	Monoclinic
FERROKESTERITE	Cu2(Fe,Zn)SnS4	Tetragonal
FERROLAUEITE	Fe++Fe+++2(PO4)2(OH)2.8H2O	Triclinic
FERRONICKELPLATINUM	Pt(Ni,Fe)	Tetragonal
FERROPARGASITE	NaCa2(Fe++,Mg)4Al(Si6Al2)O22(OH)2	Monoclinic
FERROPYROSMALITE	(Fe++,Mn)8Si6O15(Cl,OH)10	Hexagonal
FERRORICHTERITE	Na2Ca(Fe++,Mg)5Si8O22(OH)2	Monoclinic

Mineral Data

Mineral Name	Formula	Crystal System
FERROSELITE	FeSe2	Orthorhombic
FERROSILITE (ORTHOFERROSILITE)	(Fe++,Mg)2Si2O6	Orthorhombic
FERROSTRUNZITE	Fe++Fe+++2(PO4)2(OH)2.6H2O	Triclinic
FERROTANTALITE	Fe++Ta2O6	Orthorhombic
FERROTAPIOLITE	(Fe++,Mn++)(Ta,Nb)2O6	Tetragonal
FERROTSCHERMAKITE	Ca2(Fe++,Mg)3Al2(Si7Al)O22(OH)2	Monoclinic
FERROTYCHITE	Na6Fe++2(SO4)(CO3)4	Isometric
FERROWINCHITE	NaCa(Fe++,Mg)4AlSi8O22(OH)2	Monoclinic
FERROWODGINITE	(Fe++,Mn++)(Sn,Ti,Fe+++,Ta)(Ta,Nb)2O8	Monoclinic
FERROWYLLIEITE	(Na,Ca,Mn)(Fe++,Mn)(Fe++,Fe+++,Mg)Al(PO4)3	Monoclinic
FERRUCCITE	NaBF4	Orthorhombic
FERSILICITE	FeSi	Isometric
FERSMANITE	(Ca,Na)16(Ti,Nb)8Si4O12O18F6	Monoclinic
FERSMITE	(Ca,Ce,Na)(Nb,Ta,Ti)2(O,OH,F)6	Orthorhombic
FERUVITE	(Ca,Na)(Fe,Mg,Ti)3(Al,Mg,Fe)6(BO3)3Si6O18(OH)4	Trigonal
FERVANITE	Fe+++4(VO4)4.5H2O	Monoclinic
FETIASITE	(Fe++,Fe+++,Ti)3O2(As2O5)	Monoclinic
FIBROFERRITE	Fe+++(SO4)(OH).5H2O	Monoclinic
FICHTELITE	C19H34	Monoclinic
FIEDLERITE	Pb3Cl4(OH)2	Monoclinic
FILIPSTADITE	(Mn,Mg)2Sb+++++Fe+++O8	Orthorhombic
FILLOWITE	Na2Ca(Mn,Fe++)7(PO4)6	Monoclinic
FINGERITE	Cu11(VO4)6O2	Triclinic
FINNEMANITE	Pb5(As+++O3)3Cl	Hexagonal
FISCHESSERITE	Ag3AuSe2	Isometric
FIZELYITE	Pb14Ag5Sb21S48	Monoclinic
FLAGSTAFFITE	C10H22O3	Orthorhombic
FLEISCHERITE	Pb3Ge(SO4)2(OH)6.3H2O	Hexagonal
FLETCHERITE	Cu(Ni,Co)2S4	Isometric
FLINKITE	Mn++2Mn+++(AsO4)(OH)4	Orthorhombic
FLORENCITE-(Ce)	CeAl3(PO4)2(OH)6	Trigonal
FLORENCITE-(La)	(La,Ce)Al3(PO4)2(OH)6	Trigonal
FLORENCITE-(Nd)	(Nd,Ce)Al3(PO4)2(OH)6	Triclinic
FLORENSOVITE	Cu(Cr1,5Sb0,5)S4	Isometric
FLUCKITE	CaMnH2(AsO4)2.2H2O	Triclinic
FLUELLITE	Al2(PO4)F2(OH).7H2O	Orthorhombic
FLUOBORITE	Mg3(BO3)(F,OH)3	Hexagonal
FLUOCERITE-(Ce) (TYSONITE)	(Ce,La)F3	Hexagonal
FLUOCERITE-(La)	(La,Ce)F3	Hexagonal
FLUORAPATITE	Ca5(PO4)3F	Hexagonal
FLUORAPOPHYLLITE	(K,Na)Ca4Si8O20(F,OH).8H2O	Orthorhombic
FLUORELLESTADITE	Ca5(SiO4,PO4,SO4)3(F,OH,Cl)	Hexagonal
FLUORELLESTADITE (WILKEITE)	Ca5(SiO4,PO4,SO4)3(F,OH,Cl)	Hexagonal
FLUORFERROLEAKEITE	(Na,K)Na2Li(Fe++,Mn++,Mg)2Fe2+++Si8O22(O	Monoclinic

Mineral Data

Mineral Name	Formula	Crystal System
(FLUOR-FERRO-LEAKEITE)	H,F)2	
FLUORITE	CaF2	Isometric
FLUORRICHTERITE	(Na,K)(Ca,Na)2(Mg,Fe)5Si8O22(F,OH,O)2	Monoclinic
FOGGITE	CaAL(PO4)(OH)2.H2O	Orthorhombic
FOITITE	Na<0,5(Fe++,Al)3Al6Si6O18(BO3)3(OH)4	Trigonal
FOORDITE	Sn++(Nb,Ta)2O6	Monoclinic
FORMANITE-	YTaO4	Tetragonal
FORNACITE	(Pb,Cu)3[(Cr,As)O4]2(OH)	Monoclinic
FORSTERITE (OLIVINE)	Mg2SiO4	Orthorhombic
FOSHAGITE	Ca4Si3O9(OH)2	Triclinic
FOSHALLASITE	Ca3Si2O7.3H2O	Monoclinic
FOURMARIERITE	Pb(UO2)4O3(OH)4.4H2O	Orthorhombic
FRAIPONTITE	(Zn,Al)3(Si,Al)2O5(OH)4	Monoclinic
FRANCEVILLITE	(Ba,Pb)(UO2)2V2O8.5H2O	Orthorhombic
FRANCISCANITE	Mn++3V+++++(SiO4)(O,OH)7	Hexagonal
FRANCISITE	Cu3Bi(SeO3)2O2Cl	Orthorhombic
FRANCKEITE	Pb5Sn3Sb2S14	Triclinic
FRANCOANELLITE	H6(K,Na)3(Al,Fe+++)5(PO4)8.13H2O	Trigonal
FRANCOISITE-(Nd)	(Nd,Y,Sm,Ce)(UO2)3(PO4)2O(OH).6H2O	Monoclinic
FRANCONITE	Na2Nb4O11.9H2O	Monoclinic
FRANKDICKSONITE	BaF2	Isometric
FRANKHAWTHORNEITE	Cu2Te++++++O4(OH)2	Monoclinic
FRANKLINFURNACEITE	Ca2(Fe+++Al)Mn+++Mn++3Zn2Si2O10(OH)8	Monoclinic
FRANKLINITE	(Zn,Mn++,Fe++)(Fe+++,Mn+++)2O4	Isometric
FRANKLINPHILITE	(K,Na)<1(Mn++,Mg,Zn,Fe+++)8(Si,Al,Fe+++)12(O,OH)36.2-3H2O	Triclinic ps Hexagonal
FRANSOLETITE	H2Ca3Be2(PO4)4.4H2O	Monoclinic
FRANZINITE	(Na,Ca)7(Si,Al)12O24(SO4,CO3,OH,Cl)3.H2O	Hexagonal
FREBOLDITE	CoSe	Hexagonal
FREDRIKSSONITE	Mg2(Mn+++,Fe+++)BO5	Orthorhombic
FREEDITE	Pb8Cu+(As+++O3)2O3Cl5	Monoclinic
FREIBERGITE	(Ag,Cu,Fe)12(Sb,As)4S13	Isometric
FREIESLEBENITE	AgPbSbS3	Monoclinic
FRESNOITE	Ba2TiSi2O8	Tetragonal
FREUDENBERGITE	Na2(Ti,Fe)8O16	Monoclinic ps Hexagonal
FRIEDELITE	Mn8Si6O15(OH,Cl)10	Monoclinic psTrigonal
FRIEDRICHITE	Pb5Cu5Bi7S18	Orthorhombic
FRITZSCHEITE	Mn(UO2)2[(P,V)O4]2.10H2O	Tetragonal
FROHBERGITE	FeTe2	Orthorhombic
FROLOVITE	CaB2(OH)8	Triclinic
FRONDELITE	Mn++Fe+++4(PO4)3(OH)5	Orthorhombic
FROODITE	PdBi2	Monoclinic
FUENZALIDAITE	K6(Na,K)4Na6Mg10(SO4)12(IO3)12.12H2O	Trigonal
FUKALITE	Ca4Si2O6(CO3)(OH,F)2	Orthorhombic
FUKUCHILITE	Cu3FeS8	Isometric
FULOPPITE	Pb3Sb8S15	Monoclinic
FURONGITE	Al2(UO2)(PO4)3(OH)2.8H2O	Triclinic

Mineral Name	Formula	Crystal System
FURUTOBEITE	(Cu,Ag)6PbS4	Monoclinic
GABRIELSONITE	PbFe++(AsO4)(OH)	Orthorhombic
GADOLINITE-	Y2Fe++Be2Si2O10	Monoclinic
GADOLINITE-(Ce)	(Ce,La,Nd,Y)2Fe++Be2Si2O10	Monoclinic
GAGARINITE-	NaCaY(F,Cl)6	Hexagonal
GAGEITE-1A	(Mn,Mg,Zn)42Si16O54(OH)40	Triclinic
GAGEITE-2M	(Mn,Mg,Zn)42Si16O54(OH)40	Monoclinic
GAHNITE	ZnAl2O4	Isometric
GAIDONNAYITE	Na2ZrSi3O9.2H2O	Orthorhombic
GAINESITE-(NaCs)	CsNa(Be,Li)Zr2(PO4)4.2H2O	Tetragonal
GAINESITE-(NaNa)	Na(Na,K)(Be,Li)Zr2(PO4)4.1,5-2H2O	Tetragonal
GAITITE	Ca2Zn(AsO4)2.2H2O	Triclinic
GALAXITE	(Mn,Fe++,Mg)(Al,Fe+++)2O4	Isometric
GALEITE	Na15(SO4)5F4Cl	Trigonal
GALENA	PbS	Isometric
GALENOBISMUTITE	PbBi2S4	Orthorhombic
GALKHAITE	(Cs,Tl)(Hg,Cu,Zn)6(As,Sb)4S12	Isometric
GALLITE	CuGaS2	Tetragonal
GAMAGARITE	Ba2(Fe+++,Mn+++)(VO4)2(OH)	Monoclinic
GANANITE	BiF3	Isometric
GANOMALITE	Pb9Ca5Mn++Si9O33	Hexagonal
GANOPHYLLITE	(K,Na)2(Mn,Al,Mg)8(Si,Al)12O29(OH)7.8-9H2O	Monoclinic
GAOTAIITE	Ir3Te8	Isometric
GARAVELLITE	FeSbBiS4	Orthorhombic
GARRELSITE	Ba3NaSi2B7O16(OH)4	Monoclinic
GARRONITE	Na2Ca5Al12Si20O64.27H2O	Orthorhombic ps Tetragonal
GARTRELLITE	Pb(Cu++,Fe++)2(AsO4,SO4)2(CO3,H2O)0,7	Triclinic
GARYANSELLITE	(Mg,Fe+++)3(PO4)2(OH,O).1,5H2O	Orthorhombic
GASPARITE-(Ce)	CeAsO4	Monoclinic
GASPEITE	(Ni,Mg,Fe++)CO3	Trigonal
GATEHOUSEITE	Mn++5(PO4)2(OH)4	Orthorhombic
GATUMBAITE	CaAl2(PO4)2(OH)2.H2O	Monoclinic
GAUDEFROYITE	Ca4Mn+++3-x(BO3)3(CO3)(O,OH)3	Hexagonal
GAULTITE	Na4Zn2Si7O18.5H2O	Orthorhombic
GAYLUSSITE	Na2Ca(CO3)2.5H2O	Monoclinic
GEARKSUTITE	CaAl(OH)F4.H2O	Monoclinic
GEBHARDITE	Pb8(As+++2O5)2OCl6	Monoclinic
GEDRITE	(Mg,Fe++)5Al2(Si6Al2)O22(OH)2	Orthorhombic
GEERITE	Cu8S5	ps Cubic
GEFFROYITE	(Ag,Cu,Fe)9(Se,S)8	Isometric
GEHLENITE	Ca2Al(AlSi)O7	Tetragonal
GEIGERITE	Mn5(H2O)8(AsO3OH)2(AsO4)2.2H2O	Triclinic
GEIKIELITE	MgTiO3	Trigonal
GEMINITE	Cu(AsO3.OH).H2O	Triclinic
GENKINITE	(Pt,Pd)4Sb3	Tetragonal
GENTHELVITE	Zn4Be3(SiO4)3S	Isometric

Mineral Name	Formula	Crystal System
GEOCRONITE	$Pb_{14}(Sb,As)_6S_{23}$	Monoclinic
GEORGECHAOITE	$KNaZrSi_3O_9.2H_2O$	Orthorhombic
GEORGEITE	$Cu_2CO_3(OH)_2$	Amorphous
GEORGIADESITE	$Pb_{16}(AsO_4)_4Cl_{14}O_2(OH)_2$ or $Pb_{16}(AsO_4)_4Cl_{14}(OH)_6$	Monoclinic
GERASIMOVSKITE	$(Mn,Ca)(Nb,Ti)_5O_{12}.9H_2O$	Amorphous
GERDTREMMELITE	$(Zn,Fe^{++})(Al,Fe^{+++})_2(AsO_4)(OH)_5$	Triclinic
GERHARDTITE	$Cu_2(NO_3)(OH)_3$	Orthorhombic
GERMANITE	$Cu_{26}Fe_4Ge_4S_{32}$	Isometric
GERMANOCOLUSITE	$Cu_{13}V(Ge,As)_3S_{16}$	Isometric
GERSDORFFITE	$NiAsS$	Isometric
GERSTLEYITE	$Na_2(Sb,As)_8S_{13}.2H_2O$	Monoclinic
GERSTMANNITE	$(Mg,Mn)_2ZnSiO_4(OH)_2$	Orthorhombic
GETCHELLITE	$AsSbS_3$	Monoclinic
GEVERSITE	$Pt(Sb,Bi)_2$	Isometric
GIANELLAITE	$Hg_4(SO_4)N_2$	Isometric
GIBBSITE (HYDRARGILLITE)	$Al(OH)_3$	Monoclinic
GIESSENITE	$2(Cu_2Pb_{26}(Bi,Sb)_{20}S_{57})$	Monoclinic
GILALITE	$Cu_5Si_6O_{17}.7H_2O$[Ino.]	Monoclinic
GILLESPITE	$BaFe^{++}Si_4O_{10}$	Tetragonal
GILLULYITE	$Tl_2(As,Sb)_8S_{13}$	Monoclinic
GILMARITE	$Cu_3(AsO_4)(OH)_3$	Triclinic
GINIITE	$Fe^{++}Fe^{+++}{}_4(PO_4)_4(OH)_2.2H_2O$	Monoclinic
GINORITE	$Ca_2B_{14}O_{23}.8H_2O$	Monoclinic
GIORGIOSITE	$Mg_5(CO_3)_4(OH)_2.5H_2O$	
GIRAUDITE	$(Cu,Zn,Ag)_{12}(As,Sb)_4(Se,S)_{13}$	Isometric
GIRDITE	$Pb_3H_2(Te^{++++}O_3)(Te^{++++++}O_6)$	Monoclinic
GIRVASITE	$NaCa_2Mg_3(PO_4)_2[PO_2(OH)_2](CO_3)(OH)_2.4H_2O$	Monoclinic
GISMONDINE	$Ca_2Al_4Si_4O_{16}.9H_2O$	Monoclinic
GITTINSITE	$CaZrSi_2O_7$	Monoclinic
GIUSEPPETTITE	$(Na,K,Ca)_{7-8}(Si,Al)_{12}O_{24}(SO_4,Cl)_{1-2}$	Hexagonal
GLADITE	$PbCuBi_5S_9$	Orthorhombic
GLAUBERITE	$Na_2Ca(SO_4)_2$	Monoclinic
GLAUCOCERINITE	$(Zn,Cu)_{10}Al_6(SO_4)_3(OH)_{32}.18H_2O$	Hexagonal
GLAUCOCHROITE	$CaMnSiO_4$	Orthorhombic
GLAUCODOT	$(Co,Fe)AsS$	Orthorhombic ps Cubic
GLAUCONITE	$(K,Na)(Fe^{+++},Al,Mg)_2(Si,Al)_4O_{10}(OH)_2$	Monoclinic
GLAUCOPHANE	$Na_2(Mg,Fe^{++})_3Al_2Si_8O_{22}(OH)_2$	Monoclinic
GLAUKOSPHAERITE	$(Cu,Ni)_2(CO_3)(OH)_2$	Monoclinic
GLUCINE	$CaBe_4(PO_4)_2(OH)_4.0,5H_2O$	
GLUSHINSKITE	$Mg(C_2O_4).2H_2O$	Monoclinic
GMELINITE	$(Na_2,Ca)Al_2Si_4O_{12}.6H_2O$	Hexagonal
GOBBINSITE	$Na_4(Ca,Mg,K_2)Al_6Si_{10}O_{32}.12H_2O$	Orthorhombic ps Tetragonal
GODLEVSKITE	$(Ni,Fe)_7S_6$	Orthorhombic
GODOVIKOVITE	$(NH_4)(Al,Fe^{+++})(SO_4)_2$	Hexagonal

Mineral Name	Formula	Crystal System
GOEDKENITE	$(Sr,Ca)2Al(PO4)2(OH)$	Monoclinic
GOETHITE	$Fe^{+++}O(OH)$	Orthorhombic
GOLD (OR)	Au	Isometric
GOLDAMALGAM	$(Au,Ag)Hg$	Isometric
GOLDFIELDITE	$Cu12(Te,Sb,As)4S13$	Isometric
GOLDICHITE	$KFe^{+++}(SO4)2.4H2O$	Monoclinic
GOLDMANITE	$Ca3(V,Al,Fe^{+++})2(SiO4)3$	Isometric
GONNARDITE	$Na2CaAl4Si6O20.7H2O$	Orthorhombic
GONYERITE	$(Mn,Mg)5Fe^{+++}(Si3Fe^{+++})O10(OH)8$	Orthorhombic
GOOSECREEKITE	$CaAl2Si6O16.5H2O$	Monoclinic
GORCEIXITE	$BaAl3(PO4)(PO3OH)(OH)6$	Monoclinic psTrigonal
GORDONITE	$MgAl2(PO4)2(OH)2.8H2O$	Triclinic
GORGEYITE (GOERGEYITE)	$K2Ca5(SO4)6.H2O$	Monoclinic
GORMANITE	$Fe^{++}3Al4(PO4)4(OH)6.2H2O$	Triclinic
GORTDRUMITE	$(Cu,Fe)6Hg2S5$	Orthorhombic
GOSLARITE	$ZnSO4.7H2O$	Orthorhombic
GOTZENITE (GOETZENITE)	$(Ca,Na)3(Ti,Al)Si2O7(F,OH)2$	Triclinic
GOUDEYITE	$(Al,Y)Cu6(AsO4)3(OH)6.3H2O$	Hexagonal
GOWERITE	$CaB6O10.5H2O$	Monoclinic
GOYAZITE	$SrAl3(PO4)2(OH)5.H2O$	Trigonal
GRAEMITE	$CuTeO3.H2O$	Orthorhombic
GRAFTONITE	$(Fe^{++},Mn,Ca)3(PO4)2$	Monoclinic
GRANDIDIERITE	$(Mg,Fe^{++})Al3(BO4)(SiO4)O$	Orthorhombic
GRANDREEFITE	$Pb2SO4F2$	Monoclinic
GRANTSITE	$Na4CaxV^{++++}2xV^{+++++}12-2xO32.8H2O$	Monoclinic
GRAPHITE	C	Hexagonal/Trigonal
GRATONITE	$Pb9As4S15$	Trigonal
GRAVEGLIAITE	$Mn^{++}(SO3).3H2O$	Orthorhombic
GRAYITE	$(Th,Pb,Ca)PO4.H2O$	ps Hexagonal
GRECHISHCHEVITE	$Hg3S2(Br,Cl,I)2$	Tetragonal
GREENALITE	$(Fe^{++},Fe^{+++})2-3Si2O5(OH)4$	Monoclinic
GREENOCKITE	CdS	Hexagonal
GREGORYITE	$(Na2,K2,Ca)CO3$	Hexagonal
GREIGITE (MELNIKOVITE)	$Fe^{++}Fe^{+++}2S4$	Isometric
GRICEITE	LiF	Isometric
GRIMALDIITE	$Cr^{+++}O(OH)$	Trigonal
GRIMSELITE	$K3Na(UO2)(CO3)3.H2O$	Hexagonal
GRIPHITE	$Na4Ca6(Mn,Fe^{++},Mg)19Li2Al8(PO4)24(F,OH)8$	Isometric
GRISCHUNITE	$NaCa2Mn^{++}5Fe^{+++}(AsO4)6.2H2O$	Orthorhombic
GROSSITE	$CaAl4O7$	Monoclinic
GROSSULAR	$Ca3Al2(SiO4)3$	Isometric
GROUTITE	$Mn^{+++}O(OH)$	Orthorhombic
GRUMANTITE	$NaHSi2O5.H2O$	Orthorhombic
GRUNERITE	$(Fe^{++},Mg)7Si8O22(OH)2$	Monoclinic
GRUZDEVITE	$Cu6Hg3Sb4S12$	Trigonal
GUANAJUATITE	$Bi2Se3$	Orthorhombic

Mineral Name	Formula	Crystal System
GUANINE	C5H3(NH2)N4O	Monoclinic
GUARINOITE	(Zn,Co,Ni)6(SO4)(OH,Cl)10.5H2O	Hexagonal
GUDMUNDITE	FeSbS	Monoclinic
GUERINITE	Ca5H2(AsO4)4.9H2O	Monoclinic
GUETTARDITE	Pb(Sb,As)2S4	Monoclinic
GUGIAITE	Ca2BeSi2O7	Tetragonal
GUILDITE	CuFe+++(SO4)2(OH).4H2O	Monoclinic
GUILLEMINITE	Ba(UO2)3(SeO3)2O2.3H2O	Orthorhombic
GUNNINGITE	(Zn,Mn)SO4.H2O	Monoclinic
GUPEIITE	Fe3Si	Isometric
GUSTAVITE	PbAgBi3S6	Orthorhombic
GUTSEVICHITE	(Al,Fe+++)3(PO4,VO4)2(OH)3.8H2O	
GUYANAITE	CrO(OH)	Orthorhombic
GYPSUM (GYPSE)	CaSO4.2H2O	Monoclinic
GYROLITE	NaCa16Si23AlO60(OH)8.64H2O	Triclinic ps Hexagonal
GYSINITE-(Nd)	Pb(Nd,La)(CO3)2(OH).H2O	Orthorhombic
HAAPALAITE	4(Fe,Ni)S.3(Mg,Fe++)(OH)2	Hexagonal
HAFNON	HfSiO4	Tetragonal
HAGENDORFITE	NaCaMn(Fe++,Fe+++,Mg)2(PO4)3	Monoclinic
HAGGITE (HAEGGITE)	V2O2(OH)3	Monoclinic
HAIDINGERITE	CaHAsO4.H2O	Orthorhombic
HAINITE	Na2Ca4(Ti,Zr,Mn,Fe)2(Si2O7)F4	Triclinic
HAIWEEITE	Ca(UO2)2Si6O15.5H2O	Monoclinic
HAKITE	(Cu,Hg,Ag)12Sb4(Se,S)13	Isometric
HALITE (ROCKSALT,SELGEMME)	NaCl	Isometric
HALLIMONDITE	Pb2(UO2)(AsO4)2	Triclinic
HALLOYSITE	Al2Si2O5(OH)4	Monoclinic
HALOTRICHITE	Fe++Al2(SO4)4.22H2O	Monoclinic
HALURGITE	Mg2[B4O5(OH)4]2.H2O	Monoclinic
HAMBERGITE	Be2BO3(OH)	Orthorhombic
HAMMARITE	Pb2Cu2Bi4S9	Orthorhombic
HANCOCKITE	(Pb,Ca,Sr)2(Al,Fe+++)3(SiO4)3(OH)	Monoclinic
HANKSITE	KNa22(SO4)9(CO3)2Cl	Hexagonal
HANNAYITE	(NH4)2Mg3H4(PO4)4.8H2O	Triclinic
HANNEBACHITE	Ca2(SO3)2.H2O	Orthorhombic
HARADAITE	Sr2(VO2)Si4O12	Orthorhombic
HARDYSTONITE	Ca2ZnSi2O7	Tetragonal
HARKERITE	Ca24Mg8Al2(SiO4)8(BO3)6(CO3)10.2H2O	Isometric
HARMOTOME	(Ba,Na,K)1-2(Si,Al)8O16.6H2O	Monoclinic
HARRISONITE	Ca(Fe++,Mg)6(PO4)2(SiO4)2	Trigonal
HARSTIGITE	Ca6MnBe4(SiO4)2(Si2O7)2(OH)2	Orthorhombic
HARTITE	C20H34	Triclinic
HASHEMITE	Ba(Cr,S)O4	Orthorhombic
HASTINGSITE	NaCa2(Fe++,Mg)4Fe+++(Si6Al2)O22(OH)2	Monoclinic
HASTITE	CoSe2	Orthorhombic
HATCHITE	(Pb,Tl)2AgAs2S5	Triclinic

Mineral Data

Mineral Name	Formula	Crystal System
HATRURITE	Ca3SiO5	Trigonal
HAUCHECORNITE	Ni9Bi(Sb,Bi)S8	Tetragonal
HAUCKITE	(Mg,Mn++)24Zn18Fe+++3(SO4)4(CO3)2(OH)81	Hexagonal
HAUERITE	MnS2	Isometric
HAUSMANNITE	Mn++Mn+++2O4	Tetragonal
HAUYNE	(Na,Ca)4-8Al6Si6(O,S)24(SO4,Cl)1-2	Isometric
HAWLEYITE	CdS	Isometric
HAWTHORNEITE	Ba[Ti3Cr4Fe4Mg]O19	Hexagonal
HAXONITE	(FeNi)23C6	Isometric
HAYCOCKITE	Cu4Fe5S8	Orthorhombic
HAYNESITE	(UO2)3(SeO3)2(OH)2.5H2O	Orthorhombic
HEAZLEWOODITE	Ni3S2	Trigonal
HECTORFLORESITE	Na9(IO3)(SO4)4	Monoclinic
HECTORITE	Na0,3(Mg,Li)3Si4O10(F,OH)2	Monoclinic
HEDENBERGITE	CaFe++Si2O6	Monoclinic
HEDLEYITE	Bi7Te3	Trigonal
HEDYPHANE	Pb3Ca2(AsO4)3Cl	Hexagonal
HEIDEITE	(Fe,Cr)1+x(Ti,Fe)2S4	Monoclinic
HEIDORNITE	Na2Ca3B5O8(SO4)2Cl(OH)2	Monoclinic
HEINRICHITE	Ba(UO2)2(AsO4)2.10-12H2O	Tetragonal
HEJTMANITE	Ba(Mn,Fe++)2TiO(Si2O7)(OH,F)2	Monoclinic
HELIOPHYLLITE	Pb6As2O7Cl4	Orthorhombic
HELLANDITE-	(Ca,Y)6(Al,Fe+++)Si4B4O20(OH)4	Monoclinic
HELLYERITE	NiCO3.6H2O	Triclinic
HELMUTWINKLERITE	PbZn2(AsO4)2.2H2O	Triclinic
HELVITE	Mn4Be3(SiO4)3S	Isometric
HEMATITE (OLIGISTE)	Fe2O3	Trigonal
HEMATOLITE	(Mn,Mg,Al)15(AsO3)(AsO4)2(OH)23	Trigonal
HEMATOPHANITE	Pb4Fe+++3O8(OH,Cl)	Tetragonal
HEMIHEDRITE	Pb10Zn(CrO4)6(SiO4)2F2	Triclinic
HEMIMORPHITE	Zn4Si2O7(OH)2.H2O	Orthorhombic
HEMLOITE	(As,Sb)2(Ti,V,Fe,Al)12O23OH	Triclinic
HEMUSITE	Cu6SnMoS8	Isometric
HENDERSONITE	Ca2V++++V+++++8O24.8H2O	Orthorhombic
HENDRICKSITE	K(Zn,Mn)3Si3AlO10(OH)2	Monoclinic
HENEUITE	CaMg5(PO4)3(CO3)(OH)	Triclinic
HENMILITE	Ca2Cu[B(OH)4]2(OH)4	Triclinic
HENNOMARTINITE	SrMn+++2Si2O7(OH)2.H2O	Orthorhombic
HENRITERMIERITE	Ca3(Mn,Al)2(SiO4)2(OH)4	Tetragonal
HENRYITE	Cu4Ag3Te4	Isometric
HENTSCHELITE	Cu++Fe+++2(PO4)2(OH)2	Monoclinic
HENTSCHELITE (ANDREWSITE)	Cu++Fe+++2(PO4)2(OH)2	Monoclinic
HERCYNITE	Fe++Al2O4	Isometric
HERDERITE	CaBe(PO4)F	Monoclinic
HERSCHELITE	(Na,Ca,K)AlSi2O6.3H2O	Trigonal
HERZENBERGITE	SnS	Orthorhombic

Mineral Name	Formula	Crystal System
HESSITE	Ag2Te	Monoclinic
HETAEROLITE	ZnMn+++2O4	Tetragonal
HETEROGENITE-2H	Co+++O(OH)	Hexagonal
HETEROGENITE-3R	Co+++O(OH)	Trigonal
HETEROMORPHITE	Pb7Sb8S19	Monoclinic
HETEROSITE	Fe+++PO4	Orthorhombic
HEULANDITE	(Na,Ca)2-3Al3(Al,Si)2Si13O36.12H2O	Monoclinic
HEWETTITE	CaV6O16.9H2O	Monoclinic
HEXAHYDRITE	MgSO4.6H2O	Monoclinic
HEXAHYDROBORITE	Ca[B(OH)4]2.2H2O	Monoclinic
HEXATESTIBIOPANICKELITE	(Ni,Pd)(Te,Sb)	Hexagonal
HEYITE	Pb5Fe++2(VO4)2O4	Monoclinic
HEYROVSKYITE	Pb10AgBi5S18	Orthorhombic
HIBBINGITE	(Fe,Mg)2(OH)3Cl	Orthorhombic
HIBONITE	(Ca,Ce)(Al,Ti,Mg)12O19	Hexagonal
HIBSCHITE	Ca3Al2(SiO4)3-x(OH)4x(x=0,2-1,5)	Isometric
HIDALGOITE	PbAl3(AsO4)(SO4)(OH)6	Trigonal
HIERATITE	K2SiF6	Isometric
HILAIRITE	Na2ZrSi3O9.3H2O	Trigonal
HILGARDITE-1Tc	Ca2B5O9Cl.H2O	Triclinic
HILGARDITE-3Tc	Ca2B5O9Cl.H2O	Triclinic
HILGARDITE-4M	Ca2B5O9Cl.H2O	Monoclinic
HILLEBRANDITE	Ca6Si3O9(OH)6	Monoclinic ps Orthorhombic
HINGGANITE-	(Y,Yb,Er)BeSiO4(OH)	Monoclinic
HINGGANITE-(Ce)	(Ce,Y)2(_,Fe++)Be2Si2O8(OH,O)2	Monoclinic
HINGGANITE-(Yb)	(Yb,Y)BeSiO4(OH)	Monoclinic
HINSDALITE	(Pb,Sr)Al3(PO4)(SO4)(OH)6	Trigonal
HIORTDAHLITE	(Ca,Na)3(Zr,Ti,Y)Si2O7(O,OH,F)2	Triclinic
HISINGERITE	Fe+++2Si2O5(OH)4.2H2O	Monoclinic
HOCARTITE	Ag2FeSnS4	Tetragonal
HOCHELAGAITE	(Ca,Na,Sr)Nb4O11.8H2O	Monoclinic
HODGKINSONITE	MnZn2SiO4(OH)2	Monoclinic
HODRUSHITE	Cu8Bi12S22	Monoclinic
HOELITE (ANTHRAQUINONE)	(C6H4)2(CO)2	Orthorhombic
HOGBOMITE (HOEGBOMITE)	(Mg,Fe++)2(Al,Ti)5O10	Hexagonal/Trigonal
HOGTUVAITE (HOEGTUVAITE)	(Ca,Na)2(Fe++,Fe+++,Ti,Mg,Mn)6(Si,Be,Al)6O20	Triclinic
HOHMANNITE	Fe+++2(SO4)2(OH)2.7H2O	Triclinic
HOLDAWAYITE	Mn++6(CO3)2(OH)7(Cl,OH)	Monoclinic
HOLDENITE	(Mn,Mg)6Zn3(AsO4)2(SiO4)(OH)8	Orthorhombic
HOLLANDITE	Ba(Mn++++,Mn++)8O16	Monoclinic ps Tetragonal
HOLLINGWORTHITE	(Rh,Pt,Pd)AsS	Isometric
HOLMQUISTITE	Li2(Mg,Fe++)3Al2Si8O22(OH)2	Orthorhombic
HOLTEDAHLITE	Mg12(PO3OH,CO3)(PO4)5(OH,O)6	Hexagonal

Mineral Data

Mineral Name	Formula	Crystal System
HOLTITE	Al6(Al,Ta)(Si,Sb)3BO15(O,OH)2	Orthorhombic
HOMILITE	Ca2(Fe++,Mg)B2Si2O10	Monoclinic
HONESSITE	Ni6Fe+++2(SO4)(OH)16.4H2O	Trigonal
HONGQUIITE	TiO	Isometric
HONGSHIITE	PtCu	Trigonal
HOPEITE	Zn3(PO4)2.4H2O	Orthorhombic
HORNESITE (HOERNESITE)	Mg3(AsO4)2.8H2O	Monoclinic
HORSFORDITE	Cu5Sb	Cubic
HOTSONITE	Al5(PO4)(SO4)(OH)10.8H2O	Triclinic
HOWARDEVANSITE	NaCu++Fe+++2(VO4)3	Triclinic
HOWIEITE	Na(Fe++,Mn)10(Fe+++,Al)2Si12O31(OH)13	Triclinic
HOWLITE	Ca2B5SiO9(OH)5	Monoclinic
HSIANGHUALITE	Ca3Li2Be3(SiO4)3F2	Isometric
HUANGHOITE-(Ce) (HUANGHEITE)	BaCe(CO3)2F	Trigonal
HUANGITE	Ca0,5Al3(SO4)2(OH)6	Trigonal
HUBNERITE (HUEBNERITE)	MnWO4	Monoclinic
HUEMULITE	Na4MgV10O28.24H2O	Triclinic
HUGELITE (HUEGELITE)	Pb2(UO2)3(AsO4)2(OH)4.3H2O	Monoclinic
HULSITE	(Fe++,Mg)2(Fe+++,Sn)BO5	Monoclinic
HUMBERSTONITE	K3Na7Mg2(SO4)6(NO3)2.6H2O	Trigonal
HUMBOLDTINE	Fe++(C2O4).2H2O	Monoclinic
HUMITE	(Mg,Fe++)7(SiO4)3(F,OH)2	Orthorhombic
HUMMERITE	KMgV+++++5O14.8H2O	Triclinic
HUNCHUNITE	(Au,Ag)2Pb	Isometric
HUNGCHAOITE	MgB4O5(OH)4.7H2O	Triclinic ps Hexagonal
HUNTITE	CaMg3(CO3)4	Trigonal
HUREAULITE	Mn5(PO4)2[PO3(OH)]2.4H2O	Monoclinic
HURLBUTITE	CaBe2(PO4)2	Monoclinic
HUTCHINSONITE	(Pb,Tl)2As5S9	Orthorhombic
HUTTONITE	ThSiO4	Monoclinic
HYALOPHANE	(K,Ba)Al(Si,Al)3O8	Monoclinic
HYALOTEKITE	(Ba,Pb,Ca,K)6(B,Si,Al)2(Si,Be)10O28(F,Cl)	Triclinic ps Monoclinic
HYDROASTROPHYLLITE	(H3O,K,Ca)3(Fe++,Mn)5-6Ti2Si8(O,OH)31	Triclinic
HYDROBASALUMINITE	Al4(SO4)(OH)10.12-36H2O	Monoclinic
HYDROBIOTITE (BIOTITE-VERMICULITE)	K(Mg,Fe)6(Si,Al)8O20(OH)4.xH2O	Monoclinic
HYDROBORACITE	CaMgB6O8(OH)6.3H2O	Monoclinic
HYDROCALUMITE	Ca2Al(OH)6[Cl1-x(OH)x].3H2O	Monoclinic
HYDROCERUSSITE	Pb3(CO3)2(OH)2	Trigonal
HYDROCHLORBORITE	Ca2B4O4(OH)7Cl.7H2O	Monoclinic
HYDRODELHAYELITE	KCa2AlSi7O17(OH)2.6H2O	Orthorhombic
HYDRODRESSERITE	BaAl2(CO3)2(OH)4.3H2O	Triclinic
HYDROGLAUBERITE	Na4Ca(SO4)3.2H2O	Monoclinic
HYDROHALITE	NaCl.2H2O	Monoclinic
HYDROHETAEROLITE	Zn2Mn+++4O8.H2O	Tetragonal
HYDROHONESSITE	Ni6Fe+++2(SO4)(OH)16.7H2O	Hexagonal

Mineral Name	Formula	Crystal System
HYDROMAGNESITE	Mg5(CO3)4(OH)2.4H2O	Monoclinic
HYDROMBOBOMKULITE	(Ni,Cu)Al4[(NO3)2,(SO4)](OH)12.13-14H2O	Monoclinic
HYDROMOLYSITE	FeCl3.6H2O	
HYDRONIUMJAROSITE	(H3O)Fe+++3(SO4)2(OH)6	Trigonal
HYDROROMARCHITE	Sn3O2(OH)2	Tetragonal
HYDROSCARBROITE	Al14(CO3)3(OH)36.nH2O	Triclinic
HYDROTALCITE	Mg6Al2(CO3)(OH)16.4H2O	Trigonal
HYDROTUNGSTITE	H2WO4.H2O	Monoclinic
HYDROUGRANDITE	(Ca,Mg,Fe++)3(Fe+++,Al)2(SiO4)3-x(OH)4x	Isometric
HYDROXYAPOPHYLLITE	KCa4Si8O20(OH,F).8H2O	Tetragonal
HYDROXYCANCRINITE	Na4(AlSiO4)3(OH).H2O	Hexagonal
HYDROXYLAPATITE (HYDROXYAPATITE)	Ca5(PO4)3(OH)	Hexagonal
HYDROXYLBASTNAESITE-(Ce)	(Ce,La)(CO3)(OH,F)	Hexagonal
HYDROXYLBASTNAESITE-(La)	(La,Ce)(CO3)(OH,F)	Hexagonal
HYDROXYLBASTNAESITE-(Nd)	(Nd,La)(CO3)(OH,F)	Hexagonal
HYDROXYLELLESTADITE	Ca10(SiO4)3(SO4)3(OH,Cl,F)2	Monoclinic ps Hexagonal
HYDROXYLHERDERITE	CaBe(PO4)(OH)	Monoclinic
HYDROZINCITE	Zn5(CO3)2(OH)6	Monoclinic
HYPERSTHENE	(Fe,Mg)SiO3	Orthorhombic
HYPERCINNABAR	HgS	Hexagonal
HYTTSJOITE	Pb18Ba2Ca5Mn++2Fe++2(Si15O30)2Cl.6H2O	Trigonal
IANTHINITE	(UO2).5(UO3).10H2O	Orthorhombic
ICE (GLACE,EIS)	H2O	Hexagonal
IDAITE	Cu3FeS4	Hexagonal
IDRIALITE	C22H14	Orthorhombic
IIMORIITE-	Y2(SiO4)(CO3)	Triclinic
IKAITE	CaCO3.6H2O	Monoclinic
IKUNOLITE	Bi4(S,Se)3	Trigonal
ILESITE	(Mn,Zn,Fe++)SO4.4H2O	Monoclinic
ILIMAUSSITE-(Ce)	Ba2Na4CeFe+++Nb2Si8O28.5H2O	Hexagonal
ILLITE	(K,H3O)(Al,Mg,Fe)2(Si,Al)4O10[(OH)2,H2O]	Monoclinic
ILMAJOKITE	(Na,Ce,Ba)2TiSi3O5(OH)10.nH2O	Monoclinic
ILMENITE	Fe++TiO3	Trigonal
ILMENORUTILE	(Ti,Nb,Fe+++)3O6	Tetragonal
ILSEMANNITE	Mo3O8.nH2O	Amorphous
ILVAITE (LIEVRITE)	CaFe++2Fe+++Si2O7O(OH)	Orthorhombic/Monoclinic
IMANDRITE	Na12Ca3Fe+++2Si12O36	Orthorhombic
IMHOFITE	Tl6CuAs16S40	Monoclinic
IMITERITE	Ag2HgS2	Monoclinic
IMOGOLITE	Al2SiO3(OH)4	
INAGLYITE	PbCu3(Ir,Pt)8S16	Hexagonal
INCAITE	(Pb,Ag)4Sn4FeSb2S15	Monoclinic
INDERBORITE	CaMg[B3O3(OH)5]2.6H2O	Monoclinic

Mineral Data

Mineral Name	Formula	Crystal System
INDERITE	MgB3O3(OH)5.5H2O	Monoclinic
INDIALITE	Mg2Al4Si5O18	Hexagonal
INDIGIRITE	Mg2Al2(CO3)4(OH)2.15H2O	Monoclinic
INDITE	Fe++In2S4	Isometric
INDIUM	In	Tetragonal
INESITE	Ca2Mn7Si10O28(OH)2.5H2O	Triclinic
INGERSONITE	Ca3Mn++Sb+++++4O14	Hexagonal
INGODITE	Bi(S,Te)	Hexagonal
INNELITE	(Ba,K)4(Na,Ca)3Ti3(Si2O7)(SO4)O4	Triclinic
INSIZWAITE	Pt(Bi,Sb)2	Isometric
INYOITE	Ca2B6O6(OH)10.8H2O	Monoclinic
IODARGYRITE (IODYRITE)	AgI	Hexagonal
IOWAITE	Mg4Fe+++(OH)8OCl.2-4H2O	Hexagonal
IQUIQUEITE	K3Na4Mg(Cr++++++O4)B24O39(OH).12H2O	Hexagonal
IRANITE	Pb10Cu(CrO4)6(SiO4)2(F,OH)2	Triclinic
IRAQITE-(La)	K(La,Ce,Th)2(Ca,Na)4(Si,Al)16O40	Tetragonal
IRARSITE	(Ir,Ru,Rh,Pt)AsS	Isometric
IRHTEMITE	Ca4MgH2(AsO4)4.4H2O	Monoclinic
IRIDARSENITE	(Ir,Ru)As2	Monoclinic
IRIDISITE	(Ir,Cu,Rh,Ni,Pt)S2	ps Cubic
IRIDIUM	(Ir,Os,Ru)	Isometric
IRIGINITE	UO2Mo2O7.3H2O	Orthorhombic
IRON (FER,EISEN)	Fe	Isometric
IRTYSHITE	Na2(Ta,Nb)4O11	Hexagonal
ISHIKAWAITE	(U,Fe,Y,Ca)(Nb,Ta)O4	Orthorhombic
ISOCHALCOPYRITE	Cu16Fe17S32or(Fe,Cu)SorCu8Fe9S16	Isometric
ISOCLASITE	Ca2(PO4)(OH).2H2O	Monoclinic
ISOCUBANITE	CuFe2S3	Isometric
ISOFERROPLATINUM	(Pt,Pd)3(Fe,Cu)	Isometric
ISOKITE	CaMg(PO4)F	Monoclinic
ISOMERTIEITE	Pd11Sb2As2	Isometric
ITOITE	Pb3Ge(SO4)2O2(OH)2	Orthorhombic
IWAKIITE	Mn++(Fe+++,Mn+++)2O4	Tetragonal
IXIOLITE	(Ta,Nb,Sn,Fe,Mn)4O8	Monoclinic
IZOKLAKEITE	Pb27(Cu,Fe)2(Sb,Bi)19S57	Orthorhombic
JACOBSITE	(Mn++,Fe++,Mg)(Fe+++,Mn+++)2O4	Isometric
JADEITE	Na(Al,Fe+++)Si2O6	Monoclinic
JAFFEITE	Ca4(Si3O7)(OH)6	Hexagonal
JAGOITE	Pb3Fe+++Si4O12(Cl,OH)	Hexagonal
JAGOWERITE	BaAl2(PO4)2(OH)2	Triclinic
JAHNSITE-(CaMnFe)	CaMn++Fe++2Fe+++2(PO4)4(OH)2.8H2O	Monoclinic
JAHNSITE-(CaMnMg)	CaMn(Mg,Fe++)2Fe+++2(PO4)4(OH)2.8H2O	Monoclinic
JAHNSITE-(CaMnMn)	CaMnMn2Fe+++2(PO4)4(OH)2.8H2O	Monoclinic
JAHNSITE-(MnMnMn)	(Mn++,Ca)Mn++(Mn++,Fe++)2Fe+++2(PO4)4(OH)2.8H2O	Monoclinic
JAIPURITE	CoS	Hexagonal
JALPAITE	Ag3CuS2	Tetragonal

Mineral Data

Mineral Name	Formula	Crystal System
JAMBORITE	(Ni++,Ni+++,Fe)(OH)2(OH,S,H2O)	Hexagonal
JAMESITE	Pb2Zn2Fe+++5(AsO4)5O4	Triclinic
JAMESONITE	Pb4FeSb6S14	Monoclinic
JANGGUNITE	Mn++++5-x(Mn++,Fe+++)1+xO8(OH)6,x=0,2	Orthorhombic
JANHAUGITE	(Na,Ca)3(Mn++,Fe++)3(Ti++++,Zr,Nb)2(Si2O7)2 O2(OH,F)2	Monoclinic
JANKOVICITE	Tl5Sb9(As,Sb)4S22	Triclinic
JARLITE	Na2(Sr,Na)14Al12Mg2F64(OH,H2O)4	Monoclinic
JAROSEWICHITE	Mn++3Mn+++(AsO4)(OH)6	Orthorhombic
JAROSITE	KFe+++3(SO4)2(OH)6	Trigonal
JASKOLSKIITE	Pb2+xCux(Sb,Bi)2-xS5,x=0,2	Orthorhombic
JASMUNDITE	Ca11(SiO4)4O2S	Tetragonal
JEANBANDYITE	(Fe+++,Mn++)Sn++++(OH)6	Tetragonal ps Cubic
JEFFREYITE	(Ca,Na)2(Be,Al)Si2(O,OH)7	Orthorhombic ps Tetragonal
JENNITE	Ca9H2Si6O18(OH)8.6H2O	Triclinic
JENSENITE	Cu++3Te++++++O6.2H2O	Monoclinic
JEPPEITE	(K,Ba)2(Ti,Fe+++)6O13	Monoclinic
JEREMEJEVITE	Al6B5O15(F,OH)3	Hexagonal
JEROMITE	As(S,Se)2	Amorphous
JERRYGIBBSITE	Mn9(SiO4)4(OH)2	Orthorhombic
JERVISITE	(Na,Ca,Fe++)(Sc,Mg,Fe++)Si2O6	Monoclinic
JIANSHUIITE	(Mg,Mn++)Mn++++3O7.3H2O	Triclinic
JIMBOITE	Mn3B2O6	Orthorhombic
JIMTHOMPSONITE	(Mg,Fe++)5Si6O16(OH)2	Orthorhombic
JINSHAJIANGITE	(Ba,Ca)4(Na,K)5(Fe,Mn)15(Ti,Fe,Nb,Zr)8Si15O6 4(F,OH)6	Monoclinic
JIXIANITE	Pb(W,Fe+++)2(O,OH)7	Isometric
JOAQUINITE-(Ce)	Ba2NaCe2Fe++(Ti,Nb)2Si8O26(OH,F).H2O	Monoclinic
JOESMITHITE	PbCa2(Mg,Fe++,Fe+++)5Si6Be2O22(OH)2	Monoclinic
JOHACHIDOLITE	CaAlB3O7	Orthorhombic
JOHANNITE	Cu(UO2)2(SO4)2(OH)2.8H2O	Triclinic
JOHANNSENITE	CaMnSi2O6	Monoclinic
JOHILLERITE	Na(Mg,Zn)3Cu(AsO4)3	Monoclinic
JOHNBAUMITE	Ca5(AsO4)3(OH)	Hexagonal
JOHNINNESITE	Na2Mn++9(Mg,Mn++)7Si12O34(AsO4)2(OH)8	Triclinic
JOHNSOMERVILLEITE	Na2Ca(Mg,Fe++,Mn)7(PO4)6	Trigonal
JOHNWALKITE	K(Mn++,Fe+++,Fe++)2(Nb,Ta)(PO4)2O2(H2O,O H)2	Orthorhombic
JOKOKUITE	MnSO4.5H2O	Triclinic
JOLIOTITE	(UO2)(CO3).nH2O,(n=2?)	Orthorhombic
JOLLIFFEITE	(Ni,Co)AsSe	Isometric
JONESITE	Ba4(K,Na)2Ti4Al2Si10O36.6H2O	Orthorhombic
JORDANITE	Pb14(As,Sb)6S23	Monoclinic
JORDISITE	MoS2	Amorphous
JOSEITE-A	Bi4(S,Te)3	Trigonal
JOSEITE-B	Bi4(Te,S)3	Trigonal
JOURAVSKITE	Ca3Mn++++(SO4,CO3)2(OH)6.12H2O	Hexagonal

Mineral Data

Mineral Name	Formula	Crystal System
JUANITE	Ca10Mg4Al2Si11O39.4H2O	Orthorhombic
JULGOLDITE-(Fe)	Ca2Fe++(Fe+++,Al)2(SiO4)(Si2O7)(OH)2.H2O	Monoclinic
JULIENITE	Na2Co++(SCN)4.8H2O	Tetragonal
JUNGITE	Ca2Zn4Fe+++8(PO4)9(OH)9.16H2O	Orthorhombic
JUNITOITE	CaZn2Si2O7.H2O	Orthorhombic
JUNOITE	Pb3Cu2Bi8(S,Se)16	Monoclinic
JURBANITE	Al(SO4)(OH).5H2O	Monoclinic
KAATIALAITE	Fe+++As+++++3O9.6-8H2O	Monoclinic
KADYRELITE	Hg4(Br,Cl)2O	Isometric
KAERSUTITE	NaCa2(Mg,Fe++)4Ti(Si6Al2)O22(OH)2	Monoclinic
KAFEHYDROCYANITE	K4Fe++(CN)6.3H2O	Tetragonal
KAHLERITE	Fe++(UO2)2(AsO4)2.10-12H2O	Tetragonal
KAINITE	MgSO4.KCl.3H2O	Monoclinic
KAINOSITE-	Ca2(Y,Ce)2Si4O12(CO3).H2O	Orthorhombic
KALBORSITE	K6Al4Si6BO20(OH)4Cl	Tetragonal
KALIBORITE	KHMg2B12O16(OH)10.4H2O	Monoclinic
KALICINITE	KHCO3	Monoclinic
KALININITE	ZnCr2S4	Isometric
KALINITE	KAl(SO4)2.11H2O	Monoclinic
KALIOPHILITE	KAlSiO4	Hexagonal
KALIPYROCHLORE	(H2O,K)2(Nb,Ti)2O4(OH)2	Isometric
KALISTRONTITE	K2Sr(SO4)2	Trigonal
KALSILITE	KAlSiO4	Hexagonal
KALUGINITE	(Mn++,Ca)MgFe+++(PO4)2(OH).4H2O	Orthorhombic
KAMACITE	(Fe,Ni)	Isometric
KAMAISHILITE	Ca2Al2SiO6(OH)2	Tetragonal
KAMBALDAITE	NaNi4(CO3)3(OH)3.3H2O	Hexagonal
KAMCHATKITE	KCu++3OCl(SO4)2	Orthorhombic
KAMIOKITE	Fe2Mo3O8	Hexagonal
KAMITUGAITE	PbAl(UO2)5[(P,As)O4]2(OH)9.9.5H2O	Triclinic
KAMOTOITE-	(Y,Nd,Gd)2U++++++4(CO3)3.14,5H2O	Monoclinic
KAMPHAUGITE-	CaY(CO3)6(OH).1-1,5H2O	Tetragonal
KANEMITE	NaHSi2O4(OH)2.2H2O	Orthorhombic
KANKITE	Fe+++AsO4.3.5H2O	Monoclinic
KANOITE	(Mn++,Mg)2Si2O6	Monoclinic
KANONAITE	(Mn+++,Al)AlSiO5	Orthorhombic
KAOLINITE	Al2Si2O5(OH)4	Triclinic
KARASUGITE	SrCaAl(F,OH)7	Monoclinic
KARELIANITE	V2O3	Trigonal
KARIBIBITE	Fe+++2As+++4(O,OH)9	Orthorhombic
KARLITE	(Mg,Al)6(BO3)3(OH,Cl)4	Orthorhombic
KARNASURTITE-(Ce)	(Ce,La,Th)(Ti,Nb)(Al,Fe+++)(Si,P)2O7(OH)4.3H2O	Hexagonal
KARPATITE (CORONENE)	C24H12	Monoclinic
KARPINSKITE	(Mg,Ni)2Si2O5(OH)2	Monoclinic
KASHINITE	(Ir,Rh)2S3	Orthorhombic
KASOLITE	Pb(UO2)SiO4.H2O	Monoclinic

Mineral Name	Formula	Crystal System
KASSITE	CaTi2O4(OH)2	Orthorhombic
KATOITE	Ca3Al2(SiO4)3-x(OH)4xx=1,5-3	Isometric
KATOPTRITE	(Mn,Mg)13(Al,Fe+++)4Sb+++++2Si2O28	Monoclinic
KAWAZULITE	Bi2Te2Se	Trigonal
KAZAKHSTANITE	Fe+++5V++++3V+++++12O39(OH)9.9H2O	Monoclinic
KAZAKOVITE	Na6(Mn,H2)TiSi6O18	Trigonal
KECKITE	Ca(Mn,Zn)2Fe+++3(PO4)4(OH)3.2H2O	Monoclinic
KEGELITE	Pb8Al4Si8(SO4)2(CO3)4(OH)8O20	Monoclinic ps Hexagonal
KEITHCONNITE	Pd20Te7	Trigonal
KEIVIITE-	(Y,Yb)2Si2O7	Monoclinic
KEIVIITE-(Yb)	(Yb,Y)2Si2O7	Monoclinic
KELDYSHITE	Na2-xHxZrSi2O7.nH2O	Triclinic
KELLYITE	(Mn++,Mg,Al)3(Si,Al)2O5(OH)4	Hexagonal
KELYANITE	Hg36Sb3(Cl,Br)9O28	Monoclinic
KEMMLITZITE	(Sr,Ce)Al3(AsO4)(SO4)(OH)6	Trigonal
KEMPITE	Mn2Cl(OH)3	Orthorhombic
KENTROLITE	Pb2Mn+++2Si2O9	Orthorhombic
KENYAITE	Na2Si22O41(OH)8.6H2O	Monoclinic
KERMESITE	Sb2S2O	Triclinic ps Monoclinic
KERNITE	Na2B4O6(OH)2.3H2O	Monoclinic
KERSTENITE	PbSeO4	Orthorhombic
KESTERITE	Cu2(Zn,Fe)SnS4	Tetragonal
KETTNERITE	CaBi(CO3)OF	Tetragonal
KEYITE	Cd++2Cu++3(Zn,Cu)4(AsO4).2H2O	Monoclinic
KEYSTONEITE	(Ni,Mg,Fe,Mn)3Te++++3O9.5H2O	Hexagonal
KHADEMITE (ROSTITE)	Al(SO4)F.H2O	Orthorhombic
KHAMRABAEVITE	(Ti,V,Fe)C	Isometric
KHANNESHITE	(NaCa)3(Ba,Sr,Ce,Ca)3(CO3)5	Hexagonal
KHARAELAKHITE	(Pt,Cu,Pb,Fe,Ni)9S8	Orthorhombic
KHATYRKITE	(Cu,Zn)Al2	Tetragonal
KHIBINSKITE	K2ZrSi2O7	Monoclinic psTrigonal
KHINITE	PbCu++3Te++++++O6(OH)2	Orthorhombic
KHRISTOVITE-(Ce)	(Ca,Ce)(Ce,La,Nd,Dy)Mn++2AlSiO4Si2O7(OH,F,O)2	Monoclinic
KIDDCREEKITE	Cu6SnWS8	Isometric
KIDWELLITE	NaFe+++9(PO4)6(OH)10.5H2O	Monoclinic
KIEFTITE	CoSb3	Isometric
KIESERITE	MgSO4.H2O	Monoclinic
KILCHOANITE	Ca3Si2O7	Orthorhombic
KILLALAITE	2Ca3Si2O7.H2O	Monoclinic
KIMROBINSONITE	(Ta,Nb)(OH)5-2x(O,CO3)x	Isometric
KIMURAITE-	CaY2(CO3)4.6H2O	Orthorhombic
KIMZEYITE	Ca3(Zr,Ti)2(Si,Al,Fe+++)3O12	Isometric
KINGITE	Al3(PO4)2(OH,F)3.9H2O	Triclinic
KINGSMOUNTITE	(Ca,Mn++)4(Fe++,Mn++)Al4(PO4)6(OH)4.12H2O	Monoclinic
KINICHILITE	(Fe++,Mg,Zn)2(Te++++O3)3(NaxH2-x).3H2O	Hexagonal

Mineral Name	Formula	Crystal System
KINOITE	Ca2Cu2Si3O8(OH)4	Monoclinic
KINOSHITALITE	(Ba,K)(Mg,Mn,Al)3Si2Al2O10(OH)2	Monoclinic
KINTOREITE	PbFe+++3(PO4,AsO4,SO4)2(OH,H2O)6	Trigonal
KIPUSHITE	(Cu,Zn)5Zn(PO4)2(OH)6.H2O	Monoclinic
KIRKIITE	Pb10Bi3As3S19	Orthorhombic/Monoclinic
KIRSCHSTEINITE	CaFe++SiO4	Orthorhombic
KITAIBELITE	15Bi2S3.5Ag2S.PbS	
KITKAITE	NiTeSe	Trigonal
KITTATINNYITE	Ca4Mn++2Mn+++4Si4O16(OH)8.18H2O	Hexagonal
KIVUITE	(Th,Ca,Pb)H2(UO2)4(PO4)2(OH)8.7H2O	Orthorhombic
KLADNOITE	C6H4(CO)2NH	Monoclinic
KLEBELSBERGITE	Sb+++4O4(OH)2(SO4)	Orthorhombic
KLEEMANITE	ZnAl2(PO4)2(OH)2.3H2O	Monoclinic
KLEINITE	Hg2N(Cl,SO4).nH2O	Hexagonal
KLOCKMANNITE	CuSe	Hexagonal
KLYUCHEVSKITE	K3Cu3(Fe+++,Al)O2(SO4)4	Monoclinic
KNORRINGITE	Mg3Cr2(SiO4)3	Isometric
KOASHVITE	Na6(Ca,Mn)(Ti,Fe)Si6O18.H2O	Orthorhombic
KOBEITE-	(Y,U)(Ti,Nb)2(O,OH)6	Amorphous
KOBELLITE	Pb22Cu4(Bi,Sb)30S69	Orthorhombic
KOCHKARITE	PbBi4Te7	Trigonal
KOECHLINITE	Bi2MoO6	Orthorhombic
KOENENITE	Na4Mg9Al4Cl12(OH)22	Trigonal
KOGARKOITE	Na3(SO4)F	Monoclinic
KOKTAITE	(NH4)2Ca(SO4)2.H2O	Monoclinic
KOLARITE	PbTeCl2	Orthorhombic
KOLBECKITE	ScPO4.2H2O	Monoclinic
KOLFANITE	Ca2Fe+++3O2(AsO4)3.2H2O	Monoclinic
KOLICITE	Mn7Zn4(AsO4)2(SiO4)2(OH)8	Orthorhombic
KOLOVRATITE	Zn,Ni.VO4	
KOLWEZITE	(Cu,Co)2(CO3)(OH)2	Triclinic
KOLYMITE	Cu7Hg6	Isometric
KOMAROVITE	(Ca,Mn,Na)2Nb4Si4O12O2(OH,F)4.7H2O	Orthorhombic
KOMBATITE	Pb14(VO4)2O9Cl4	Monoclinic
KOMKOVITE	BaZrSi3O9.3H2O	Trigonal
KONDERITE (KONDORITE)	PbCu3(Rh,Pt,Ir)8S16	Hexagonal
KONINCKITE	Fe+++PO4.3H2O	Tetragonal
KONYAITE	Na2Mg(SO4)2.5H2O	Monoclinic
KORAGOITE	(Mn++,Mn+++)3(Nb,Mn++)2(Nb,Ta)3W2O20	Monoclinic
KORITNIGITE	Zn(As+++++O3)(OH).H2O	Triclinic
KORNELITE	Fe+++2(SO4)3.7H2O	Monoclinic
KORNERUPINE	Mg3-4(Al,Fe+++)5,5-6(SiO4,BO4)5(O,OH)2-3	Orthorhombic
KORNITE	(K,Na)(Na,Li)2(Mg,Mn+++,Fe+++,Li)5Si8O22(OH,F)2	Monoclinic
KORSHUNOVSKITE	Mg2Cl(OH)3.3,5-4H2O	Triclinic
KORZHINSKITE	CaB2O4.H2O	Monoclinic
KOSMOCHLOR	NaCr+++Si2O6	Monoclinic

Mineral Name	Formula	Crystal System
KOSNARITE	KZr++++2(PO4)3	Trigonal ps Cubic
KOSTOVITE	CuAuTe4	Orthorhombic
KOSTYLEVITE	K2ZrSi3O9.H2O	Monoclinic
KOTOITE	Mg3B2O6	Orthorhombic
KOTTIGITE (KOETTIGITE)	Zn3(AsO4)2.8H2O	Monoclinic
KOTULSKITE	Pd(Te,Bi)	Hexagonal
KOUTEKITE	Cu5As2	Hexagonal
KOVDORSKITE	Mg2(PO4)(OH).3H2O	Monoclinic
KOZULITE	Na3Mn4(Fe+++,Al)Si8O22(OH,F)2	Monoclinic
KRAISSLITE	(Mn++,Mg)24Zn3Fe+++(As+++O3)2(As+++++O4)3(SiO4)6(OH)18	Hexagonal
KRASNOVITE	Ba(Al,Mg)[(OH)2/(PO4,CO3)].H2O	Orthorhombic
KRATOCHVILITE	(C6H4)2CH2	Orthorhombic
KRAUSITE	KFe+++(SO4)2.H2O	Monoclinic
KRAUSKOPFITE	BaSi2O4(OH)2.2H2O	Monoclinic
KRAUTITE	MnAs+++++O3(OH).H2O	Monoclinic
KREMERSITE	(NH4,K)2Fe+++Cl5.H2O	Orthorhombic
KRENNERITE	AuTe2	Orthorhombic
KRIBERGITE	Al5(PO4)3(SO4)(OH)4.2H2O	Triclinic
KRINOVITE	NaMg2CrSi3O10	Triclinic
KROHNKITE (KROEHNKITE)	Na2Cu(SO4)2.2H2O	Monoclinic
KRUPKAITE	PbCuBi3S6	Orthorhombic
KRUTAITE	CuSe2	Isometric
KRUTOVITE	NiAs2	Isometric
KRYZHANOVSKITE	MnFe+++2(PO4)2(OH)2.H2O	Orthorhombic
KTENASITE	(Cu,Zn)5(SO4)2(OH)6.6H2O	Monoclinic
KUKISVUMITE	Na6ZnTi4Si8O28.4H2O	Orthorhombic
KUKSITE	(Pb,Ca)3Zn3Te++++++O6(PO4,VO4)2	Orthorhombic
KULANITE	Ba(Fe++,Mn,Mg)2Al2(PO4)3(OH)3	Triclinic
KULIOKITE-	(Y,Yb,Er,Dy,Lu,Gd,Tm,Ho)4Al(SiO4)2(OH)2F5	Triclinic
KULKEITE	Na0,35Mg8Al(AlSi7)O20(OH)10	Monoclinic
KULLERUDITE	NiSe2	Orthorhombic
KUPLETSKITE	(K,Na)3(Mn,Fe++)7(Ti,Nb)2Si8O24(O,OH)7	Triclinic
KURAMITE	Cu3SnS4	Tetragonal
KURANAKHITE	PbMn++++Te++++++O6	Orthorhombic
KURCHATOVITE	Ca(Mg,Mn,Fe++)B2O5	Orthorhombic
KURILITE	(Ag,Au)2(Te,Se,S)	Isometric
KURNAKOVITE	MgB3O3(OH)5.5H2O	Triclinic
KURUMSAKITE	(Zn,Ni,Cu)8Al8V2Si5O35.27H2O	Orthorhombic
KUSACHIITE	CuBi2O4	Tetragonal
KUTINAITE	Cu14Ag6As7	Isometric
KUTNOHORITE (KUTNAHORITE)	Ca(Mn,Mg,Fe++)(CO3)2	Trigonal
KUZMINITE	Hg2(Br,Cl)2	Tetragonal
KUZNETSOVITE	Hg3Cl(AsO4)	Isometric
KVANEFJELDITE	Na4(Ca,Mn)Si6O14(OH)2	Orthorhombic
KYANITE (DISTHENE)	Al2SiO5=Al[6]Al[6]OSiO4	Triclinic

Mineral Name	Formula	Crystal System
KYZYLKUMITE	$V^{+++}_2Ti_3O_9$	Monoclinic
LABRADORITE	$(Ca,Na)(Si,Al)_4O_8$	Triclinic
LABUNTSOVITE	$(K,Na)_8(Ti^{++++},Nb,Fe)_9(Si_4O_{12})_4O_8.16H_2O$	Monoclinic
LACROIXITE	$NaAl(PO_4)F$	Monoclinic
LAFFITTITE	$AgHgAsS_3$	Monoclinic
LAIHUNITE	$Fe^{++}Fe^{+++}_2(SiO_4)_2$	Monoclinic
LAITAKARITE	$Bi_4(Se,S)_3$	Trigonal
LAMMERITE	$Cu_3[(As,P)O_4]_2$	Monoclinic
LAMPROPHYLLITE	$Na_2(Sr,Ba)_2Ti_3(SiO_4)_4(OH,F)_2$	Monoclinic
LANARKITE	$Pb_2(SO_4)O$	Monoclinic
LANDAUITE	$NaMnZn_2(Ti,Fe^{+++})_6Ti_{12}O_{38}$	Monoclinic psTrigonal
LANDESITE	$(Mn,Mg)_9Fe^{+++}_3(PO_4)_8(OH)_3.9H_2O$	Orthorhombic
LANGBANITE	$(Mn,Ca,Fe,Mg)^{++}_4(Mn,Fe)_9Sb^{+++++}[O_{16}(SiO_4)_2]$	Monoclinic
LANGBEINITE	$K_2Mg_2(SO_4)_3$	Isometric
LANGISITE	$(Co,Ni)As$	Hexagonal
LANGITE	$Cu_4(SO_4)(OH)_6.2H_2O$	Monoclinic
LANNONITE	$HCa_4Mg_2Al_4(SO_4)_8F_9.3H_2O$	Tetragonal
LANSFORDITE	$MgCO_3.5H_2O$	Monoclinic
LANTHANITE-(Ce)	$(Ce,La)_2(CO_3)_3.8H_2O$	Orthorhombic
LANTHANITE-(La)	$(La,Ce)_2(CO_3)_3.8H_2O$	Orthorhombic
LANTHANITE-(Nd)	$(Nd,La)_2(CO_3)_3.8H_2O$	Orthorhombic
LAPHAMITE	$As_2(Se,S)_3$	Monoclinic
LAPIEITE	$CuNiSbS_3$	Orthorhombic
LAPLANDITE-(Ce)	$Na_4CeTiPSi_7O_{22}.5H_2O$	Orthorhombic
LARDERELLITE	$(NH_4)B_5O_6(OH)_4$	Monoclinic
LARNITE	Ca_2SiO_4	Monoclinic
LAROSITE	$(Cu,Ag)_{21}(Pb,Bi)_2S_{13}$	Orthorhombic
LARSENITE	$PbZnSiO_4$	Orthorhombic
LATIUMITE	$(Ca,K)_8(Al,Mg,Fe)(Si,Al)_{10}O_{25}(SO_4)$	Monoclinic
LATRAPPITE	$(Ca,Na)(Nb,Ti,Fe)O_3$	Orthorhombic
LAUBMANNITE	$Fe^{++}_3Fe^{+++}_6(PO_4)_4(OH)_{12}$	Orthorhombic
LAUEITE	$Mn^{++}Fe^{+++}_2(PO_4)_2(OH)_2.8H_2O$	Triclinic
LAUMONTITE	$CaAl_2Si_4O_{12}.4H_2O$	Monoclinic
LAUNAYITE	$Pb_{22}Sb_{26}S_{61}$	Monoclinic
LAURELITE	$Pb(F,Cl,OH)_2$	Hexagonal
LAURIONITE	$PbCl(OH)$	Orthorhombic
LAURITE	RuS_2	Isometric
LAUSENITE	$Fe^{+++}_2(SO_4)_3.6H_2O$	Monoclinic
LAUTARITE	$Ca(IO_3)_2$	Monoclinic
LAUTENTHALITE	$PbCu_4(SO_4)_2(OH)_6.3H_2O$	Monoclinic
LAUTITE	$CuAsS$	Orthorhombic
LAVENDULAN	$NaCaCu_5(AsO_4)_4Cl.5H_2O$	Orthorhombic
LAVENITE	$(Na,Ca)_2(Mn,Fe^{++})(Zr,Ti)Si_2O_7(O,OH,F)$	Monoclinic
LAVRENTIEVITE	$Hg_3S_2(Cl,Br)_2$	Monoclinic/Triclinic
LAWRENCITE	$(Fe^{++},Ni)Cl_2$	Trigonal
LAWSONBAUERITE	$(Mn,Mg)_9Zn_4(SO_4)_2(OH)_{22}.8H_2O$	Monoclinic

Mineral Data

Mineral Name	Formula	Crystal System
LAWSONITE	CaAl2Si2O7(OH)2.H2O	Orthorhombic
LAZARENKOITE	(Ca,Fe++)Fe+++As+++3O7.3H2O	Orthorhombic
LAZULITE	MgAl2(PO4)2(OH)2	Monoclinic
LAZURITE	(Na,Ca)7-8(Al,Si)12(O,S)24[(SO4),Cl2,(OH)2]	
LEAD (PLOMB,BLEI)	Pb	Isometric
LEADAMALGAM	HgPb2	Tetragonal
LEADHILLITE	Pb4(SO4)(CO3)2(OH)2	
LEAKEITE	NaNa2(Mg2Fe+++2Li)Si8O22(OH)2	Monoclinic
LECHATELIERITE	SiO2	
LECONTITE	(NH4,K)Na(SO4).2H2O	Orthorhombic
LEGRANDITE	Zn2(AsO4)(OH).H2O	Monoclinic
LEHNERITE	Mn[UO2/PO4]2.8H2O	Monoclinic
LEIFITE	Na2(Si,Al,Be)7(O,OH,F)14	Trigonal
LEIGHTONITE	K2Ca2Cu(SO4)4.2H2O	Triclinic ps Orthorhombic
LEITEITE	ZnAs+++2O4	Monoclinic
LEMOYNITE	(Na,K)2CaZr2Si10O26.5-6H2O	Monoclinic
LENGENBACHITE	Pb6(Ag,Cu)2As4S13	Triclinic
LENINGRADITE	PbCu++3(VO4)Cl2	Orthorhombic
LENNILENAPEITE	K6-7(Mg,Mn,Fe++,Fe+++,Zn)48(Si,Al)72(O,OH)216.16H2O	Triclinic
LENOBLITE	V2O4.2H2O	Orthorhombic
LEONITE	K2Mg(SO4)2.4H2O	Monoclinic
LEPERSONNITE-(Gd)	CaO.(Gd,Dy)2O3.24UO3.8CO2.4SiO2.60H2O	Orthorhombic
LEPIDOLITE	K(Li,Al)3(Si,Al)4O10(F,OH)2	Monoclinic
LERMONTOVITE	U++++(PO4)(OH).H2O	Orthorhombic
LESSINGITE-(Ce)	(Ce,Ca)5(SiO4)3(F,OH)	Monoclinic
LETOVICITE	(NH4)3H(SO4)2	Triclinic
LEUCITE	KAlSi2O6	Tetragonal
LEUCOPHANITE	(Na,Ca)2BeSi2(O,OH,F)7	Triclinic ps Orthorhombic
LEUCOPHOENICITE	Mn7(SiO4)3(OH)2	Monoclinic
LEUCOPHOSPHITE	KFe+++2(PO4)2(OH).2H2O	Monoclinic
LEUCOSPHENITE	BaNa4Ti2B2Si10O30	Monoclinic
LEVYCLAUDITE	Pb8Sn7Cu3(Bi,Sb)3S28	
LEVYNE (LEVYNITE)	(Ca,Na2,K2)Al2Si4O12.6H2O	Trigonal
LEWISITE	(Ca,Fe++,Na)2(Sb,Ti)2O7	Isometric
LIANDRATITE	U++++++(Nb,Ta)2O8	Hexagonal
LIBERITE	Li2BeSiO4	Monoclinic
LIBETHENITE	Cu2(PO4)(OH)	Orthorhombic
LIDDICOATITE	Ca(Li,Al)3Al6(BO3)3Si6O18(O,OH,F)4	Trigonal
LIEBAUITE	Ca3Cu5Si9O26	Monoclinic
LIEBENBERGITE	(Ni,Mg)2SiO4	Orthorhombic
LIEBIGITE	Ca2(UO2)(CO3)3.11H2O	Orthorhombic
LIKASITE	Cu3(NO3)(OH)5.2H2O	Orthorhombic
LILLIANITE	Pb3Bi2S6	Orthorhombic
LIME (CALCIUMOXIDE,CHAUX)	CaO	Isometric

Mineral Name	Formula	Crystal System
LINARITE	PbCu(SO4)(OH)2	Monoclinic
LINDACKERITE	(Cu,Co)5(AsO4)2(AsO3OH)2.10H2O	Triclinic
LINDGRENITE	Cu3(MoO4)2(OH)2	Monoclinic
LINDQVISTITE	Pb2(Mn++,Mg)2Fe+++15-16O27	Hexagonal
LINDSLEYITE	(BaSr)(Ti,Cr,Fe,Mg,Zr)21O38	Trigonal
LINDSTROMITE (LINDSTROEMITE)	Pb3Cu3Bi7S15	Orthorhombic
LINNAEITE (LINNEITE)	Co++Co+++2S4	Isometric
LINTISITE	Na3LiTi2Si4O14.2H2O	Monoclinic
LIOTTITE	(Ca,Na,K)8(Si,Al)12O24[(SO4),(CO3),Cl,OH]4.H2O	Hexagonal
LIPSCOMBITE	(Fe++,Mn)Fe+++2(PO4)2(OH)2	Tetragonal
LIROCONITE	Cu2Al(AsO4)(OH)4.4H2O	Monoclinic
LISETITE	Na2CaAl4Si4O16	Orthorhombic
LISHIZHENITE	ZnFe+++2(SO4)4.14H2O	Triclinic
LISKEARDITE	(Al,Fe+++)3(AsO4)(OH)6.5H2O	Orthorhombic
LITHARGE	PbO	Tetragonal
LITHIOMARSTURITE	LiCa2Mn2HSi5O15	Triclinic
LITHIOPHILITE	LiMnPO4	Orthorhombic
LITHIOPHORITE	Li6Al14Mn++3Mn++++18O42(OH)42	Hexagonalrthorhombic Monoclinic
LITHIOPHOSPHATE	Li3PO4	Orthorhombic
LITHIOTANTITE	Li(Ta,Nb)3O8	Monoclinic
LITHIOWODGINITE	(Li,Mn,Fe)(Ta,Nb,Sn)3O8	Monoclinic
LITHOSITE	K6Al4Si8O25.2H2O	Monoclinic ps Orthorhombic
LITIDIONITE	KNaCuSi4O10	Triclinic
LIVEINGITE	Pb9As13S28	Monoclinic
LIVINGSTONITE	HgSb4S8	Monoclinic
LIZARDITE	Mg3Si2O5(OH)4	Trigonal/H
LOKKAITE-	CaY4(CO3)7.9H2O	Orthorhombic
LOLLINGITE (LOELLINGITE)	FeAs2	Orthorhombic
LOMONOSOVITE	Na8Mn++Ti++++3Si4O12(PO4)2O4	Triclinic
LONECREEKITE	(NH4)(Fe+++,Al)(SO4)2.12H2O	Isometric
LONSDALEITE	C	Hexagonal
LOPARITE-(Ce)	(Ce,Na,Ca)2(Ti,Nb)2O6	Isometric
LOPEZITE	K2Cr2O7	Triclinic
LORANDITE	TlAsS2	Monoclinic
LORANSKITE-	(Y,Ce,Ca)ZrTaO6	Orthorhombic
LORENZENITE (RAMSAYITE)	Na2Ti2Si2O9	Orthorhombic
LOSEYITE	(Mn,Zn)7(CO3)2(OH)10	Monoclinic
LOTHARMEYERITE	CaZnMn+++(As+++++O3OH)2(OH)3	Monoclinic
LOUDOUNITE	NaCa5Zr4Si16O40(OH)11.8H2O	
LOUGHLINITE	Na2Mg3Si6O16.8H2O	Orthorhombic
LOURENSWALSITE	(K,Ba)2(Ti,Mg,Ca,Fe)4(Si,Al,Fe)6O14(OH)12	Hexagonal
LOVDARITE	K2Na6Be4Si14O36.9H2O	Orthorhombic
LOVERINGITE	(Ca,Ce)(Ti,Fe+++,Cr,Mg)21O38	Trigonal

Mineral Name	Formula	Crystal System
LOVOZERITE	Na2Ca(Zr,Ti)Si6(O,OH)18	Trigonal
LOWEITE (LOEWEITE)	Na12Mg7(SO4)13.15H2O	Trigonal
LUANHEITE	Ag3Hg	Hexagonal
LUBEROITE	Pt5Se4	Monoclinic
LUCASITE-(Ce)	(Ce,La)Ti2(O,OH)6	Monoclinic
LUDDENITE	Pb2Cu2Si5O14.14H2O	Monoclinic
LUDJIBAITE	Cu5(PO4)2(OH)4	Triclinic
LUDLAMITE	(Fe++,Mg,Mn)3(PO4)2.4H2O	Monoclinic
LUDLOCKITE	PbFe+++4(As+++5O11)2	Triclinic
LUDWIGITE	Mg2Fe+++BO5	Orthorhombic
LUESHITE	NaNbO3	Monoclinic ps Cubic
LUETHEITE	Cu2Al2(AsO4)2(OH)4.H2O	Monoclinic
LUNEBURGITE (LUENEBURGITE)	Mg3B2(PO4)2(OH)6.5H2O	Monoclinic
LUNIJIANLAITE	Li0,72Al6(Si7AlO20)(OH,O)10	Monoclinic
LUNOKITE	(Mn,Ca)(Mg,Fe++,Mn)Al(PO4)2(OH).4H2O	Orthorhombic
LUZONITE	Cu3AsS4	Tetragonal
LYONSITE	Cu++3Fe+++4(VO4)6	Orthorhombic
MACAULAYITE	(Fe+++,Al)24Si4O43(OH)2	Monoclinic
MACDONALDITE	BaCa4Si16O36(OH)2.10H2O	Orthorhombic
MACEDONITE	PbTiO3	Tetragonal
MACFALLITE	Ca2Mn+++3(SiO4)(Si2O7)(OH)3	Monoclinic
MACHATSCHKIITE	(Ca,Na)6(As+++++O4)(As+++++O3OH)3(PO4,SO4).15H2O	Trigonal
MACKAYITE	Fe+++Te2O5(OH)	Tetragonal
MACKINAWITE	(Fe,Ni)9S8	Tetragonal
MACPHERSONITE	Pb4(SO4)(CO3)2(OH)2	Orthorhombic
MACQUARTITE	Pb7Cu2(CrO4)4(SiO4)2(OH)2.H2O	Monoclinic
MADOCITE	Pb17(Sb,As)16S41	Orthorhombic
MAGADIITE	NaSi7O13(OH)3.4H2O	Monoclinic
MAGBASITE	KBa(Al,Sc)(Mg,Fe++)6Si6O20F2	
MAGHAGENDORFITE	NaMgMn(Fe++,Fe+++)2(PO4)3	Monoclinic
MAGHEMITE	Fe2.67O4	
MAGNESIOALUMINOKATOPHORITE	Na2Ca(Mg,Fe++)4Al(Si7Al)O22(OH)2	Monoclinic
MAGNESIOANTHOPHYLLITE	(Mg,Fe++)7Si8O22(OH)2	Orthorhombic
MAGNESIOARFVEDSONITE	Na3(Mg,Fe++)4Fe+++Si8O22(OH)2	Monoclinic
MAGNESIOAUBERTITE	(Mg,Cu)Al(SO4)2Cl.14H2O	Triclinic
MAGNESIOAXINITE	Ca2MgAl2BO3Si4O12(OH)	Triclinic
MAGNESIOCARPHOLITE	MgAl2Si2O6(OH)4	Orthorhombic
MAGNESIOCHLORITOID	MgAl2SiO5(OH)2	Triclinic
MAGNESIOCHROMITE	MgCr2O4	Isometric
MAGNESIOCLINOHOLMQUISTITE	Li2(Mg,Fe++)3Al2Si8O22(OH)2	Monoclinic
MAGNESIOCOPIAPITE	MgFe+++4(SO4)6(OH)2.20H2O	Triclinic
MAGNESIOCOULSONITE	Mg++V+++2O4	Isometric
MAGNESIOCUMMINGTONIT	(Mg,Fe++)7Si8O22(OH)2	Monoclinic

Mineral Data

Mineral Name	Formula	Crystal System
E		
MAGNESIODUMORTIERITE	(Mg,Ti++++)<1(Al,Mg)2Al4Si3O18-y(OH)yB(y=2-3)	Orthorhombic
MAGNESIOFERRIKATOPHORITE	Na2Ca(Mg,Fe++)4Fe+++Si7AlO22(OH)2	Monoclinic
MAGNESIOFERRITE	MgFe+++2O4	Isometric
MAGNESIOGEDRITE	(Mg,Fe++)5Al2Si6Al2O22(OH)2	Orthorhombic
MAGNESIOHASTINGSITE	NaCa2(Mg,Fe++)4Fe+++(Si6Al2)O22(OH)2	Monoclinic
MAGNESIOHOLMQUISTITE	Li2(Mg,Fe++)3Al2Si8O22(OH)2	Orthorhombic
MAGNESIOHORNBLENDE	Ca2(Mg,Fe++)4Al(Si7Al)O22(OH,F)2	Monoclinic
MAGNESIOHULSITE	(Mg,Fe++)2(Fe+++,Sn++++,Mg)BO5	Monoclinic
MAGNESIORIEBECKITE	Na2(Mg,Fe++)3Fe+++2Si8O22(OH)2	Monoclinic
MAGNESIOSADANAGAITE	(K,Na)Ca2(Mg,Fe++,Al,Fe+++,Ti)5(Si,Al)8O22(OH)2	Monoclinic
MAGNESIOTARAMITE	Na2Ca(Mg,Fe++)3Al2Si6Al2O22(OH)2	Monoclinic
MAGNESITE (GIOBERTITE)	MgCO3	Trigonal
MAGNESIUMASTROPHYLLITE	(Na,K)4Mg2(Fe++,Fe+++,Mn)5Ti2Si8O24(O,OH,F)7	Monoclinic
MAGNESIUMCHLOROPHOENICITE	(Mg,Mn)3Zn2(AsO4)(OH,O)6	Monoclinic
MAGNESIUMZIPPEITE	Mg++0,5(UO2)2(SO4)(OH)3.H2O	Monoclinic
MAGNETITE	Fe++Fe+++2O4	Isometric
MAGNETOPLUMBITE	Pb(Fe+++,Mn+++)12O19	Hexagonal
MAGNIOTRIPLITE	(Mg,Fe++,Mn)2(PO4)F	Monoclinic
MAGNOCOLUMBITE	(Mg,Fe++,Mn)(Nb,Ta)2O6	Orthorhombic
MAGNOLITE	Hg+2Te++++O3	Orthorhombic
MAGNUSSONITE	Mn++5As+++3O9(OH,Cl)=Mn10(AsO3)6(OH,Cl)2	Cubic/Tetragonal
MAHLMOODITE	FeZr(PO4)2.4H2O	Monoclinic
MAJAKITE (MAYAKITE)	PdNiAs	Hexagonal
MAJORITE	Mg3(Fe,Al,Si)2(SiO4)3	Isometric
MAKATITE	Na2Si4O8(OH)2.4H2O	Monoclinic
MAKINENITE (MAEKINENITE)	NiSe	Trigonal
MAKOVICKYITE	(Ag,Cu)1,5(Bi,Pb)5,5S9	Monoclinic
MALACHITE	Cu2(CO3)(OH)2	Monoclinic
MALANITE	Cu(Pt,Ir)2S4	Isometric
MALAYAITE	CaSnSiO5=CaSnOSiO4	Monoclinic
MALDONITE	Au2Bi	Isometric
MALLADRITE	Na2SiF6	Trigonal
MALLARDITE	Mn++SO4.7H2O	Monoclinic
MAMMOTHITE	Pb6Cu4AlSb+++++O2(SO4)2Cl4(OH)16	Monoclinic
MANAKSITE	KNaMn++Si4O10	Triclinic
MANANDONITE	Li2Al4[(Si2AlB)O10](OH)8	Orthorhombic
MANASSEITE	Mg6Al2(CO3)(OH)16.4H2O	Hexagonal
MANDARINOITE	Fe+++2Se3O9.6H2O	Monoclinic
MANGANARSITE	Mn++3As+++2O4(OH)4	Trigonal
MANGANAXINITE	Ca2Mn++Al2BO3Si4O12(OH)	Triclinic

Mineral Name	Formula	Crystal System
MANGANBABINGTONITE	Ca2(Mn,Fe++)Fe+++Si5O14(OH)	Triclinic
MANGANBELYANKINITE	(Mn,Ca)(Ti,Nb)5O12.9H2O	Amorphous
MANGANBERZELIITE	(Ca,Na)3(Mn,Mg)2(AsO4)3	Isometric
MANGANESEHORNESITE	(Mn,Mg)3(AsO4)2.8H2O	Monoclinic
MANGANESESHADLUNITE	(Mn,Pb,Cd)(Cu,Fe)8S8	Isometric
MANGANGORDONITE	(Mn++,Fe++,Mg)Al2(PO4)2(OH)2.8H2O	Triclinic
MANGANHUMITE	(Mn,Mg)7(SiO4)3(OH)2	Orthorhombic
MANGANITE	MnO(OH)	Monoclinic
MANGANNEPTUNITE	KNa2Li(Mn,Fe++)2Ti2Si8O24	Monoclinic
MANGANOCHROMITE	(Mn,Fe++)(Cr,V)2O4	Isometric
MANGANOCOLUMBITE	(Mn,Fe++)(Nb,Ta)2O6	Orthorhombic
MANGANOLANGBEINITE	K2Mn2(SO4)3	Isometric
MANGANOSEGELERITE	(Mn,Ca)(Mn,Fe++,Mg)Fe+++(PO4)2(OH).4H2O	Orthorhombic
MANGANOSITE	MnO	Isometric
MANGANOSTIBITE	(Mn++,Fe++)7(SbO4)(AsO4,SiO4)O4	Orthorhombic
MANGANOTANTALITE	MnTa2O6	Orthorhombic
MANGANOTAPIOLITE	(Mn++,Fe++)(Ta,Nb)2O6	Tetragonal
MANGANOTYCHITE	Na6(Mn++,Fe++,Mg)2(SO4)(CO3)4	Isometric
MANGANPYROSMALITE	(Mn,Fe++)8Si6O15(OH,Cl)10	Hexagonal
MANJIROITE	(Na,K)(Mn++++,Mn++)8O16.nH2O	Tetragonal
MANNARDITE	BaTi6V+++2O16.H2O	Tetragonal
MANSFIELDITE	AlAsO4.2H2O	Orthorhombic
MANTIENNEITE	KMg2Al2Ti(PO4)4(OH)3.15H2O	Orthorhombic
MAPIMITE	Zn2Fe+++3(AsO4)3(OH)4.10H2O	Monoclinic
MARCASITE	FeS2	Orthorhombic
MARGARITASITE	(Cs,K,H3O)2(UO2)2V2O8.H2O	Monoclinic
MARGARITE	CaAl2(Al2Si2)O10(OH)2	Monoclinic
MARGAROSANITE	Pb(Ca,Mn++)2Si3O9	Triclinic
MARIALITE	Na4Al3Si9O24Cl	Tetragonal
MARICITE	NaFe++PO4	Orthorhombic
MARICOPAITE	Pb7Ca2(Si,Al)48O100.32H2O	Orthorhombic
MAROKITE	CaMn+++2O4	Orthorhombic
MARRITE	PbAgAsS3	Monoclinic
MARSHITE	CuI	Isometric
MARSTURITE	NaCaMn3Si5O14(OH)	Triclinic
MARTHOZITE	Cu(UO2)3(SeO3)3(OH)2.7H2O	Orthorhombic
MASCAGNITE	(NH4)2SO4	Orthorhombic
MASLOVITE	PtBiTe	Isometric
MASSICOT	PbO	Orthorhombic
MASUTOMILITE	K(Li,Al,Mn++)3(Si,Al)4O10(F,OH)2	Monoclinic
MASUYITE	Pb3U++++++8O27.10H2O	Orthorhombic
MATHEWROGERSITE	Pb7(Fe,Cu)Al3GeSi12O36.(OH,H2O)6	Trigonal
MATHIASITE	(K,Ca,Sr)(Ti,Cr,Fe,Mg)21O38	Trigonal
MATILDITE	AgBiS2	Hexagonal
MATLOCKITE	PbFCl	Tetragonal
MATRAITE	ZnS	Trigonal
MATTAGAMITE	CoTe2	Orthorhombic

Mineral Name	Formula	Crystal System
MATTEUCCITE	NaHSO4.H2O	Monoclinic
MATTHEDDLEITE	Pb20(SiO4)7(SO4)4Cl4	Hexagonal
MATULAITE	CaAl18(PO4)12(OH)20.28H2O	Monoclinic
MAUCHERITE	Ni11As8	Tetragonal
MAUFITE	(Mg,Ni)Al4Si3O13.4H2O	
MAWBYITE	Pb(Fe+++Zn)2(AsO4)2(OH,H2O)2	Monoclinic
MAWSONITE	Cu+6Fe+++2Sn++++S8	Tetragonal
MAXWELLITE	(Na,Ca)(Fe+++,Al,Mg)(AsO4)(F,O)	Monoclinic
MAYENITE	Ca12Al14O33	Isometric
MAYINGITE	IrBiTe	Isometric
MAZZITE	K2CaMg2(Al,Si)36O72.28H2O	Hexagonal
MBOBOMKULITE	(Ni,Cu)Al4[(NO3)2,(SO4)](OH)12.3H2O	Monoclinic
MCALLISTERITE	Mg2B12O14(OH)12.9H2O	Trigonal
MCALPINEITE	Cu++3Te++++++O6.H2O	Isometric
MCAUSLANITE	HFe++3Al2(PO4)4F.18H2O	Triclinic
MCBIRNEYITE	Cu3(VO4)2	Triclinic
MCCONNELLITE	CuCrO2	Trigonal
MCCRILLISITE	NaCs(Be,Li)Zr2(PO4)4.1-2H2O	Tetragonal
MCGILLITE	(Mn,Fe++)8Si6O15(OH)8Cl2	Monoclinic psTrigonal
MCGOVERNITE	(Mn,Mg,Zn)22(AsO3)(AsO4)3(SiO4)3(OH)21	Trigonal
MCGUINNESSITE	(Mg,Cu)2(CO3)(OH)2	Monoclinic
MCKELVEYITE-	Ba3Na(Ca,U)Y(CO3)6.3H2O	Triclinic psTrigonal
MCKELVEYITE-(Nd)	Na(Ba,Sr)3Ca(Nd,Ce,La)(CO3)6.3H2O	Triclinic
MCKINSTRYITE	(Ag,Cu)2S	Orthorhombic
MCNEARITE	NaCa5H4(AsO4)5.4H2O	Triclinic
MEDAITE	(Mn,Ca)6(V+++++,As)Si5O18(OH)	Monoclinic
MEGACYCLITE	K2Na16Si18O36(OH)18.32H2O	Monoclinic
MEIONITE	Ca4Al6Si6O24CO3	Tetragonal
MEIXNERITE	Mg6Al2(OH)18.4H2O	Trigonal
MELANOCERITE-(Ce)	(Ce,Ca)5(Si,B)3O12(OH,F).nH2O	Hexagonal
MELANOPHLOGITE	SiO2	Cubic/Tetragonal
MELANOSTIBITE	Mn(Sb+++++,Fe+++)O3	Trigonal
MELANOTEKITE	Pb2Fe+++2Si2O9	Orthorhombic
MELANOTHALLITE	Cu2OCl2	Orthorhombic
MELANOVANADITE	CaV++++2V+++++2O10.5H2O	Triclinic
MELANTERITE	Fe++SO4.7H2O	Monoclinic
MELILITE	(Ca,Na)2(Al,Mg,Fe++)(Si,Al)2O7	Tetragonal
MELIPHANITE (MELINOPHANE)	(Ca,Na)2Be(Si,Al)2(O,OH,F)7	Tetragonal
MELKOVITE	CaFe+++H6(MoO4)4(PO4).6H2O	Monoclinic
MELLITE	Al2[C6(COO)6].16H2O	Tetragonal
MELONITE	NiTe2	Trigonal
MELONJOSEPHITE	CaFe++Fe+++(PO4)2(OH)	Orthorhombic
MENDIPITE	Pb3Cl2O2	Orthorhombic
MENDOZAVILITE	Na(Ca,Mg)2Fe+++6(PO4)2(P+++++Mo++++++11O39)(OH,Cl)10.33H2O	Monoclinic/Triclinic
MENDOZITE	NaAl(SO4)2.11H2O	Monoclinic

Mineral Name	Formula	Crystal System
MENEGHINITE	Pb13CuSb7S24	Orthorhombic
MENGXIANMINITE	(Ca,Na)3(Fe++,Mn++)2Mg2(Sn++++,Zn)5Al8O29	Orthorhombic
MERCALLITE	KHSO4	Orthorhombic
MERCURY (MERCURE,QUECKSILBER)	Hg	Trigonal
MEREITERITE	K2Fe++(SO4)2.4H2O	Monoclinic
MERENSKYITE	(Pd,Pt)(Te,Bi)2	Trigonal
MERLINOITE	(K,Ca,Na,Ba)7(Si25Al9)O64.23H2O	Orthorhombic
MERRIHUEITE	(K,Na)2(Fe++,Mg)5Si12O30	Hexagonal
MERTIEITE-I	Pd11(Sb,As)4	ps Hexagonal
MERTIEITE-II	Pd8(Sb,As)3	Trigonal
MERWINITE	Ca3Mg(SiO4)2	Monoclinic
MESOLITE	Na2Ca2Al6Si9O30.8H2O	Monoclinic
MESSELITE	Ca2(Fe++,Mn)(PO4)2.2H2O	Triclinic
METAALUMINITE	Al2(SO4)(OH)4.5H2O	Monoclinic
METAALUNOGEN	Al4(SO4)6.27H2O	Monoclinic
METAANKOLEITE	K2(UO2)2(PO4)2.6H2O	Tetragonal
METAAUTUNITE (METAAUTUNITE)	Ca(UO2)2(PO4)2.2-6H2O	Tetragonal
METABORITE	HBO2	Isometric
METACALCIOURANOITE	(Ca,Na,Ba)U2O7.2H2O	
METACINNABAR	HgS	Isometric
METADELRIOITE	CaSrV2O6(OH)2	Triclinic
METAHAIWEEITE	Ca(UO2)2Si6O15.nH2O	Monoclinic
METAHEINRICHITE	Ba(UO2)2(AsO4)2.8H2O	Tetragonal
METAHEWETTITE	CaV6O16.3H2O	Monoclinic
METAHOHMANNITE	Fe+++2(SO4)2(OH)2.3H2O	
METAKAHLERITE	Fe++(UO2)2(AsO4)2.8H2O	Tetragonal
METAKIRCHHEIMERITE	Co(UO2)2(AsO4)2.8H2O	Tetragonal
METAKOTTIGITE (METAKOETTIGITE)	(Zn,Fe+++,Fe++)3(AsO4)2.8(H2O,OH)	Triclinic
METALODEVITE	Zn(UO2)2(AsO4)2.10H2O	Tetragonal
METAMUNIRITE	Na2[V+++++2O6]orNaVO3	Orthorhombic
METANOVACEKITE	Mg(UO2)2(AsO4)2.4-8H2O	Tetragonal
METAROSSITE	CaV2O6.2H2O	Triclinic
METASCHODERITE	Al2(PO4)(VO4).6H2O	Monoclinic
METASCHOEPITE	UO3.nH2O(n<2)	Orthorhombic
METASIDERONATRITE	Na2Fe+++(SO4)2(OH).H2O	Orthorhombic
METASTIBNITE	Sb2S3	Amorphous
METASTUDTITE	UO4.2H2O	Orthorhombic
METASWITZERITE	Mn3(PO4)2.4H2O	Monoclinic
METATORBERNITE	Cu(UO2)2(PO4)2.8H2O	Tetragonal
METATYUYAMUNITE	Ca(UO2)2V2O8.3H2O	Orthorhombic
METAURANOCIRCITE	Ba(UO2)2(PO4)2.6-8H2O	Monoclinic
METAURANOPILITE	(UO2)6(SO4)(OH)10.5H2O	Monoclinic
METAURANOSPINITE	Ca(UO2)2(AsO4)2.8H2O	Tetragonal
METAVANDENDRIESSCHEI	PbU7O22.nH2O(n<12)	Orthorhombic

Mineral Data

Mineral Name	Formula	Crystal System
TE		
METAVANMEERSSCHEITE	U++++++(UO2)3(PO4)2(OH)6.2H2O	Orthorhombic
METAVANURALITE	Al(UO2)2V2O8(OH).8H2O	Triclinic
METAVARISCITE	AlPO4.2H2O	Monoclinic
METAVAUXITE	Fe++Al2(PO4)2(OH)2.8H2O	Monoclinic
METAVIVIANITE	Fe++3-xFe+++x(PO4)2(OH)x.(8-x)H2O	Triclinic
METAVOLTINE	K2Na6Fe++Fe+++6(SO4)12O2.18H2O	Hexagonal
METAZELLERITE	Ca(UO2)(CO3)2.3H2O	Orthorhombic
METAZEUNERITE	Cu(UO2)2(AsO4)2.8H2O	Tetragonal
MEYERHOFFERITE	Ca2B6O6(OH)10.2H2O	Triclinic
MEYMACITE	WO3.2H2O	Amorphous
MGRIITE	Cu3AsS3	Isometric
MIARGYRITE	AgSbS2	Monoclinic
MICHENERITE	(Pd,Pt)BiTe	Isometric
MICROCLINE	KAlSi3O8	Triclinic
MICROLITE	(Na,Ca)2Ta2O6(O,OH,F)	Isometric
MICROSOMMITE	(Na,Ca,K)7-8(Si,Al)12O24(Cl,SO4,CO3)2-3	Hexagonal
MIERSITE	(Ag,Cu)I	Isometric
MIHARAITE	PbCu4FeBiS6	Orthorhombic
MIKASAITE	(Fe+++,Al)2(SO4)3	Trigonal
MILARITE	K2Ca4Al2Be4Si24O60.H2O	Hexagonal
MILLERITE	NiS	Trigonal
MILLISITE	(Na,K)CaAl6(PO4)4(OH)9.3H2O	Tetragonal
MILLOSEVICHITE	(Al,Fe+++)2(SO4)3	Trigonal
MIMETITE	Pb5(AsO4)3Cl	Hexagonal
MINAMIITE	(Na,Ca,K)Al3(SO4)2(OH)6	Trigonal
MINASGERAISITE-	CaY2Be2Si2O10	Monoclinic
MINASRAGRITE	VO(SO4).5H2O	Monoclinic
MINEEVITE-	Na25Ba(Y,Gd,Dy)2(HCO3)4(CO3)11(SO4)2ClF2	Hexagonal
MINEHILLITE	(K,Na)2-3Ca28(Zn4Al4Si40)O112(OH)16	Hexagonal
MINGUZZITE	K3Fe+++(C2O4)3.3H2O	Monoclinic
MINIUM	Pb++2Pb++++O4	Tetragonal
MINNESOTAITE	(Fe++,Mg)3Si4O10(OH)2	Monoclinic
MINRECORDITE	CaZn(CO3)2	Trigonal
MINYULITE	KAl2(PO4)2(OH,F).4H2O	Orthorhombic
MIRABILITE	Na2SO4.10H2O	Monoclinic
MISENITE	K2SO4.6KHSO4	Monoclinic
MISERITE	K(Ca,Ce)6Si8O22(OH,F)2	Triclinic
MITRIDATITE	Ca2Fe+++3(PO4)3O2.3H2O	Monoclinic
MITSCHERLICHITE	K2CuCl4.2H2O	Tetragonal
MIXITE	BiCu6(AsO4)3(OH)6.3H2O	Hexagonal
MOCTEZUMITE	Pb(UO2)(TeO3)2	Monoclinic
MODDERITE	(Co,Fe)As	Orthorhombic
MOHITE	Cu2SnS3	Triclinic
MOHRITE	(NH4)2Fe++(SO4)2.6H2O	Monoclinic
MOISSANITE-15R (CARBORUNDUM)	SiC	Trigonal

Mineral Name	Formula	Crystal System
MOISSANITE-33R (CARBORUNDUM)	SiC	Trigonal
MOISSANITE-5H (CARBORUNDUM)	SiC	Hexagonal
MOISSANITE-6H (CARBORUNDUM)	SiC	Hexagonal
MOLURANITE	H4U++++(UO2)3(MoO4)7.18H2O	Amorphous
MOLYBDENITE (MOLYBDENITE-2H)	MoS2	Hexagonal
MOLYBDENITE-3R	MoS2	Trigonal
MOLYBDITE	MoO3	Orthorhombic
MOLYBDOFORNACITE	Pb2Cu[(As,P)O4][(Mo,Cr)O4](OH)	Monoclinic
MOLYBDOMENITE	PbSeO3	Monoclinic
MOLYBDOPHYLLITE	Pb2Mg2Si2O7(OH)2	Trigonal
MOLYSITE	Fe+++Cl3	Hexagonal
MONAZITE-(Ce)	(Ce,La,Nd,Th)PO4	Monoclinic
MONAZITE-(La)	(La,Ce,Nd)PO4	Monoclinic
MONAZITE-(Nd)	(Nd,Ce,La,Pr,Sm,Gd)(P,Si)O4	Monoclinic
MONCHEITE	(Pt,Pd)(Te,Bi)2	Trigonal
MONETITE	CaHPO4	Triclinic
MONGOLITE	Ca4Nb6Si5O24(OH)10.5-6H2O	Tetragonal
MONGSHANITE	(Mg,Cr,Fe++)2(Ti,Zr,Cr,FE+++)5O12	Hexagonal
MONIMOLITE	(Pb,Ca)3Sb2O8	Isometric
MONOHYDROCALCITE	CaCO3.H2O	Hexagonal
MONTANITE	Bi2Te++++++O6.2H2O	Monoclinic
MONTBRAYITE	(Au,Sb)2Te3	Triclinic
MONTEBRASITE	LiAl(PO4)(OH,F)	Triclinic
MONTEPONITE	CdO	Isometric
MONTEREGIANITE-	(Na,K)6(Y,Ca)2Si16O38.10H2O	Monoclinic
MONTESOMMAITE	(K,Na)9Al9Si23O64.10H2O	Orthorhombic ps Tetragonal
MONTGOMERYITE	Ca4MgAl4(PO4)6(OH)4.12H2O	Monoclinic
MONTICELLITE	CaMgSiO4	Orthorhombic
MONTMORILLONITE	(Na,Ca)0,3(Al,Mg)2Si4O10(OH)2.nH2O	Monoclinic
MONTROSEITE	(V+++,Fe+++)O(OH)	Orthorhombic
MONTROYALITE	Sr4Al8(CO3)3(OH,F)26.10-11H2O	Triclinic
MONTROYDITE	HgO	Orthorhombic
MOOIHOEKITE	Cu9Fe9S16	Tetragonal
MOOLOOITE	Cu++(C2O4).0,5H2O	Orthorhombic
MOOREITE	(Mg,Zn,Mn)15(SO4)2(OH)26.8H2O	Monoclinic
MOORHOUSEITE	(Co,Ni,Mn)SO4.6H2O	Monoclinic
MOPUNGITE	NaSb(OH)6	Tetragonal
MORAESITE	Be2(PO4)(OH).4H2O	Monoclinic
MORDENITE	(Ca,Na2,K2)Al2Si10O24.7H2O	Orthorhombic
MOREAUITE	Al3(UO2)(PO4)3(OH)2.13H2O	Monoclinic
MORELANDITE	(Ba,Ca,Pb)5(AsO4,PO4)3Cl	Hexagonal
MORENOSITE	NiSO4.7H2O	Orthorhombic
MORIMOTOITE	Ca3TiFe++(SiO4)3	Isometric

Mineral Data

Mineral Name	Formula	Crystal System
MORINITE	NaCa2Al2(PO4)2(F,OH)5.2H2O	Monoclinic
MOROZEVICZITE	(Pb,Fe)3Ge1-xS4	Isometric
MOSANDRITE	Na2Ca4(Ce,Y)(Ti,Zr)(Si2O7)2OF3	Monoclinic
MOSCHELITE	Hg2I2	Tetragonal
MOSCHELLANDSBERGITE	Ag2Hg3	Isometric
MOSESITE	Hg2N(Cl,SO4,MoO4,CO3).H2O	Isometric
MOTTRAMITE	PbCu(VO4)(OH)	Orthorhombic
MOTUKOREAITE	Na2Mg38Al24(CO3)13(SO4)8(OH)108.56H2O	Hexagonal
MOUNANAITE	PbFe+++2(VO4)2(OH)2	Triclinic
MOUNTAINITE	(Ca,Na2,K2)2Si4O10.3H2O	Monoclinic
MOUNTKEITHITE	(Mg,Ni)11(Fe+++,Cr,Al)3(OH)24(SO4,CO3)3.5.11H2O	Hexagonal
MOURITE	U++++Mo++++++5O12(OH)10	Monoclinic
MOYDITE-	YB(OH)4(CO3)	Orthorhombic
MOZARTITE	CaMn+++SiO4(OH)	Orthorhombic
MPOROROITE	W++++++AlO3(OH)3.2H2O	Triclinic
MRAZEKITE	Bi+++2Cu++3(PO4)2O2(OH)2.2H2O	Monoclinic
MROSEITE	CaTe++++(CO3)O2	Orthorhombic
MUCKEITE (MUECKEITE)	CuNiBiS3	Orthorhombic
MUIRITE	Ba10Ca2MnTiSi10O30(OH,Cl,F)10	Tetragonal
MUKHINITE	Ca2Al2V+++(SiO4)3(OH)	Monoclinic
MULLITE	Al6Si2O13	Orthorhombic
MUMMEITE	Ag3CuPbBi6S13orAg2Cu2Pb2Bi6S13	Monoclinic
MUNDITE	Al(UO2)3(PO4)2(OH)3,5.5H2O	Orthorhombic
MUNDRABILLAITE	(NH4)2Ca(HPO4)2.H2O	Monoclinic
MUNIRITE	NaVO3.(2-x)H2O	Monoclinic
MURATAITE-	(Na,Y)4(Zn,Fe++)3(Ti,Nb)6O18(F,OH)4	Isometric
MURDOCHITE	PbCu6O8-x(Cl,Br)2x	Isometric
MURMANITE	Na3(Ti,Nb)4(Si4O12)O4.4H2O	Triclinic
MURUNSKITE	K2Cu3FeS4	Tetragonal
MUSCOVITE	KAl2(Si3Al)O10(OH,F)2	Monoclinic ps Hexagonal
MUSGRAVITE	(Mg,Fe++,Zn)2Al6BeO12	Trigonal
MUSHISTONITE	(Cu,Zn,Fe)Sn++++(OH)6	Isometric
MUSKOXITE	Mg7Fe+++4O13.10H2O	Trigonal
MUTHMANNITE	(Ag,Au)Te	Orthorhombic
NABAPHITE	NaBaPO4.9H2O	Isometric
NABOKOITE	Cu++7Te++++O4(SO4)5.KCl	Tetragonal
NACAPHITE	Na2Ca(PO4)F	Orthorhombic
NACARENIOBSITE-(Ce)	NbNa3Ca3(Ce,La,Pr,Nd)(Si2O7)2OF3	Monoclinic
NACRITE	Al2Si2O5(OH)4	Monoclinic
NADORITE	PbSbO2Cl	Orthorhombic
NAFERTISITE	(Na,K)3(Fe++,Fe+++,Mg)9-10Ti2(Si,Fe+++,Al)12O34(O,OH)9	Monoclinic
NAGASHIMALITE	Ba4(V+++,Ti)4Si8B2O27Cl(O,OH)2	Orthorhombic
NAGELSCHMIDTITE	Ca3(PO4)2.2(a-Ca2SiO4)	Hexagonal
NAGYAGITE	AuPb(Sb,Bi)Te2-3S6	Orthorhombic
NAHCOLITE	NaHCO3	Monoclinic

Mineral Data

Mineral Name	Formula	Crystal System
NAHPOITE	Na2HPO4	Monoclinic
NAKAURIITE	Cu8(SO4)4(CO3)(OH)6.48H2O	Orthorhombic
NALIPOITE	NaLi2PO4	Orthorhombic
NAMANSILITE	NaMn+++(Si2O6)	Monoclinic
NAMBULITE	(Li,Na)(Mn,Ca)4Si5O14(OH)	Triclinic
NAMIBITE	Cu++(BiO)2VO4OH	Monoclinic
NAMUWITE	(Zn,Cu)4(SO4)(OH)6.4H2O	Hexagonal
NANLINGITE	CaMg4(AsO3)2F4	Trigonal
NANPINGITE	CsAl2(Si,Al)4O10(OH,F)2	Monoclinic
NANTOKITE	CuCl	Isometric
NARSARSUKITE	Na2(Ti,Fe+++)Si4(O,F)11	Tetragonal
NASINITE	Na2B5O8(OH).2H2O	Orthorhombic
NASLEDOVITE	PbMn3Al4(CO3)4(SO4)O5.5H2O	
NASONITE	Pb6Ca4Si6O21Cl2	Hexagonal
NASTROPHITE	Na(Sr,Ba)(PO4).9H2O	Isometric
NATALYITE	Na(V+++,Cr+3)Si2O6	Monoclinic
NATANITE	Fe++Sn++++(OH)6	Isometric
NATISITE	Na2(TiO)SiO4	Tetragonal
NATRITE	Na2CO3	Monoclinic
NATROALUNITE	NaAl3(SO4)2(OH)6	Trigonal
NATROAPOPHYLLITE	NaCa4Si8O20F.8H2O	Orthorhombic
NATROBISTANTITE	(Na,Cs)Bi(Ta,Nb,Sb)4O12	Isometric
NATROCHALCITE	NaCu2(SO4)2(OH).H2O	Monoclinic
NATRODUFRENITE	Na(Fe+++,Fe++)(Fe+++,Al)5(PO4)4(OH)6.2H2O	Monoclinic
NATROFAIRCHILDITE	Na2Ca(CO3)2	Orthorhombic
NATROJAROSITE	NaFe+++3(SO4)2(OH)6	Trigonal
NATROLITE	Na2Al2Si3O10.2H2O	Orthorhombic
NATROMONTEBRASITE	(Na,Li)Al(PO4)(OH,F)	Triclinic
NATRON (SODA)	Na2CO3.10H2O	Monoclinic
NATRONAMBULITE	(Na,Li)Mn4Si5O14(OH)	Triclinic
NATRONIOBITE	NaNbO3	Monoclinic
NATROPHILITE	NaMnPO4	Orthorhombic
NATROPHOSPHATE	Na7(PO4)2F.19H2O	Isometric
NATROSILITE	Na2Si2O5	Monoclinic
NATROTANTITE	NaTa3O8	Monoclinic
NAUJAKASITE	Na6(Fe++,Mn)Al4Si8O26	Monoclinic
NAUMANNITE	Ag2Se	Orthorhombic ps Cubic
NAVAJOITE	V2O5.3H2O	Monoclinic
NCHWANINGITE	(Mn++,Mg)2Si2O6(OH)4.2H2O	Orthorhombic
NEALITE	Pb4Fe++(As+++++O3)2Cl4.2H2O	Triclinic
NEFEDOVITE	Na5Ca4(PO4)4F	Triclinic
NEIGHBORITE	NaMgF3	Orthorhombic
NEKOITE	Ca3Si6O15.7H2O	Triclinic
NEKRASOVITE	Cu13V(Sn,As,Sb)3S16	Isometric
NELENITE (FERROSCHALLERITE)	(Mn,Fe++)16Si12As+++3O36(OH)17	Monoclinic
NELTNERITE	CaMn+++6SiO12	Tetragonal

Mineral Name	Formula	Crystal System
NENADKEVICHITE	(Na,Ca,K)(Nb,Ti)Si2O6(O,OH).2H2O	Orthorhombic
NEOTOCITE	(Mn,Fe++)SiO3.H2O	Amorphous
NEPHELINE	(Na,K)AlSiO4	Hexagonal
NEPOUITE	Ni3Si2O5(OH)4	Monoclinic
NEPTUNITE	KNa2Li(Fe++,Mn)2Ti2Si8O24	Monoclinic
NESQUEHONITE	Mg(HCO3)(OH).2H2O	Monoclinic
NEVSKITE	Bi(Se,S)	Trigonal
NEWBERYITE	MgHPO4.3H2O	Orthorhombic
NEYITE	Pb7(Cu,Ag)2Bi6S17	Monoclinic
NIAHITE	(NH4)(Mn++,Mg,Ca)PO4.H2O	Orthorhombic
NICHROMITE	(Ni,Co,Fe++)(Cr,Fe+++,Al)2O4	Isometric
NICKEL	Ni	Isometric
NICKELALUMITE	(Ni,Cu)Al4[(SO4),(NO3)2](OH)12.3H2O	Monoclinic
NICKELAUSTINITE	Ca(Ni,Zn)(AsO4)(OH)	Orthorhombic
NICKELBISCHOFITE	NiCl2.6H2O	Monoclinic
NICKELBLODITE (NICKELBLOEDITE)	Na2(Ni,Mg)(SO4)2.4H2O	Monoclinic
NICKELBOUSSINGAULTITE	(NH4)2(Ni,Mg)(SO4)2.6H2O	Monoclinic
NICKELHEXAHYDRITE	(Ni,Mg,Fe++)(SO4).6H2O	Monoclinic
NICKELINE (NICCOLITE)	NiAs	Hexagonal
NICKELSKUTTERUDITE	(Ni,Co)As3-x	Isometric
NICKELZIPPEITE	Ni++0,5(UO2)2(SO4)(OH)3.H2O	Monoclinic
NICKENICHITE	Na0,8Ca0,4(Mg,Fe+++,Al)3Cu0,4(AsO4)3	Monoclinic
NIERITE	Si3N4	Trigonal
NIFONTOVITE	Ca3B6O6(OH)12.2H2O	Monoclinic
NIGERITE	(Zn,Mg,Fe++)(Sn,Zn)2(Al,Fe+++)12O22(OH)2	Trigonal
NIGGLIITE	PtSn	Hexagonal
NIMITE	(Ni,Mg,Fe++)5Al(Si3Al)O10(OH)8	Monoclinic
NINGYOITE	(U,Ca,Ce)2(PO4)2.1-2H2O	Orthorhombic ps Hexagonal
NININGERITE	(Mg,Fe++,Mn)S	Isometric
NIOBOAESCHYNITE-(Ce)	(Ce,Ca,Th)(Nb,Ti)2(O,OH)6	Orthorhombic
NIOBOAESCHYNITE-(Nd)	(Nd,Ce)(Nb,Ti)2(O,OH)6	Orthorhombic
NIOBOPHYLLITE	(K,Na)3(Fe++,Mn)6(Nb,Ti)2Si8(O,OH,F)31	Triclinic
NIOCALITE	Ca14Nb2(Si2O7)4O6F2	Monoclinic
NISBITE	NiSb2	Orthorhombic
NISSONITE	Cu2Mg2(PO4)2(OH)2.5H2O	Monoclinic
NITER (SALPETER,NITRE)	KNO3	Orthorhombic
NITRATINE (SODANITER,NITRONATRITE)	NaNO3	Trigonal
NITROBARITE	Ba(NO3)2	Isometric
NITROCALCITE	Ca(NO3)2.4H2O	Monoclinic
NITROMAGNESITE	Mg(NO3)2.6H2O	Monoclinic
NOBLEITE	CaB6O9(OH)2.3H2O	Monoclinic
NOELBENSONITE	BaMn+++2Si2O7(OH)2.H2O	Orthorhombic
NOLANITE	(V+++,Fe++,Fe+++,Ti)10O14(OH)2	Hexagonal
NONTRONITE	Na0,3Fe+++2(Si,Al)4O10(OH)2.nH2O	Monoclinic

Mineral Name	Formula	Crystal System
NORBERGITE	Mg3(SiO4)(F,OH)2	Orthorhombic
NORDENSKIOLDINE (NORDENSKIOELDINE)	CaSnB2O6	Trigonal
NORDITE-(Ce)	(Ce,La)(Sr,Ca)Na2(Na,Mn)(Zn,Mg)Si6O17	Orthorhombic
NORDITE-(La)	(La,Ce)(Sr,Ca)Na2(Na,Mn)(Zn,Mg)Si6O17	Orthorhombic
NORDSTRANDITE	Al(OH)3	Triclinic
NORDSTROMITE (NORDSTROEMITE)	Pb3CuBi7(S10Se4)	Monoclinic
NORMANDITE	NaCa(Mn++,Fe++)(Ti,Nb,Zr)Si2O7(O,F)2	Monoclinic
NORRISHITE	K(Mn+++2Li)Si4O12	Monoclinic
NORSETHITE	BaMg(CO3)2	Trigonal
NORTHUPITE	Na3Mg(CO3)2Cl	Isometric
NOSEAN	Na8Al6Si6O24(SO4)	Isometric
NOVACEKITE	Mg(UO2)2(AsO4)2.12H2O	Tetragonal
NOVAKITE	(Cu,Ag)21As10	Monoclinic ps Tetragonal
NOWACKIITE	Cu6Zn3As4S12	Trigonal
NSUTITE	(g-MnO2)Mn++xMn++++1-xO2-2x(OH)2x(xissmall)	Hexagonal
NUFFIELDITE	Pb2Cu(Pb,Bi)Bi2S7	Orthorhombic
NUKUNDAMITE	(Cu,Fe)4S4	Hexagonal
NULLAGINITE	Ni2(CO3)(OH)2	Monoclinic
NYBOITE	NaNa2Mg3Al2(Si7Al)O22(OH)2	Monoclinic
NYEREREITE	Na2Ca(CO3)2	Orthorhombic ps Hexagonal
OBOYERITE	Pb6H6(Te++++O3)3(Te++++++O6)2.2H2O	Triclinic
OBRADOVICITE	H4(K,Na)Cu++Fe+++2(AsO4)(MoO4)5.12H2O	Orthorhombic
ODANIELITE	Na(Zn,Mg)3H2(AsO4)3	Monoclinic
ODINITE	(Fe+++,Mg,Al,Fe++,Ti,Mn)2.4(Si1,8Al0,2)O5(OH)4	Monoclinic/Trigonal
ODINTSOVITE	K2(Na,Li)4Ca3Be4Ti++++2Si6O20	Orthorhombic
OFFRETITE	(K2,Ca)5Al10Si26O72.30H2O	Hexagonal
OGDENSBURGITE	Ca2(Zn,Mn)Fe+++4(AsO4)4(OH)6.6H2O	Orthorhombic ps Hexagonal
OHMILITE	Sr3(Ti,Fe+++)(Si2O6)2(O,OH).2-3H2O	Monoclinic
OJUELAITE	ZnFe+++2(AsO4)2(OH)2.4H2O	Monoclinic
OKANOGANITE-	(Na,Ca)3(Y,Ce)12Si6B2O27F14	Trigonal
OKENITE	Ca5Si9O23.9H2O	Triclinic
OKHOTSKITE	Ca8(Mn++,Mg)4(Mn+++,Al,Fe+++)8Si12O46(OH)12	Monoclinic
OLDHAMITE	(Ca,Mn)S	Isometric
OLEKMINSKITE	Sr(Sr,Ca,Ba)(CO3)2	Trigonal
OLENITE	NaAl3Al6(BO3)3(Si6O18)(O,OH)4	Trigonal
OLGITE	Na(Sr,Ba)PO4	Hexagonal
OLIGOCLASE	(Na,Ca)(Si,Al)4O8	Triclinic
OLIVENITE	Cu2AsO4(OH)	Orthorhombic
OLKHONSKITE	(Cr,V+++)2Ti3O9	Monoclinic
OLMSTEADITE	KFe++2(Nb,Ta)(PO4)2O2.2H2O	Orthorhombic
OLSACHERITE	Pb2(SeO4)(SO4)	Orthorhombic

Mineral Name	Formula	Crystal System
OLSHANSKYITE	Ca3B4(OH)18	Monoclinic
OLYMPITE	Na5Li(PO4)2	Orthorhombic
OMEIITE	(Os,Ru)As2	Orthorhombic
OMPHACITE	(Ca,Na)(Mg,Fe++,Fe+++,Al)Si2O6	Monoclinic
ONORATOITE	Sb8O11Cl2	Monoclinic
OOSTERBOSCHITE	(Pd,Cu)7Se3	Orthorhombic
OPAL	SiO2.nH2O	Amorphous
ORCELITE	Ni5As2	Hexagonal
ORDONEZITE	ZnSb2O6	Tetragonal
OREBROITE (OEREBROITE)	Mn++3(Sb+++++,Fe+++)Si(O,OH)7	Hexagonal
OREGONITE	Ni2FeAs2	Hexagonal
ORICKITE	2CuFeS2.H2O	Hexagonal
ORIENTITE	Ca2Mn++Mn+++2Si3O10(OH)4	Orthorhombic
ORLYMANITE	Ca4Mn++3Si8O20(OH)6.2H2O	Trigonal
ORPHEITE	PbAl3(PO4,SO4)2(OH)6	Trigonal
ORPIMENT	As2S3	Monoclinic
ORSCHALLITE	Ca3(SO3)2(SO4).12H2O	Trigonal
ORTHOBRANNERITE	U++++U++++++Ti4O12(OH)2	Orthorhombic
ORTHOCHAMOSITE	(Fe++,Mg,Fe+++)5Al(Si3Al)O10(OH,O)8	Orthorhombic
ORTHOCHRYSOTILE	Mg3Si2O5(OH)4	Orthorhombic
ORTHOCLASE (ORTHOSE)	KAlSi3O8	Monoclinic
ORTHOERICSSONITE	BaMn2(Fe+++O)Si2O7(OH)	Orthorhombic
ORTHOJOAQUINITE-(Ce)	Ba2NaCe2Fe++Ti2Si8O26(O,OH).H2O	Orthorhombic
ORTHOPINAKIOLITE	(Mg,Mn++)2Mn+++BO5	Orthorhombic
ORTHOSERPIERITE	Ca(Cu,Zn)4(SO4)2(OH)6.3H2O	Orthorhombic
ORTHOWALPURGITE	Bi(UO2)(AsO4)2O4	Orthorhombic
OSARIZAWAITE	PbCuAl2(SO4)2(OH)6	Trigonal
OSARSITE	(Os,Ru)AsS	Monoclinic
OSBORNITE	TiN	Isometric
OSMIUM	(Os,Ir)	Hexagonal
OSUMILITE	(K,Na)(Fe++,Mg)2(Al,Fe+++)3(Si,Al)12O30	Hexagonal
OSUMILITE-(Mg)	(K,Na)(Mg,Fe++)2(Al,Fe+++)3(Si,Al)12O30	Hexagonal
OTAVITE	CdCO3	Trigonal
OTJISUMEITE	PbGe4O9	Triclinic ps Hexagonal
OTTEMANNITE	Sn2S3	Orthorhombic
OTTRELITE	(Mn,Fe++,Mg)2Al4Si2O10(OH)4	Monoclinic/Triclinic
OTWAYITE	Ni2(CO3)(OH)2.H2O	Orthorhombic
OULANKAITE	(Pd,Pt)5(Cu,Fe)4SnTe2S2	Tetragonal
OURAYITE	Ag25Pb30Bi41S104	Orthorhombic
OURAYITE-P	Pb14Ag18Bi28S65	Orthorhombic
OURSINITE	(Co,Mg)(UO2)2Si2O7.6H2O	Orthorhombic
OVERITE	CaMgAl(PO4)2(OH).4H2O	Orthorhombic
OWENSITE	(Ba,Pb)6(Cu,Fe,Ni)25S27	Isometric
OWYHEEITE	Pb7Ag2(Sb,Bi)8S20	Orthorhombic
OXAMMITE	(NH4)2(C2O4).H2O	Orthorhombic
OXIBERAUNITE	(Fe+++,Mn+++)Fe+++5(PO4)4O(OH)4.6H2O	Monoclinic
OYELITE	Ca10B2Si8O29.12H2O	Orthorhombic

Mineral Name	Formula	Crystal System
PAAKKONENITE	Sb2AsS2	Monoclinic
PABSTITE	Ba(Sn,Ti)Si3O9	Hexagonal
PACHNOLITE	NaCaAlF6.H2O	Monoclinic
PADERAITE	AgPb2Cu6Bi11S22	Monoclinic
PADMAITE	PdBiSe	Isometric
PAHASAPAITE	(Ca,Li,K,Na)27Li16Be48(PO4)48.76H2O	Isometric
PAINITE	CaZrBAl9O18	Hexagonal
PALARSTANIDE	Pd8(Sn,As)3	Hexagonal
PALENZONAITE	(Ca,Na)3Mn++(V+++++,As+++++,Si)3O12	Isometric
PALERMOITE	(Sr,Ca)(Li,Na)2Al4(PO4)4(OH)4	Orthorhombic
PALLADIUM	Pdor(Pd,Hg)	Isometric
PALLADOARSENIDE	Pd2As	Monoclinic
PALLADOBISMUTHARSENIDE	Pd2(As,Bi)	Orthorhombic
PALLADSEITE	Pd17Se15	Isometric
PALMIERITE	(K,Na)2Pb(SO4)2	Trigonal
PALYGORSKITE	(Mg,Al)2Si4O10(OH).4H2O	Monoclinic
PANASQUEIRAITE	CaMg(PO4)(OH,F)	Monoclinic
PANETHITE	(Na,Ca,K)2(Mg,Fe++,Mn)2(PO4)2	Monoclinic
PANUNZITE	(K,Na)AlSiO4	Hexagonal
PAOLOVITE	Pd2Sn	Orthorhombic
PAPAGOITE	CaCuAlSi2O6(OH)3	Monoclinic
PARAALUMOHYDROCALCITE	CaAl2(CO3)2(OH)4.6H2O	
PARABARIOMICROLITE	BaTa4O10(OH)2.2H20	Trigonal
PARABRANDTITE	Ca2Mn++(AsO4).2H2O	Triclinic
PARABUTLERITE	Fe+++(SO4)(OH).2H2O	Orthorhombic
PARACELSIAN	BaAl2Si2O8	Monoclinic
PARACHRYSOTILE	Mg3Si2O5(OH)4	Orthorhombic
PARACOQUIMBITE	Fe+++2(SO4)3.9H2O	Trigonal
PARACOSTIBITE	CoSbS	Orthorhombic
PARADAMITE	Zn2(AsO4)(OH)	Triclinic
PARADOCRASITE	Sb2(Sb,As)2	Monoclinic
PARAFRANSOLEITE	Ca3Be2(PO4)2(PO3OH)2.4H2O	Triclinic
PARAGONITE	NaAl2(Si3Al)O10(OH)2	Monoclinic
PARAGUANAJUATITE	Bi2(Se,S)3	Trigonal
PARAHOPEITE	Zn3(PO4)2.4H2O	Triclinic
PARAJAMESONITE	Pb4FeSb6S14	Orthorhombic
PARAKELDYSHITE	Na2ZrSi2O7	Triclinic
PARAKHINITE	PbCu++3Te++++++O6(OH)2	Trigonal ps Hexagonal
PARALAURIONITE	PbCl(OH)	Monoclinic
PARALSTONITE	BaCa(CO3)2	Trigonal
PARAMELACONITE	Cu+2Cu++2O3	Tetragonal
PARAMENDOZAVILITE	NaAl4Fe+++7(PO4)5(P+++++Mo++++++12O40)(OH)16.56H2O	Monoclinic/Triclinic
PARAMONTROSEITE	VO2	Orthorhombic
PARANATISITE	Na2(TiO)SiO4	Orthorhombic

Mineral Data

Mineral Name	Formula	Crystal System
PARANATROLITE	Na2Al2Si3O10.3H2O	Monoclinic ps Tetragonal
PARANIITE-	Ca2Y(AsO4)(WO4)2	Tetragonal
PARAOTWAYITE	Ni(OH)2-x(SO4,CO3)0,5x	Monoclinic
PARAPIERROTITE	Tl(Sb,As)5S8	Monoclinic
PARARAMMELSBERGITE	NiAs2	Orthorhombic
PARAREALGAR	AsS	Monoclinic
PARAROBERTSITE	Ca2Mn+++3(PO4)3O2.3H2O	Monoclinic
PARASCHACHNERITE	Ag3Hg2	Orthorhombic
PARASCHOEPITE	UO3.2H2O	Orthorhombic
PARASCHOLZITE	CaZn2(PO4)2.2H2O	Monoclinic
PARASPURRITE	Ca5(SiO4)2(CO3)	Monoclinic
PARASYMPLESITE	Fe++3(AsO4)2.8H2O	Monoclinic
PARATACAMITE	(Cu,Zn)2(OH)3Cl	Trigonal
PARATELLURITE	TeO2	Tetragonal
PARAUMBITE	K3Zr2HSi6O18.nH2O	Orthorhombic
PARAVAUXITE	Fe++Al2(PO4)2(OH)2.8H2O	Triclinic
PARGASITE	NaCa2(Mg,Fe++)4Al(Si6Al2)O22(OH)2	Monoclinic
PARISITE-(Ce)	Ca(Ce,La)2(CO3)3F2	Trigonal
PARISITE-(Nd)	Ca(Nd,Ce,La)2(CO3)3F2	Trigonal
PARKERITE	Ni3(Bi,Pb)2S2	Monoclinic
PARKINSONITE	(Pb,Mo)O8Cl2	Tetragonal
PARNAUITE	Cu9(AsO4)2(SO4)(OH)10.7H2O	Orthorhombic
PARSETTENSITE	(K,Na,Ca)(Mn,Al)7Si8O20(OH)8.2H2O	Monoclinic ps Hexagonal
PARSONSITE	Pb2(UO2)(PO4)2.2H2O	Triclinic
PARTHEITE	Ca2Al4Si4O15(OH)2.4H2O	Monoclinic
PARTZITE	Cu2Sb2(O,OH)7	Cubic
PARWELITE	(Mn,Mg)5Sb(As,Si)2O12	Monoclinic
PASCOITE	Ca3V10O28.17H2O	Monoclinic
PATRONITE	VS4	Monoclinic
PAULINGITE	(K2,Ca,Na2,Ba)5Al10Si35O90.45H2O	Isometric
PAULKELLERITE	Bi2Fe+++(PO4)O2(OH)2	Monoclinic
PAULKERRITE	K(Mg,Mn)2(Fe+++,Al)2Ti(PO4)4(OH)3.15H2O	Orthorhombic
PAULMOOREITE	Pb2As+++2O5	Monoclinic
PAVONITE	(Ag,Cu)(Bi,Pb)3S5	Monoclinic
PAXITE	Cu2As3	Monoclinic ps Orthorhombic
PEARCEITE	Ag16As2S11	Monoclinic
PECORAITE	Ni3Si2O5(OH)4	Monoclinic
PECTOLITE	NaCa2Si3O8(OH)	Triclinic
PECTOLITE-M2abc	NaCa2Si3O8(OH)	Monoclinic
PEHRMANITE	(Fe++,Zn,Mg)2Al6BeO12	Trigonal
PEISLEYITE	Na3Al16(SO4)2(PO4)10(OH)17.20H2O	Monoclinic
PEKOITE	PbCuBi11(S,Se)18	Orthorhombic
PELLYITE	Ba2Ca(Fe++,Mg)2Si6O17	Orthorhombic
PENFIELDITE	Pb2Cl3(OH)	Hexagonal
PENGZHIZHONGITE-24R	(Mg,Zn,Fe+++,Al)4(Sn,Fe+++)2Al10O22(OH)2	Trigonal
PENGZHIZHONGITE-6H	(Mg,Zn,Fe+++,Al)4(Sn,Fe+++)2Al10O22(OH)2	Trigonal

Mineral Name	Formula	Crystal System
PENIKISITE	Ba(Mg,Fe++)2Al2(PO4)3(OH)3	Triclinic
PENKVILKSITE	Na4(Ti++++,Zr)Si8O22.4H2O	Monoclinic
PENNANTITE	Mn5Al(Si3Al)O10(OH)8	Monoclinic
PENOBSQUISITE	Ca2(Fe++,Mg)B9O13Cl(OH)6.4H2O	Monoclinic
PENROSEITE	(Ni,Co,Cu)Se2	Isometric
PENTAGONITE	Ca(VO)Si4O10.4H2O	Orthorhombic
PENTAHYDRITE	MgSO4.5H2O	Triclinic
PENTAHYDROBORITE	CaB2O(OH)6.2H2O	Triclinic
PENTLANDITE	(Fe,Ni)9S8	Isometric
PENZHINITE	(Ag,Cu)4Au(S,Se)4	Hexagonal
PEPROSSIITE-(Ce)	(Ce,La)Al2(BO3)3	Hexagonal
PERETAITE	CaSb+++4O4(OH)2(SO4)2.2H2O	Monoclinic
PERHAMITE	Ca3Al7(SiO4)3(PO4)(OH)3.16.5H2O	Hexagonal
PERICLASE	MgO	Isometric
PERITE	PbBiO2Cl	Orthorhombic
PERLIALITE	K8Tl4Al12Si24O72.20H2O	Hexagonal
PERLOFFITE	Ba(Mn,Fe++)2Fe+++2(PO4)3(OH)3	Monoclinic
PERMINGEATITE	Cu3SbSe4	Tetragonal
PEROVSKITE	CaTiO3	Orthorhombic ps Cubic
PERRAULTITE	Na2KBaMn++8(Ti,Nb)4Si8O32(OH,F,H2O)7	Monoclinic
PERRIERITE	(Ce,Ca,La,Nd,Th)4(Fe++,Mg)2(Ti,Al,Zr,Fe+++)2 Ti2(Si2O7)2O8	Monoclinic
PERROUDITE	Hg5Ag4S5(Cl,I,Br)4	Orthorhombic
PERRYITE	(Ni,Fe)8(Si,P)3	Hexagonal
PETALITE	LiAlSi4O10	Monoclinic
PETARASITE	Na5Zr2Si6O18(Cl,OH).2H2O	Monoclinic
PETEDUNNITE	Ca(Zn,Mn++,Fe++,Mg)Si2O6	Monoclinic
PETERBAYLISSITE	Hg+3(CO3)(OH).2H2O	Orthorhombic
PETERSENITE-(Ce)	(Na,Ca)4(Ce,La,Nd,Sr,Pr,Sm,Ba)2(CO3)5	Monoclinic
PETERSITE-	(Y,Ce,Nd,Ca)Cu6(PO4)3(OH)6.3H2O	Hexagonal
PETITJEANITE	(Bi,Pb)3(PO4,VO4,AsO4)2(O,OH)2	Triclinic
PETROVICITE	PbHgCu3BiSe5	Orthorhombic
PETROVSKAITE	AuAg(S,Se)	Monoclinic
PETRUKITE	(Cu,Fe,Zn)2(Sn,In)S4	Orthorhombic
PETSCHECKITE	U++++Fe++(Nb,Ta)2O8	Hexagonal
PETZITE	Ag3AuTe2	Isometric
PHARMACOLITE	CaHAsO4.2H2O	Monoclinic
PHARMACOSIDERITE	KFe+++4(AsO4)3(OH)4.6-7H2O	Cubic/Tetragonal
PHAUNOUXITE	Ca3(AsO4).11H2O	Triclinic
PHENAKITE	Be2SiO4	Trigonal
PHILIPSBORNITE	PbAl3(AsO4)2(OH)5.H2O	Trigonal
PHILIPSBURGITE	(Cu,Zn)6(AsO4,PO4)2(OH)6.H2O	Monoclinic
PHILLIPSITE	(K,Na,Ca)1-2(Si,Al)8O16.6H2O	Monoclinic
PHLOGOPITE	KMg3(Si3Al)O10(F,OH)2	Monoclinic
PHOENICOCHROITE	Pb2(CrO4)O	Monoclinic
PHOSGENITE	Pb2(CO3)Cl2	Tetragonal
PHOSINAITE	Na3(Ca,Ce)PSiO7	Orthorhombic

Mineral Name	Formula	Crystal System
PHOSPHAMMITE	(NH4)2HPO4	Monoclinic
PHOSPHOFERRITE	(Fe++,Mn)3(PO4)2.3H2O	Orthorhombic
PHOSPHOFIBRITE	KCuFe+++15(PO4)12(OH)12.12H2O	Orthorhombic
PHOSPHOPHYLLITE	Zn2(Fe++,Mn)(PO4)2.4H2O	Monoclinic
PHOSPHORROSSLERITE (PHOSPHORROESSLERITE)	MgHPO4.7H2O	Monoclinic
PHOSPHOSIDERITE	Fe+++PO4.2H2O	Monoclinic
PHOSPHURANYLITE	KCa(H3O)3(UO2)7(PO4)4O4.8H2O	Orthorhombic
PHURALUMITE	Al2(UO2)3(PO4)2(OH)6.10H2O	Monoclinic
PHURCALITE	Ca2(UO2)3O2(PO4)2.7H2O	Orthorhombic
PHYLLOTUNGSTITE	CaFe+++3H(WO4)6.10H2O	Orthorhombic
PIANLINITE	Al2Si2O6(OH)2	Amorphous
PICKERINGITE	MgAl2(SO4)4.22H2O	Monoclinic
PICOTPAULITE	TlFe2S3	Orthorhombic
PICROMERITE	K2Mg(SO4)2.6H2O	Monoclinic
PICROPHARMACOLITE	H2Ca4Mg(AsO4)4.11H2O	Triclinic
PIEMONTITE	Ca2(Al,Mn,Fe)3(SiO4)3(OH)=Ca2(Mn,Fe)Al2(SiO4)(Si2O7)O(OH)	Monoclinic
PIERROTITE	Tl2Sb6As4S16	Orthorhombic
PIGEONITE	(Mg,Fe++,Ca)(Mg,Fe++)Si2O6	Monoclinic
PILSENITE	Bi4Te3	Trigonal
PIMELITE	(Ni,Mg)3Si4O10(OH)2.H2O	Monoclinic
PINAKIOLITE	(Mg,Mn++)2(Mn+++,Sb+++)BO5	Monoclinic
PINALITE	Pb3WO5Cl2	Orthorhombic
PINCHITE	Hg++5O4Cl2	Orthorhombic
PINGGUITE	Bi+++6Te++++2O13	Orthorhombic
PINNOITE	MgB2O4.3H2O	Tetragonal
PINTADOITE	Ca2V2O7.9H2O	
PIRQUITASITE	Ag2ZnSnS4	Tetragonal
PIRSSONITE	Na2Ca(CO3)2.2H2O	Orthorhombic
PITIGLIANOITE	Na6K2Si6Al6O24(SO4).2H2O	Hexagonal
PITTICITE	Fe,AsO4,SO4,H2O	Amorphous
PIYPITE	K2Cu2(SO4)2O	Tetragonal
PLAGIONITE	Pb5Sb8S17	Monoclinic
PLANCHEITE	Cu8Si8O22(OH)4.H2O	Orthorhombic
PLANERITE	Al6(PO4)2(PO3OH)2(OH)8.4H2O	Triclinic
PLATARSITE	(Pt,Rh,Ru)AsS	Isometric
PLATINUM	Pt	Isometric
PLATTNERITE	PbO2	Tetragonal
PLATYNITE	(Bi,Pb)3(Se,S)4	Trigonal
PLAYFAIRITE	Pb16Sb18S43	Monoclinic
PLOMBIERITE	Ca5H2Si6O18.6H2O	Orthorhombic
PLUMALSITE	Pb4Al2(SiO3)7	Orthorhombic
PLUMBOBETAFITE	(Pb,U,Ca)(Ti,Nb)2O6(OH,F)	Isometric
PLUMBOFERRITE	Pb2(Fe,Mn++,Mg)11O19	Hexagonal
PLUMBOGUMMITE	PbAl3(PO4)2(OH)5.H2O	Trigonal
PLUMBOJAROSITE	PbFe+++6(SO4)4(OH)12	Trigonal

Mineral Name	Formula	Crystal System
PLUMBOMICROLITE	(Pb,Ca,U)2Ta2O6(OH)	Isometric
PLUMBONACRITE	Pb10(CO3)6O(OH)6	Hexagonal
PLUMBOPALLADINITE	Pd3Pb2	Hexagonal
PLUMBOPYROCHLORE	(Pb,Y,U,Ca)2-xNb2O6(OH)	Isometric
PLUMBOTELLURITE	PbTe++++O3	Orthorhombic
PLUMBOTSUMITE	Pb5Si4O8(OH)10	Orthorhombic
PLUMOSITE	Pb2Sb2S5	Monoclinic
POITEVINITE	(Cu,Fe++,Zn)SO4.H2O	Monoclinic
POKROVSKITE	Mg2(CO3)(OH)2.0,5H2O	Monoclinic
POLARITE	Pd2PbBi	Orthorhombic
POLDERVAARTITE	(Ca,Mn++)2SiO3(OH)2	Orthorhombic
POLHEMUSITE	(Zn,Hg)S	Tetragonal ps Cubic
POLKOVICITE	(Fe,Pb)3(Ge,Fe)1-xS4	Isometric
POLLUCITE	(Cs,Na)2Al2Si4O12.H2O	Isometric
POLYBASITE	(Ag,Cu)16Sb2S11	Monoclinic ps Hexagonal
POLYCRASE-	(Y,Ca,Ce,U,Th)(Ti,Nb,Ta)2O6	Orthorhombic
POLYDYMITE	NiNi2S4	Isometric
POLYHALITE	K2Ca2Mg(SO4)4.2H2O	Triclinic
POLYLITHIONITE	KLi2AlSi4O10(F,OH)2	Monoclinic
POLYPHITE	Na17Ca3(Mg,Mn)(Ti,Mn,Zr,Nb)4Si4O12(PO4)6O4F6	Triclinic
PONOMAREVITE	K4Cu++4OCl10	Monoclinic
PORTLANDITE	Ca(OH)2	Hexagonal
POSNJAKITE	Cu4(SO4)(OH)6.H2O	Monoclinic
POTARITE	PdHg	Tetragonal
POTASSIUMALUM	KAl(SO4)2.12H2O	Isometric
POTASSIUMFLUORRICHTERITE	(K,Na)(Ca,Na)2(Mg,Fe)5Si8O22(F,OH)2	Monoclinic
POTASSIUMRICHTERITE	(K,Na)(Ca,Na)2(Mg,Fe)5Si8O22(OH,F)2	Monoclinic
POTOSIITE	Pb48Sn++++18Fe++7Sb+++16S115	Triclinic
POTTSITE	PbBiH(VO4)2.2H2O	Tetragonal
POUBAITE	PbBi2Se2(Te,S)2	Trigonal
POUDRETTEITE	KNa2B3Si12O30	Hexagonal
POUGHITE	Fe+++2(TeO3)2(SO4).3H2O	Orthorhombic
POVONDRAITE	(Na,K)(Fe+++,Fe++)3(Fe,Mg,Al)6(BO3)3Si6O18(OH)4	Trigonal
POWELLITE	CaMoO4	Tetragonal
POYARKOVITE	Hg3ClO	Monoclinic
PRASSOITE	Rh17S15	Isometric
PREHNITE	Ca2Al2Si3O10(OH)2	Orthorhombic
PREISINGERITE	Bi3(AsO4)2O(OH)	Triclinic
PREISWERKITE	NaMg2Al3Si2O10(OH)2	Monoclinic
PREOBRAZHENSKITE	Mg3B11O15(OH)9	Orthorhombic
PRICEITE	Ca4B10O19.7H2O	Triclinic
PRIDERITE	(K,Ba)(Ti,Fe+++)8O16	Tetragonal
PRINGLEITE	Ca9B26O34(OH)24Cl4.13H2O	Triclinic
PROBERTITE	NaCaB5O7(OH)4.3H2O	Monoclinic

Mineral Name	Formula	Crystal System
PROSOPITE	CaAL2(F,OH)8	Monoclinic
PROSPERITE	CaZn2(AsO4)2.H2O	Monoclinic
PROTASITE	Ba(UO2)3O3(OH)2.3H2O	Monoclinic ps Hexagonal
PROTOJOSEITE	Bi4TeS2	Hexagonal
PROUDITE	(Pb,Cu)8Bi9-10(S,Se)22	Monoclinic
PROUSTITE	Ag3AsS3	Trigonal
PRZHEVALSKITE	Pb(UO2)2(PO4)2.4H2O	Orthorhombic
PSEUDOAUTUNITE	(H3O)4Ca2(UO2)2(PO4)4.5H2O	Tetragonal
PSEUDOBOLEITE	Pb5Cu4Cl10(OH)8.2H2O	Tetragonal
PSEUDOBROOKITE	(Fe+++,Fe++)2(Fe++,Ti)O5	Orthorhombic
PSEUDOCOTUNNITE	K2PbCl4	Orthorhombic
PSEUDOGRANDREEFITE	Pb6SO4F1O	Orthorhombic ps Tetragonal
PSEUDOLAUEITE	Mn++Fe+++2(PO4)2(OH).7-8H2O	Monoclinic
PSEUDOMALACHITE	Cu5(PO4)2(OH)4	Monoclinic
PSEUDORUTILE	Fe+++2Ti3O9	Hexagonal
PUCHERITE	BiVO4	Orthorhombic
PUMPELLYITE-(Fe) (FERROPUMPELLYITE)	Ca2Fe++Al2(SiO4)(Si2O7)(OH)2.H2O	Monoclinic
PUMPELLYITE-(Mg)	Ca2MgAl2(SiO4)(Si2O7)(OH)2.H2O	Monoclinic
PUMPELLYITE-(Mn)	Ca2(Mn++,Mg)(Al,Mn+++,Fe)2(SiO4)(Si2O7(OH)2.H2O	Monoclinic
PURPURITE	Mn+++PO4	Orthorhombic
PUTORANITE	Cu16-18(Fe,Ni)18-19S32	Isometric
PYRARGYRITE	Ag3SbS3	Trigonal
PYRITE	FeS2	Isometric
PYROAURITE	Mg6Fe+++2(CO3)(OH)16.4H2O	Trigonal
PYROBELONITE	PbMn(VO4)(OH)	Orthorhombic
PYROCHLORE	(Na,Ca)2Nb2O6(OH,F)	Isometric
PYROCHROITE	Mn(OH)2	Trigonal
PYROLUSITE	MnO2	Tetragonal
PYROMORPHITE	Pb5(PO4)3Cl	Hexagonal
PYROPE	Mg3Al2(SiO4)3	Isometric
PYROPHANITE	MnTiO3	Trigonal
PYROPHYLLITE	Al2Si4O10(OH)2	Monoclinic
PYROSTILPNITE	Ag3SbS3	Monoclinic
PYROXFERROITE	(Fe++,Mn,Ca)SiO3	Triclinic
PYROXMANGITE	MnSiO3	Triclinic
PYRRHOTITE	Fe1-xS(x=0-0,17)	Monoclinic/Hexagonal
QANDILITE	(Mg,Fe++)2(Ti,Fe+++,Al)O4	Isometric
QILIANSHANITE	NaH4(CO3)(BO3).2H2O	Monoclinic
QINGHEIITE	Na2NaMn2Mg2(Al,Fe+++)2(PO4)6	Monoclinic
QITIANLINGITE	(Fe++,Mn)2(Nb,Ta)2W++++++O10	Orthorhombic
QUADRIDAVYNE	(Na,K)6Ca2Al6Si6O24Cl4	Hexagonal
QUADRUPHITE	Na14Ca(Mg,Mn)(Ti,Mn,Zr,Nb)4Si4O12(PO4)4O6 F2	Triclinic
QUARTZ	SiO2	Trigonal
QUEITITE	Pb4Zn2(SiO4)(Si2O7)(SO4)	Monoclinic

Mineral Name	Formula	Crystal System
QUENSELITE	PbMn+++O2(OH)	Monoclinic
QUENSTEDTITE	Fe+++2(SO4)3.10H2O	Triclinic
QUETZALCOATLITE	Zn8Cu4(TeO3)3(OH)18	Hexagonal
RABBITTITE	Ca3Mg3(UO2)2(CO3)6(OH)4.18H2O	Monoclinic
RABEJACITE	Ca(UO2)4(SO4)2(OH)6.6H2O	Orthorhombic
RADHAKRISHNAITE	PbTe3(Cl,S)2	Tetragonal
RADTKEITE	Hg3S2ClI	Orthorhombic
RAGUINITE	TlFeS2	Orthorhombic ps Hexagonal
RAITE	Na4Mn3Si8(O,OH)24.9H2O	Orthorhombic
RAJITE	CuTe++++2O5	Monoclinic
RALSTONITE	NaxMgxAl2-x(F,OH)6.H2O	Isometric
RAMDOHRITE	Ag3Pb6Sb11S24	Monoclinic
RAMEAUITE	K2CaU++++++6O20.9H2O	Monoclinic
RAMMELSBERGITE	NiAs2	Orthorhombic
RAMSBECKITE	(Cu,Zn)15(SO4)4(OH)22.6H2O	Monoclinic
RAMSDELLITE	MnO2	Orthorhombic
RANCIEITE	(Ca,Mn++)Mn++++4O9.3H2O	Hexagonal
RANKACHITE	CaFe++V+++++4W++++++8O36.12H2O	Orthorhombic
RANKAMAITE	(Na,K,Pb,Li)3(Ta,Nb,Al)11(O,OH)30	Orthorhombic
RANKINITE	Ca3Si2O7	Monoclinic
RANSOMITE	CuFe+++2(SO4)4.6H2O	Monoclinic
RANUNCULITE	HAl(UO2)PO4(OH)3.4H2O	Monoclinic ps Orthorhombic
RAPIDCREEKITE	Ca2(SO4)(CO3).4H2O	Orthorhombic
RASPITE	PbWO4	Monoclinic
RASVUMITE	KFe2S3	Orthorhombic
RATHITE	(Pb,Tl)3As5S10	Monoclinic
RAUENTHALITE	Ca3(AsO4)2.10H2O	Monoclinic/Triclinic
RAUVITE	Ca(UO2)2V+++++10O28.16H2O	
RAVATITE	C14H10	Orthorhombic
RAYITE	(Ag,Tl)2Pb8Sb8S21	Monoclinic
REALGAR	AsS	Monoclinic
REBULITE	Tl5Sb5As8S22	Monoclinic
RECTORITE (MICA-SMECTITE)	(Na,Ca)Al4(Si,Al)8O20(OH)4.2H2O	Monoclinic
REDDINGITE	Mn++3(PO4)2.3H2O	Orthorhombic
REDINGTONITE	(Fe++,Mg,Ni)(Cr,Al)2(SO4)4.22H2O	Monoclinic
REDLEDGEITE	BaTi6Cr+++2O16.H2O	Tetragonal
REEDERITE-	(Na,Al,Mn,Ca)15(Y,Ce,Nd,La)2(CO3)9(SO3F)(Cl,F)	Hexagonal
REEDMERGNERITE	NaBSi3O8	Triclinic
REEVESITE	Ni6Fe+++2(CO3)(OH)16.4H2O	Trigonal
REFIKITE	C19H31COOH	Orthorhombic
REICHENBACHITE	Cu++5(PO4)2(OH)4	Monoclinic
REINERITE	Zn3(As+++O3)2	Orthorhombic
REINHARDBRAUNSITE	Ca5(SiO4)2(OH,F)2	Monoclinic
REMONDITE-(Ce)	Na3(Ce,La,Ca,Na,Sr)3(CO3)5	Monoclinic ps Hexagonal

Mineral Data

Mineral Name	Formula	Crystal System
RENIERITE	(Cu,Zn)11(Ge,As)2Fe4S16	Tetragonal ps Cubic
REPPIAITE	Mn++[(V,As)O4(OH)2]2	Monoclinic
RETGERSITE	NiSO4.6H2O	Tetragonal
RETZIAN-(Ce)	Mn2Ce(AsO4)(OH)4	Orthorhombic
RETZIAN-(La)	(Mn,Mg)2(La,Ce,Nd)(AsO4)(OH)4	Orthorhombic
RETZIAN-(Nd)	Mn2(Nd,Ce,La)(AsO4)(OH)4	Orthorhombic
REVDITE	Na2Si2O5.5H2O	Triclinic
REYERITE	(Na,K)4Ca14Si22Al2O58(OH)8.6H2O	Trigonal
RHABDOPHANE-(Ce)	(Ce,La)PO4.H2O	Hexagonal
RHABDOPHANE-(La)	(La,Ce)PO4.H2O	Hexagonal
RHABDOPHANE-(Nd)	(Nd,Ce,La)PO4.H2O	Hexagonal
RHENIITE	ReS2	Hexagonal
RHENIUM	Re	Hexagonal
RHODESITE	(Ca,Na2,K2)8Si16O40.11H2O	Orthorhombic
RHODIUM	(Rh,Pt)	Isometric
RHODIZITE	(K,Cs)Al4Be4(B,Be)12O28	Isometric
RHODOCHROSITE (DIALOGITE)	MnCO3	Trigonal
RHODONITE	(Mn++,Fe++,Mg,Ca)SiO3	Triclinic
RHODOSTANNITE	Cu2FeSn3S8	HexagonalTetragonal
RHODPLUMSITE	Pb2Rh3S2	Trigonal
RHOMBOCLASE	HFe+++(SO4)2.4H2O	Orthorhombic
RHONITE (RHOENITE)	Ca2(Mg,Fe++,Fe+++,Ti)6(Si,Al)6O20	Triclinic
RIBBEITE	(Mn++,Mg)5(SiO4)2(OH)2	Orthorhombic
RICHELLITE	(Ca,Fe++)(Fe+++,Al)(PO4)2(OH,F)2	Tetragonal
RICHELSDORFITE	Ca2Cu5Sb(AsO4)4Cl(OH)6.6H2O	Monoclinic
RICHETITE	PbU++++++4O13.4H2O	Triclinic
RICHTERITE	Na2Ca(Mg,Fe++)5Si8O22(OH)2	Monoclinic
RICKARDITE	Cu7Te5	Orthorhombic ps Tetragonal
RIEBECKITE	Na2(Fe++,Mg)3Fe+++2Si8O22(OH)2	Monoclinic
RILANDITE	(Cr,Al)6SiO11.5H2O	
RIMKOROLGITE	(Mg,Mn)5(Ba,Sr,Ca)(PO4)4.8H2O	Orthorhombic
RINGWOODITE	(Mg,Fe++)2SiO4	Isometric
RINKITE (RINKOLITE)	(Na,Ca)3(Ca,Ce)4Ti4(Si2O7)2(O,F)4	Monoclinic
RINNEITE	K3NaFe++Cl6	Trigonal
RITTMANNITE	Mn++Mn++Fe++Al2(OH)2(PO4)4.8H2O	Monoclinic
RIVADAVITE	Na6MgB24O40.22H2O	Monoclinic
RIVERSIDEITE	Ca5Si6O16(OH)2.2H2O	Orthorhombic
ROALDITE	Fe4N	Isometric
ROBERTSITE	Ca3Mn+++4(PO4)4(OH)6.3H2O	Monoclinic
ROBINSONITE	Pb4Sb6S13	Triclinic
ROCKBRIDGEITE	(Fe++,Mn)Fe+++4(PO4)3(OH)5	Orthorhombic
RODALQUILARITE	H3Fe+++2(Te++++O3)4Cl	Triclinic
ROEBLINGITE	Pb2Ca6(Si6O18)(SO4)2(OH)2.4H2O	Monoclinic
ROEDDERITE	(Na,K)2(Mg,Fe++)5Si12O30	Hexagonal
ROGGIANITE	Ca2[Be(OH)2Al2Si4O13].<2.5H2O	Tetragonal

Mineral Name	Formula	Crystal System
ROHAITE	Tl2Cu8,5Sb2S4	Orthorhombic
ROKUHNITE (ROKUEHNITE)	Fe++Cl2.H2O	Monoclinic
ROMANECHITE (PSILOMELANE)	(Ba,H2O)(Mn++++,Mn+++)5O10	Monoclinic
ROMARCHITE	SnO	Tetragonal
ROMEITE	(Ca,Fe++,Mn,Na)2(Sb,Ti)2O6(O,OH,F)	Isometric
ROMERITE (ROEMERITE)	Fe++Fe+++2(SO4)4.14H2O	Triclinic
RONTGENITE-(Ce) (ROENTGENITE-(Ce))	Ca2(Ce,La)3(CO3)5F3	Trigonal
ROOSEVELTITE	BiAsO4	Monoclinic
ROQUESITE	CuInS2	Tetragonal
RORISITE	(Ca,Mg)FCl	Tetragonal
ROSASITE	(Cu,Zn)2(CO3)(OH)2	Monoclinic
ROSCHERITE	Ca(Mn++,Fe++)5Be4(PO4)6(OH)4.6H2O	Monoclinic/Triclinic
ROSCOELITE	K(V,Al,Mg)2AlSi3O10(OH)2	Monoclinic
ROSELITE	Ca2(Co,Mg)(AsO4)2.2H2O	Monoclinic
ROSELITEBETA (ROSELITEBETA)	Ca2Co(AsO4)2.2H2O	Triclinic
ROSEMARYITE	(Na,Ca,Mn++)(Mn++,Fe++)(Fe+++,Fe++,Mg)Al(PO4)3	Monoclinic
ROSENBEGITE	AlF3.3H2O	Tetragonal
ROSENBUSCHITE	(Ca,Na)3(Zr,Ti)Si2O8F	Triclinic
ROSENHAHNITE	Ca3Si3O8(OH)2	Triclinic
ROSHCHINITE	Ag19Pb10Sb51S96orPb(Ag,Cu)2(Sb,As)5S10	Orthorhombic
ROSICKYITE	S	Monoclinic
ROSIERESITE	Pb,Cu,Al,PO4,H2O	Amorphous
ROSSITE	CaV2O6.4H2O	Triclinic
ROSSLERITE (ROESSLERITE)	MgHAsO4.7H2O	Monoclinic
ROUBAULTITE	Cu2(UO2)3(CO3)2O2(OH)2.4H2O	Triclinic
ROUSEITE	Pb2Mn++(As+++O3)2.2H2O	Triclinic
ROUTHIERITE	TlHgAsS3	Tetragonal
ROUVILLEITE	Na3(Ca,Mn++)2(CO3)3(F,OH)	Monoclinic
ROWEITE	Ca2Mn++2B4O7(OH)6	Orthorhombic
ROWLANDITE-	Y4Fe++Si4O14F2	Amorphous
ROXBYITE	Cu9S5	Monoclinic
ROZENITE	Fe++SO4.4H2O	Monoclinic
RUARSITE	RuAsS	Monoclinic
RUCKLIDGEITE	(Bi,Pb)3Te4	Trigonal
RUITENBERGITE	Ca9B26O34(OH)24Cl4.13H2O	Monoclinic
RUIZITE	CaMn+++Si2O6(OH).2H2O	Monoclinic
RUSAKOVITE	(Fe+++,Al)5(VO4,PO4)2(OH)9.3H2O	
RUSSELLITE	Bi2WO6	Tetragonal
RUSTENBURGITE	(Pt,Pd)3Sn	Isometric
RUSTUMITE	Ca10(Si2O7)2(SiO4)Cl2(OH)2	Monoclinic
RUTHENARSENITE	(Ru,Ni)As	Orthorhombic
RUTHENIRIDOSMINE	(Ir,Os,Ru)	Hexagonal
RUTHENIUM	(Ru,Ir,Os)	Hexagonal

Mineral Data

Mineral Name	Formula	Crystal System
RUTHERFORDINE	UO2(CO3)	Orthorhombic
RUTILE	TiO2	Tetragonal
RYNERSONITE	Ca(Ta,Nb)2O6	Orthorhombic
SABATIERITE	Cu6TlSe4	Orthorhombic
SABELLIITE	(Cu,Zn)2Zn[(As,Sb)O4](OH)3	Trigonal
SABIEITE	(NH4)Fe+++(SO4)2	Trigonal
SABINAITE	Na4Zr2TiO4(CO3)4	Monoclinic
SABUGALITE	HAl(UO2)4(PO4)4.16H2O	Monoclinic ps Tetragonal
SACROFANITE	(Na,Ca,K)9(Si,Al)12O24[(OH)2,(SO4),(CO3),Cl2)]3.nH2O	Hexagonal
SADANAGAITE	(K,Na)Ca2(Fe++,Mg,Al,Fe+++,Ti)5(Si,Al)8O22(OH)2	Monoclinic
SAFFLORITE	CoAs2	Orthorhombic
SAHAMALITE-(Ce)	(Mg,Fe++)Ce2(CO3)4	Monoclinic
SAHLINITE	Pb14(AsO4)2O9Cl4	Monoclinic
SAINFELDITE	Ca5(AsO4)[AsO3(OH)]2.4H2O	Monoclinic
SAKHAITE	Ca3Mg(BO3)2(CO3).nH2O,(n<1)	Isometric
SAKHAROVAITE	(Pb,Fe)(Bi,Sb)2S4	Orthorhombic
SAKURAIITE	(Cu,Zn,Fe,In,Sn)S	Cubic or Tetragonal
SALAMMONIAC (SALMIAC)	NH4Cl	Isometric
SALEEITE	Mg(UO2)2(PO4)2.10H2O	Monoclinic ps Tetragonal
SALESITE	Cu(IO3)(OH)	Orthorhombic
SALIOTITE	Na0,5Li0,5Al3AlSi3O10(OH)5	Monoclinic
SAMARSKITE-	(Y,Ce,U,Fe+++)3(Nb,Ta,Ti)5O16	Monoclinic
SAMFOWLERITE	Ca14Mn++3Zn2(Be,Zn)2Be6(SiO4)6(Si2O7)4(OH,F)6	Monoclinic
SAMPLEITE	NaCaCu5(PO4)4Cl.5H2O	Orthorhombic
SAMSONITE	Ag4MnSb2S6	Monoclinic
SAMUELSONITE	(Ca,Ba)Ca8(Fe++,Mn)4Al2(PO4)10(OH)2	Monoclinic
SANBORNITE	BaSi2O5	Orthorhombic
SANDERITE	MgSO4.2H2O	
SANEROITE	Na2(Mn++,Mn+++)10Si11VO34(OH)4	Triclinic
SANIDINE	(K,Na)(Si,Al)4O8	Monoclinic
SANJUANITE	Al2(PO4)(SO4)(OH).9H2O	Monoclinic
SANMARTINITE	(Zn,Fe++)WO4	Monoclinic
SANTACLARAITE	CaMn++4Si5O14(OH)2.H2O	Triclinic
SANTAFEITE	(Mn,Fe,Al,Mg)8(Mn,Mn)8(Ca,Sr,Na)12(VO4,AsO4)16(OH)20.8H2O	Orthorhombic
SANTANAITE	9PbO.2PbO2.CrO3	Hexagonal
SANTITE	KB5O6(OH)4.2H2O	Orthorhombic
SAPONITE	(Ca/2,Na)0,3(Mg,Fe++)3(Si,Al)4O10(OH)2.4H2O	Monoclinic
SAPPHIRINE-1Tc	(Mg,Al)8(Al,Si)6O20	Triclinic
SAPPHIRINE-2M	(Mg,Al)8(Al,Si)6O20	Monoclinic
SARABAUITE	CaSb10O10S6	Monoclinic
SARCOLITE	Na,Ca6Al4Si6O24F	Tetragonal
SARCOPSIDE	(Fe++,Mn,Mg)3(PO4)2	Monoclinic
SARKINITE	Mn++2(AsO4)(OH)	Monoclinic
SARMIENTITE	Fe+++2(AsO4)(SO4)(OH).5H2O	Monoclinic

Mineral Data

Mineral Name	Formula	Crystal System
SARTORITE	PbAs2S4	Monoclinic
SARYARKITE-	Ca(Y,Th)Al5(SiO4)2(PO4,SO4)2(OH)7.6H2O	Hexagonal
SASAITE	(Al,Fe+++)6(PO4,SO4)5(OH)3.36H2O	Orthorhombic
SASSOLITE (BORICACID)	H3BO3	Triclinic
SATIMOLITE	KNa2Al4B6O15Cl3.13H2O	Orthorhombic
SATPAEVITE	Al12V++++2V+++++6O37.30H2O	Orthorhombic
SATTERLYITE	(Fe++,Mg,Fe+++)2(PO4)(OH)	Hexagonal
SAUCONITE	Na0,3Zn3(Si,Al)4O10(OH)2.4H2O	Monoclinic
SAYRITE	Pb2(UO2)5O6(OH)2.4H2O	Monoclinic
SAZHINITE-(Ce)	Na2CeSi6O14(OH).nH2O,(nca.5)	Orthorhombic
SAZYKINAITE-	(Na,K)5Y(Zr,Ti)Si6O18.6H2O	Trigonal
SBORGITE	NaB5O6(OH)4.3H2O	Monoclinic
SCACCHITE	MnCl2	Trigonal
SCARBROITE	Al5(OH)13(CO3).5H2O	Hexagonal
SCAWTITE	Ca7Si6(CO3)O18.2H2O	Monoclinic
SCHACHNERITE	Ag1,1Hg0,9	Hexagonal
SCHAFARZIKITE	Fe++Sb+++2O4	Tetragonal
SCHAIRERITE	Na21(SO4)7F6Cl	Trigonal
SCHALLERITE	(Mn++,Fe++)16Si12As+++3O36(OH)17	Trigonal
SCHAPBACHITE	AgBiS2	Isometric
SCHAURTEITE	Ca3Ge++++(SO4)2(OH)6.3H2O	Hexagonal
SCHEELITE	CaWO4	Tetragonal
SCHERTELITE	(NH4)2MgH2(PO4)2.4H2O	Orthorhombic
SCHETELIGITE	(Ca,Y,Sb,Mn)2(Ti,Ta,Nb,W)2O6(O,OH)	Orthorhombic
SCHIEFFELINITE	Pb(Te++++++,S)O4.H2O	Orthorhombic
SCHIRMERITE	Ag3Pb3Bi9S18toAg3Pb6Bi7S18	Orthorhombic
SCHLOSSMACHERITE	(H3O,Ca)Al3(AsO4,SO4)2(OH)6	Trigonal
SCHMIEDERITE (SCHMEIDERITE)	Pb2Cu++2(Se++++O3)(Se++++++O4)(OH)4	Monoclinic
SCHMITTERITE	(UO2)TeO3	Orthorhombic
SCHNEIDERHOHNITE (SCHNEIDERHOEHNITE)	Fe++Fe+++3As+++5O13	Triclinic
SCHODERITE	Al2(PO4)(VO4).8H2O	Monoclinic
SCHOENFLIESITE	MgSn++++(OH)6	Isometric
SCHOEPITE	UO3.2H2O	Orthorhombic
SCHOLLHORNITE (SCHOELLHORNITE)	Na0,3CrS2.H2O	Trigonal
SCHOLZITE	CaZn2(PO4)2.2H2O	Orthorhombic
SCHOONERITE	Fe++2ZnMnFe+++(PO4)3(OH)2.9H2O	Orthorhombic
SCHORL	NaFe++3Al6(BO3)3Si6O18(OH)4	Trigonal
SCHORLOMITE	Ca3Ti++++2(Fe+++2Si)O12	Isometric
SCHREIBERSITE (RHABDITE)	(Fe,Ni)3P	Tetragonal
SCHREYERITE	V+++2Ti3O9	Monoclinic
SCHROCKINGERITE (SCHROECKINGERITE)	NaCa3(UO2)(CO3)3(SO4)F.10H2O	Triclinic
SCHUBNELITE	Fe+++2-x(V+++++,V++++)2O4(OH)4	Triclinic
SCHUETTEITE	Hg3(SO4)O2	Hexagonal

Mineral Data

Mineral Name	Formula	Crystal System
SCHUILINGITE-(Nd)	PbCu(Nd,Gd,Sm,Y)(CO3)3(OH).1,5H2O	Orthorhombic
SCHULENBERGITE	(Cu,Zn)7(SO4,CO3)2(OH)10.3H2O	Trigonal
SCHULTENITE	PbHAsO4	Monoclinic
SCHUMACHERITE	Bi3[(V,As,P)O4]2O(OH)	Triclinic
SCHWARTZEMBERGITE	Pb6(IO3)2Cl4O2(OH)2	Orthorhombic ps Tetragonal
SCHWERTMANNITE	Fe+++16O16(OH)12(SO4)2	Tetragonal
SCLARITE	(Zn,Mg,Mn++)4Zn3(CO3)2(OH)10	Monoclinic
SCOLECITE	CaAl2Si3O10.3H2O	Monoclinic
SCORODITE	Fe+++AsO4.2H2O	Orthorhombic
SCORZALITE	(Fe++,Mg)Al2(PO4)2(OH)2	Monoclinic
SCOTLANDITE	PbSO3	Monoclinic
SCRUTINYITE	àPbO2	Orthorhombic
SEAMANITE	Mn3(PO4)B(OH)6	Orthorhombic
SEARLESITE	NaBSi2O5(OH)2	Monoclinic
SEDERHOLMITE	NiSe	Hexagonal
SEDOVITE	U(MoO4)2	Orthorhombic
SEELIGERITE	Pb3Cl3(IO3)O	Orthorhombic
SEELITE	Mg(UO2)2(AsO4,AsO3)2.4-7H2O	Monoclinic
SEGELERITE	CaMgFe+++(PO4)2(OH).4H2O	Orthorhombic
SEGNITITE	PbFe+++3H(AsO4)2(OH)6	Trigonal
SEIDOZERITE	(Na,Ca)2(Zr,Ti,Mn)2Si2O7(O,F)2	Monoclinic
SEINAJOKITE	(Fe,Ni)(Sb,As)2	Orthorhombic
SEKANINAITE	(Fe++,Mg)2Al4Si5O18	Orthorhombic
SELENIUM	Se	Trigonal
SELENOSTEPHANITE	Ag5Sb(Se,S)4	Orthorhombic
SELENTELLURIUM	(Se,Te)	Trigonal
SELIGMANNITE	PbCuAsS3	Orthorhombic
SELLAITE	MgF2	Tetragonal
SELWYNITE	NaK(Be,Al)Zr2(PO4)4.2H2O	Tetragonal
SEMENOVITE-(Ce)	(Ca,Ce,La,Na)10-12(Fe++,Mn)(Si,Be)20(O,OH,F)48	Orthorhombic
SEMSEYITE	Pb9Sb8S21	Monoclinic
SENAITE	Pb(Ti,Fe,Mn)21O38	Trigonal
SENARMONTITE	Sb2O3	Isometric
SENEGALITE	Al2(PO4)(OH)3.H2O	Orthorhombic
SENGIERITE	Cu2(UO2)2V2O8.6H2O	Monoclinic
SEPIOLITE	Mg4Si6O15(OH)2.6H2O	Orthorhombic
SERANDITE	Na(Mn++,Ca)2Si3O8(OH)	Triclinic
SERENDIBITE	Ca2(Mg,Al)6(Si,Al,B)6O20	Triclinic
SERGEEVITE	Ca2Mg11(CO3)27(HCO3)12(OH)4.6H2O	Trigonal
SERPIERITE	Ca(Cu,Zn)4(SO4)2(OH)6.3H2O	Monoclinic
SHABAITE-(Nd)	Ca(Nd,Y)2(UO2)(CO3)4(OH)2.6H2O	Monoclinic
SHABYNITE	Mg5(BO3)2(Cl,OH)2(OH)5.4H2O	Monoclinic
SHADLUNITE	(Pb,Cd)(Fe,Cu)8S8	Isometric
SHAFRANOVSKITE	(Na,K)6(Mn++,Fe++)3Si9O24.6H2O	Trigonal
SHAKHOVITE (SHAHOVITE)	Hg+4Sb+++++O3(OH)3	Monoclinic

Mineral Name	Formula	Crystal System
SHANDITE	Pb2Ni3S2	Trigonal ps Cubic
SHANNONITE	Pb2CO3O	Orthorhombic
SHARPITE	Ca(UO2)6(CO3)5(OH)4.6H2O	Orthorhombic
SHATTUCKITE	Cu5(SiO3)4(OH)2	Orthorhombic
SHCHERBAKOVITE	(K,Na,Ba)3(Ti,Nb)2Si4O14	Orthorhombic
SHCHERBINAITE	V2O5	Orthorhombic
SHERWOODITE	Ca9Al2V++++4V+++++24O80.56H2O	Tetragonal
SHIGAITE	NaMn++6Al3(SO4)2(OH)18.12H2O	Trigonal
SHOMIOKITE-	Na3Y(CO3)3.3H2O	Orthorhombic
SHORTITE	Na2Ca2(CO3)3	Orthorhombic
SHUANGFENGITE	IrTe2	Trigonal
SHUBNIKOVITE	Ca2Cu8(AsO4)6Cl(OH).7H2O	Orthorhombic
SHUISKITE	Ca2(Mg,Al)(Cr,Al)2(SiO4)(Si2O7)(OH)2.H2O	Monoclinic
SIBIRSKITE	CaHBO3	Monoclinic
SICKLERITE	Li(Mn++,Fe+++)PO4	Orthorhombic
SIDERAZOT	Fe5N2	Hexagonal
SIDERITE	Fe++CO3	Trigonal
SIDERONATRITE	Na2Fe+++(SO4)2(OH).3H2O	Orthorhombic
SIDEROPHYLLITE	KFe++2Al(Al2Si2)O10(F,OH)2	Monoclinic
SIDEROTIL	Fe++SO4.5H2O	Triclinic
SIDORENKITE	Na3Mn(PO4)(CO3)	Monoclinic ps Orthorhombic
SIDWILLITE	MoO3.2H2O	Monoclinic
SIEGENITE	(Ni,Co)3S4	Isometric
SIELECKIITE	Cu3Al4(PO4)2(OH)12.2H2O	Triclinic
SIGLOITE	Fe+++Al2(PO4)2(OH)3.5H2O	Triclinic
SILHYDRITE	3SiO2.H2O	Orthorhombic
SILICON (SILICIUM)	Si	Isometric
SILINAITE	NaLiSi2O5.2H2O	Monoclinic
SILLENITE	Bi12SiO20	Isometric
SILLIMANITE	Al2SiO5=Al[6]Al[4]OSiO4	Orthorhombic
SILVER-2H	Ag	Hexagonal
SILVER-3C (ARGENT)	Ag	Isometric
SILVER-4H	Ag	Hexagonal
SIMFERITE	Li0,5(Mg,Mn+++)5(PO4)3	Orthorhombic
SIMONELLITE	C19H24	Orthorhombic
SIMONITE	TlHgAs3S6	Monoclinic
SIMONKOLLEITE	Zn5(OH)8Cl2.H2O	Hexagonal
SIMPLOTITE	CaV++++4O9.5H2O	Monoclinic
SIMPSONITE	Al4(Ta,Nb)3(O,OH,F)14	Trigonal
SINCOSITE	CaV++++2(PO4)2(OH)4.3H2O	Tetragonal
SINHALITE	MgAlBO4	Orthorhombic
SINJARITE	CaCl2.2H2O	Tetragonal
SINKANKASITE	H2MnAl(PO4)2(OH).6H2O	Triclinic
SINNERITE	Cu6As4S9	Triclinic
SINOITE	Si2N2O	Orthorhombic
SITINAKITE	Na2K(Ti,Nb)4O4(SiO4)2(O,OH).4H2O	Tetragonal

Mineral Name	Formula	Crystal System
SJOGRENITE (SJOEGRENITE)	$Mg_6Fe^{++}{}_2(CO_3)(OH)_{14}.5H_2O$	Hexagonal
SKINNERITE	Cu_3SbS_3	Monoclinic
SKIPPENITE	$Bi_2Se_2(Te,S)$	Trigonal
SKLODOWSKITE	$(H_3O)_2Mg(UO_2)_2(SiO_4)_2.2H_2O$	Monoclinic
SKUTTERUDITE	$(Co,Ni)As_{3-x}$	Isometric
SLAVIKITE	$NaMg_2Fe^{+++}{}_5(SO_4)_7(OH)_6.33H_2O$	Trigonal
SLAWSONITE	$(Sr,Ca)Al_2Si_2O_8$	Monoclinic
SMIRNITE	$Bi_2Te^{++++}O_5$	Orthorhombic
SMITHITE	$AgAsS_2$	Monoclinic
SMITHSONITE	$ZnCO_3$	Trigonal
SMOLIANINOVITE	$(Co,Ni,Mg,Ca)_3(Fe^{+++},Al)_2(AsO_4)_4.11H_2O$	Orthorhombic
SMRKOVECITE	$Bi_2(PO_4)O(OH)$	Monoclinic
SMYTHITE	$(Fe,Ni)_9S_{11}$ or $(Fe,Ni)_{13}S_{16}$	Trigonal
SOBOLEVITE	$Na_{11}(Na,Ca)_4(Mg,Mn)Ti^{++++}{}_4(Si_4O_{12})(PO_4)_4$ O_5F_3	Triclinic
SOBOLEVSKITE	$PdBi$	Hexagonal/Monoclinic
SODALITE	$Na_8Al_6Si_6O_{24}Cl_2$	Isometric
SODDYITE	$(UO_2)_2SiO_4.2H_2O$	Orthorhombic
SODIUMALUM	$NaAl(SO_4)_2.12H_2O$	Isometric
SODIUMANTHOPHYLLITE	$Na(Mg,Fe^{++})_7(Si_7Al)O_{22}(OH)_2$	Orthorhombic
SODIUMAUTUNITE	$Na_2(UO_2)_2(PO_4)_2.8H_2O$	Tetragonal
SODIUMBETPAKDALITE	$(Na,Ca)_3Fe^{+++}{}_2(As_2O_4)(MoO_4)_6.15H_2O$	Monoclinic
SODIUMBOLTWOODITE	$(H_3O)(Na,K)(UO_2)SiO_4.H_2O$	Orthorhombic
SODIUMDACHIARDITE	$(Na_2,Ca,K_2)_{4-5}Al_8Si_{40}O_{96}.26H_2O$	Monoclinic
SODIUMGEDRITE	$Na(Mg,Fe^{++})_6Al(Si_6Al_2)O_{22}(OH)_2$	Orthorhombic
SODIUMKOMAROVITE	$(Na,Ca,H)_4Nb_4Si_4O_{12}O_8(OH,F)_4.2H_2O$	Orthorhombic
SODIUMPHARMACOSIDERITE	$(Na,K)_2Fe^{+++}{}_4(AsO_4)_3(OH)_5.7H_2O$	Tetragonal ps Cubic
SODIUMPHLOGOPITE	$NaMg_3Si_3AlO_{10}(OH)_2$	Monoclinic
SODIUMURANOSPINITE	$(Na_2,Ca)(UO_2)_2(AsO_4)_2.5H_2O$	Tetragonal
SODIUMZIPPEITE	$Na(UO_2)_2(SO_4)(OH)_3.H_2O$	Monoclinic
SOFIITE	$Zn_2(Se^{++++}O_3)Cl_2$	Orthorhombic ps Hexagonal
SOGDIANITE	$(K,Na)_2(Li,Fe^{+++},Al)_3ZrSi_{12}O_{30}$	Hexagonal
SOHNGEITE (SOEHNGEITE)	$Ga(OH)_3$	Isometric
SOLONGOITE	$Ca_2B_3O_4(OH)_4Cl$	Monoclinic
SONOLITE	$Mn_9(SiO_4)_4(OH,F)_2$	Monoclinic
SONORAITE	$Fe^{+++}Te^{++++}O_3(OH).H_2O$	Monoclinic
SOPCHEITE	$Ag_4Pd_3Te_4$	Orthorhombic
SORBYITE	$Pb_{19}(Sb,As)_{20}S_{49}$	Monoclinic
SORENSENITE	$Na_4SnBe_2Si_6O_{18}.2H_2O$	Monoclinic
SOSEDKOITE	$(K,Na)_5Al_2(Ta,Nb)_{22}O_{60}$	Orthorhombic
SOUCEKITE	$PbCuBi(S,Se)_3$	Orthorhombic
SOUZALITE	$(Mg,Fe^{++})_3(Al,Fe^{+++})_4(PO_4)_4(OH)_6.2H_2O$	Monoclinic
SPADAITE	$MgSiO_2(OH)_2.H_2O$	
SPANGOLITE	$Cu_6Al(SO_4)(OH)_{12}Cl.3H_2O$	Trigonal
SPENCERITE	$Zn_4(PO_4)_2(OH)_2.3H_2O$	Monoclinic

Mineral Name	Formula	Crystal System
SPERRYLITE	PtAs2	Isometric
SPERTINIITE	Cu(OH)2	Orthorhombic
SPESSARTINE (SPESSARTITE)	Mn++3Al2(SiO4)3	Isometric
SPHAEROBISMOITE	Bi2O3	Tetragonal
SPHAEROCOBALTITE (COBALTOCALCITE)	CoCO3	Trigonal
SPHALERITE (BLENDE)	(Zn,Fe)S	Isometric
SPHENISCIDITE	(NH4,K)(Fe+++,Al)2(PO4)2(OH).2H2O	Monoclinic
SPINEL	MgAl2O4	Isometric
SPIONKOPITE	Cu39S28	Hexagonal
SPIROFFITE	(Mn,Zn)2Te3O8	Monoclinic
SPODIOPHYLLITE	(Na,K)4(Mg,Fe++)3(Fe+++,Al)2(Si8O24)	Monoclinic
SPODIOSITE	Ca2(PO4)F	Orthorhombic
SPODUMENE	LiAlSi2O6	Monoclinic
SPURRITE	Ca5(SiO4)2(CO3)	Monoclinic
SQUAWCREEKITE	(Fe+++,Sb+++++,Sn++++,Ti)O2	Tetragonal
SREBRODOLSKITE	Ca2Fe+++2O5	Orthorhombic
SRILANKITE	(Ti,Zr)O2	Orthorhombic
STALDERITE	(Tl,Cu)(Zn,Fe,Hg)AsS3	Tetragonal
STANFIELDITE	Ca4(Mg,Fe++,Mn)5(PO4)6	Monoclinic
STANLEYITE	VOSO4.6H2O	Orthorhombic
STANNITE	Cu2FeSnS4	Tetragonal
STANNOIDITE	Cu8(Fe,Zn)3Sn2S12	Orthorhombic
STANNOMICROLITE (SUKULAITE)	(Sn++,Fe++)(Ta,Nb,Sn++++)2(O,OH)7	Isometric
STANNOPALLADINITE	(Pd,Cu)3Sn2	Hexagonal
STARKEYITE	MgSO4.4H2O	Monoclinic
STAUROLITE (STAUROTIDE)	(Fe++,Mg,Zn)2Al9(Si,Al)4O22(OH)2	Monoclinic ps Orthorhombic
STEACYITE	Th(Ca,Na)2K1-xSi8O20	Tetragonal
STEENSTRUPINE-(Ce)	Na14Ce6Mn++Mn+++Fe++2(Zr,Th)(Si6O18)2(PO4)7.3H2O	Trigonal
STEIGERITE	AlVO4.3H2O	Monoclinic
STELLERITE	CaAl2Si7O18.7H2O	Orthorhombic
STENHUGGARITE	CaFe+++(As+++O2)(As+++Sb+++O5)	Tetragonal
STENONITE	(Sr,Ba,Na)2Al(CO3)F5	Monoclinic
STEPANOVITE	NaMgFe+++(C2O4)3.8-9H2O	Trigonal
STEPHANITE	Ag5SbS4	Orthorhombic
STERCORITE	H(NH4)Na(PO4).4H2O	Triclinic
STERLINGHILLITE	Mn++3(AsO4)2.4H2O	
STERNBERGITE	AgFe2S3	Orthorhombic
STERRYITE	Ag2Pb10(Sb,As)12S29	Orthorhombic
STETEFELDTITE	Ag2Sb2(O,OH)7	Isometric
STEVENSITE	(Ca/2)0,3Mg3Si4O10(OH)2	Monoclinic
STEWARTITE	Mn++Fe+++2(PO4)2(OH)2.8H2O	Triclinic
STIBARSEN	SbAs	Trigonal
STIBICONITE	Sb+++Sb+++++2O6(OH)	Isometric

Mineral Data

Mineral Name	Formula	Crystal System
STIBIOBETAFITE	(Sb+++,Ca)2(Ti,Nb,Ta)2(O,OH)7	Isometric
STIBIOCOLUMBITE	SbNbO4	Orthorhombic
STIBIOCOLUSITE	Cu13V(Sb,As,Sn)3S16	Isometric
STIBIOENARGITE	Cu3(Sb,As)S4	Orthorhombic
STIBIOMICROLITE	(Sb,Ca,Na)2(Ta,Nb)2(O,OH)7	
STIBIOPALLADINITE	Pd5Sb2	Hexagonal
STIBIOTANTALITE	SbTaO4	Orthorhombic
STIBIVANITE-2M (STIBIVANITE)	Sb+++2V++++O5	Monoclinic
STIBIVANITE-2O	Sb+++2V++++O5	Orthorhombic
STIBNITE (STIBINE,ANTIMONITE)	Sb2S3	Orthorhombic
STICHTITE	Mg6Cr2(CO3)(OH)16.4H2O	Trigonal
STILBITE (DESMINE)	NaCa2Al5Si13O36.14H2O	Monoclinic
STILLEITE	ZnSe	Isometric
STILLWATERITE	Pd8As3	Hexagonal
STILLWELLITE-(Ce)	(Ce,La,Ca)BSiO5	Trigonal
STILPNOMELANE	K(Fe++,Mg,Fe+++,Al)8(Si,Al)12(O,OH)27.2H2O	Monoclinic/Triclinic
STISHOVITE	SiO2	Tetragonal
STISTAITE	SnSb	Isometric
STOIBERITE	Cu5V+++++2O10	Monoclinic
STOKESITE	CaSnSi3O9.2H2O	Orthorhombic
STOLZITE	PbWO4	Tetragonal
STOTTITE	Fe++Ge(OH)6	Tetragonal
STRACZEKITE	(Ca,K,Ba,Na)(V++++,V+++++)8O20.3H2O	Monoclinic
STRAKHOVITE	NaBa3(Mn++,Mn+++)4Si6O19(OH)3	Monoclinic
STRANSKIITE	Zn2Cu(AsO4)2	Triclinic
STRASHIMIRITE	Cu8(AsO4)4(OH)4.5H2O	Monoclinic
STRATLINGITE (STRAETLINGITE)	Ca2Al[(OH)6AlSiO2-3(OH)4-3].2,5H2O	Trigonal
STRELKINITE	Na2(UO2)2V2O8.6H2O	Orthorhombic
STRENGITE	Fe+++PO4.2H2O	Orthorhombic
STRINGHAMITE	CaCuSiO4.2H2O	Monoclinic
STROMEYERITE	AgCuS	Orthorhombic
STRONALSITE	SrNa2Al4Si4O16	Orthorhombic
STRONTIANITE	SrCO3	Orthorhombic
STRONTIOBORITE	SrB8O11(OH)4	Monoclinic
STRONTIOCHEVKINITE	(Sr,La,Ce,Ca)4(Fe++,Fe+++)(Ti+++,Zr)3Si4O12 O10	Monoclinic
STRONTIODRESSERITE	(Sr,Ca)Al2(CO3)2(OH)4.H2O	Orthorhombic
STRONTIOGINORITE	(Sr,Ca)2B14O23.8H2O	Monoclinic
STRONTIOJOAQUINITE	Sr2Ba2(Na,Fe++)2Ti2Si8O24(O,OH)2.H2O	Monoclinic
STRONTIOORTHOJOAQUIN ITE	Sr2Ba2(Na,Fe++)2Ti2Si8O24(O,OH)2.H2O	Orthorhombic
STRONTIOPIEMONTITE	(Ca,Mn++)(Sr,Ca)Mn+++(Al,Fe+++)2(SiO4)(Si2 O7)O(OH)	Monoclinic
STRONTIOPYROCHLORE	Sr2Nb2(O,OH)7	Isometric
STRONTIOWHITLOCKITE	Sr7(Mg,Ca)3(PO4)6[PO3(OH)]	Trigonal

Mineral Data

Mineral Name	Formula	Crystal System
STRONTIUMAPATITE	(Sr,Ca)5(PO4)3(OH,F)	Hexagonal
STRUNZITE	Mn++Fe+++2(PO4)2(OH)2.6H2O	Triclinic ps Monoclinic
STRUVERITE (STRUEVERITE)	(Ti,Ta,Fe+++)3O6	Tetragonal
STRUVITE	(NH4)MgPO4.6H2O	Orthorhombic
STUDENITSITE	NaCa2B9O14(OH)4.2H2O	Monoclinic
STUDTITE	UO4.4H2O	Monoclinic
STUMPFLITE	Pt(Sb,Bi)	Hexagonal
STURMANITE	Ca6(Fe+++,Al,Mn++)2(SO4)2[B(OH)4](OH)12.25 H2O	Hexagonal
STURTITE =HYDROUSMnSILICATE	Fe+++(Mn++,Ca,Mg)Si4O10(OH)3.10H2O	Amorphous
STUTZITE (STUETZITE)	Ag7Te4	Hexagonal
SUANITE	Mg2B2O5	Monoclinic
SUDBURYITE	(Pd,Ni)Sb	Hexagonal
SUDOITE	Mg2(Al,Fe+++)3Si3AlO10(OH)8	Monoclinic
SUESSITE	(Fe,Ni)3Si	Isometric
SUGILITE	KNa2(Fe++,Mn++,Al)2Li3Si12O30	Hexagonal
SULFOBORITE	Mg3B2(SO4)(OH)8(OH,F)2	Orthorhombic
SULFUR (SOUFRE,SCHWEFEL)	S	Orthorhombic
SULPHOHALITE	Na6(SO4)2FCl	Isometric
SULPHOTSUMOITE	Bi(Te,S)	Trigonal
SULVANITE	Cu3VS4	Isometric
SUNDIUSITE	Pb10(SO4)Cl2O8	Monoclinic
SUOLUNITE	Ca2Si2O5(OH)2.H2O	Orthorhombic
SURINAMITE	(Mg,Fe++)3Al4BeSi3O16	Monoclinic
SURITE	Pb(Pb,Ca)(Al,Fe+++,Mg)2(Si,Al)4O10(OH)2(CO3)2	Monoclinic
SURSASSITE	Mn++2Al3(SiO4)(Si2O7)(OH)3	Monoclinic
SUSANNITE	Pb4(SO4)(CO3)2(OH)2	Trigonal
SUSSEXITE	MnBO2(OH)	Monoclinic
SUZUKIITE	Ba2(VO2)Si4O12	Orthorhombic
SVABITE	Ca5(AsO4)3F	Hexagonal
SVANBERGITE	SrAl3(PO4)(SO4)(OH)6	Trigonal
SVEITE	KAl7(NO3)4Cl2(OH)16.8H2O	Monoclinic
SVERIGEITE	NaMnMgSn++++Be2Si3O12(OH)	Orthorhombic
SVYATOSLAVITE	CaAl2Si2O8	Orthorhombic
SVYAZHINITE	(Mg,Mn)(Al,Fe+++)(SO4)2F.14H2O	Triclinic
SWAKNOITE	(NH4)2Ca(HPO4)2.H2O	Orthorhombic
SWAMBOITE	U++++++H6(UO2)6(SiO4)6.30H2O	Monoclinic
SWARTZITE	CaMg(UO2)(CO3)3.12H2O	Monoclinic
SWEDENBORGITE	NaBe4SbO7	Hexagonal
SWEETITE	Zn(OH)2	Tetragonal
SWINEFORDITE	(Ca,Na)0,3(Al,Li,Mg)2(Si,Al)4O10(OH,F)2.2H2O	Monoclinic
SWITZERITE	(Mn++,Fe++)3(PO4)2.7H2O	Monoclinic
SYLVANITE	(Au,Ag)2Te4	Monoclinic
SYLVITE	KCl	Isometric

Mineral Data

Mineral Name	Formula	Crystal System
SYMPLESITE	Fe++3(AsO4)2.8H2O	Triclinic
SYNADELPHITE	(Mn,Mg,Ca,Pb)9(As+++O3)(As++++O4)2(OH)9 .2H2O	Triclinic ps Orthorhombic
SYNCHYSITE- (DOVERITE)	(Y,Ce)Ca(CO3)2F	Monoclinic ps Hexagonal
SYNCHYSITE-(Ce)	Ca(Ce,La)(CO3)2F	Monoclinic ps Hexagonal
SYNCHYSITE-(Nd)	Ca(Nd,La)(CO3)2F	Monoclinic ps Hexagonal
SYNGENITE	K2Ca(SO4)2.H2O	Monoclinic
SZAIBELYITE	MgBO2(OH)	Monoclinic
SZENICSITE	Cu3MoO4(OH)4	Orthorhombic
SZMIKITE	MnSO4.H2O	Monoclinic
SZOMOLNOKITE	Fe++SO4.H2O	Monoclinic
SZTROKAYITE	Bi3TeS2	
SZYMANSKIITE	Hg+16(Ni,Mg)6(CO3)12(OH)12(H3O8).3H2O	Hexagonal
TAAFFEITE	Mg3Al8BeO16	Hexagonal
TACHARANITE	Ca12Al2Si18O51.18H2O	Monoclinic
TACHYHYDRITE	CaMg2Cl6.12H2O	Trigonal
TADZHIKITE-(Ce)	Ca3(Ce,Y)2(Ti,Al,Fe+++)B4Si4O22	Monoclinic
TAENIOLITE	KLiMg2Si4O10F2	Monoclinic
TAENITE	(Ni,Fe)	Isometric
TAIKANITE	(Ba,Sr)2Mn+++2Si4O12	Monoclinic
TAIMYRITE	(Pd,Cu,Pt)3Sn	Orthorhombic
TAKANELITE	(Mn++,Ca)Mn++++4O8.H2O	Hexagonal
TAKEDAITE	Ca3B2O6	Trigonal
TAKEUCHIITE	(Mg,Mn++)2(Mn+++,Fe+++)BO5	Orthorhombic
TAKOVITE	Ni6Al2(OH)16(CO3,OH).4H2O	Trigonal
TALC (STEATITE)	Mg3Si4O10(OH)2	Monoclinic/Triclinic
TALMESSITE	Ca2Mg(AsO4)2.2H2O	Triclinic
TALNAKHITE	Cu9(Fe,Ni)8S16	Isometric
TAMARUGITE	NaAl(SO4)2.6H2O	Monoclinic
TANCOITE	HNa2LiAl(PO4)2(OH)	Orthorhombic
TANEYAMALITE	Na(Mn++,Mg,Fe++)12Si12(O,OH)44	Triclinic
TANGEITE	CaCu(VO4)(OH)	Orthorhombic
TANTALAESCHYNITE-	(Y,Ce,Ca)(Ta,Ti,Nb)2O6	Orthorhombic
TANTALCARBIDE	TaC	Isometric
TANTEUXENITE-	(Y,Ce,Ca)(Ta,Nb,Ti)2(O,OH)6	Orthorhombic
TANTITE	Ta2O5	Triclinic
TARAMELLITE	Ba4(Fe+++,Ti,Fe++,Mg)4(B2Si2O27)O2Clx	Orthorhombic
TARAMITE	Na2Ca(Fe++,Mg)3Al2(Si6Al2)O22(OH)2	Monoclinic
TARANAKITE	H7K2(Al,Fe+++)5(PO4)8.20H2O	Trigonal
TARAPACAITE	K2CrO4	Orthorhombic
TARBUTTITE	Zn2(PO4)(OH)	Triclinic
TATARSKITE	Ca6Mg2(SO4)2(CO3)2Cl4(OH)4.7H2O	Orthorhombic
TAUSONITE	SrTiO3	Isometric
TAVORITE	LiFe+++(PO4)(OH)	Triclinic
TAZHERANITE	(Zr,Ti,Ca)O2	Isometric
TEALLITE	PbSnS2	Orthorhombic
TEEPLEITE	Na2B(OH)4Cl	Tetragonal

Mineral Name	Formula	Crystal System
TEINEITE	$CuTeO_3.2H_2O$	Orthorhombic
TELARGPALITE	$(Pd,Ag)_3Te$	Isometric
TELLURANTIMONY	Sb_2Te_3	Trigonal
TELLURITE	TeO_2	Orthorhombic
TELLURIUM	Te	Trigonal
TELLUROBISMUTHITE	Bi_2Te_3	Trigonal
TELLUROHAUCHECORNITE	Ni_9BiTeS_8	Tetragonal
TELLUROPALLADINITE	Pd_9Te_4	Monoclinic
TEMAGAMITE	Pd_3HgTe_3	Orthorhombic
TENGCHONGITE	$CaU^{++++++}6Mo^{++++++}2O25.12H_2O$	Orthorhombic
TENGERITE-	$Y_3(CO_3)4.2-3H_2O$	Orthorhombic
TENNANTITE	$(Cu,Fe)_{12}As_4S_{13}$	Isometric
TENORITE	CuO	Monoclinic
TEPHROITE	Mn_2SiO_4	Orthorhombic
TERLINGUAITE	$Hg^+Hg^{++}ClO$	Monoclinic
TERSKITE	$Na_4Zr(H_4Si_6O_{18})$	Orthorhombic
TERTSCHITE	$Ca_4B_{10}O_{19}.20H_2O$	Monoclinic
TERUGGITE	$Ca_4MgAs_2B_{12}O_{22}(OH)12.12H_2O$	Monoclinic
TESCHEMACHERITE	$(NH_4)HCO_3$	Orthorhombic
TESTIBIOPALLADITE	$PdTe(Sb,Te)$	Isometric
TETRAAURICUPRIDE	$AuCu$	Tetragonal
TETRADYMITE	Bi_2Te_2S	Trigonal
TETRAFERROPLATINUM	$PtFe$	Tetragonal
TETRAHEDRITE	$(Cu,Fe)_{12}Sb_4S_{13}$	Isometric
TETRANATROLITE	$Na_2Al_2Si_3O_{10}.2H_2O$	Tetragonal
TETRAROOSEVELTITE	$BiAsO_4$	Tetragonal
TETRATAENITE	$FeNi$	Tetragonal
TETRAWICKMANITE	$Mn^{++}Sn^{++++}(OH)_6$	Tetragonal
THADEUITE	$(Ca,Mn^{++})(Mg,Fe^{++},Mn^{+++})3(PO_4)2(OH,F)2$	Orthorhombic
THALCUSITE	$TlCu_3FeS_4$	Tetragonal
THALENITE-	$Y_3Si_3O_{10}(F,OH)$	Monoclinic
THALFENISITE	$(Tl,K)6Na(Fe,Ni,Cu)25S26Cl$	Isometric
THAUMASITE	$Ca_3Si(CO_3)(SO_4)(OH)6.12H_2O$	Hexagonal
THEISITE	$Cu_5Zn_5(As^{+++++},Sb^{+++++})2O8(OH)14$	Orthorhombic
THENARDITE	Na_2SO_4	Orthorhombic
THEOPHRASTITE	$Ni(OH)_2$	Trigonal
THERESMAGNANITE	$(Co,Zn,Ni)6(SO_4)(OH,Cl)10.8H_2O$	Hexagonal
THERMONATRITE	$Na_2CO_3.H_2O$	Orthorhombic
THOMETZEKITE	$Pb(Cu,Zn)2(AsO_4)2.2H_2O$	Monoclinic/Triclinic
THOMSENOLITE	$NaCaAlF_6.H_2O$	Monoclinic
THOMSONITE	$NaCa_2Al_5Si_5O_{20}.6H_2O$	Orthorhombic
THORBASTNASITE (THORBASTNAESITE)	$Th(Ca,Ce)(CO_3)2F_2.3H_2O$	Hexagonal
THOREAULITE	$Sn^{++}Ta_2O_6$	Monoclinic
THORIANITE	ThO_2	Isometric
THORIKOSITE	$Pb_3(Sb^{+++},As^{+++})O_3(OH)Cl_2$	Tetragonal
THORITE	$(Th,U)SiO_4$	Tetragonal

Mineral Name	Formula	Crystal System
THORNASITE	(Na,K)ThSi11(O,F,OH)25.8H2O	Trigonal
THOROGUMMITE	Th(SiO4)1-x(OH)4x	Tetragonal
THOROSTEENSTRUPINE	(Ca,Th,Mn)3Si4O11F.6H2O	Amorphous
THORTVEITITE	(Sc,Y)2Si2O7	Monoclinic
THORUTITE	(Th,U,Ca)Ti2(O,OH)6	Monoclinic
THREADGOLDITE	Al(UO2)2(PO4)2(OH).8H2O	Monoclinic
TIEMANNITE	HgSe	Isometric
TIENSHANITE	BaNa2MnTiB2Si6O20	Hexagonal
TIETTAITE	(Na,K)17Fe+++TiSi16O29(OH)30.2H2O	Orthorhombic
TIKHONENKOVITE	SrAlF4(OH).H2O	Monoclinic
TILASITE	CaMg(AsO4)F	Monoclinic
TILLEYITE	Ca5Si2O7(CO3)2	Monoclinic
TIN (ETAIN,ZINN)	Sn	Tetragonal
TINAKSITE	K2Na(Ca,Fe++,Mn++,Mg)2(Ti,Fe)Si7O19(OH)	Triclinic
TINCALCONITE	Na2B4O5(OH)4.3H2O	Trigonal
TINNUNKULITE	C10H12N6O8	
TINSLEYITE	KAl2(PO4)2(OH).2H2O	Monoclinic
TINTICITE	Fe+++4(PO4)3(OH)3.5H2O	Monoclinic
TINTINAITE	Pb22Cu+4(Sb,Bi)30S69	Orthorhombic
TINZENITE	(Ca,Mn,Fe)3Al2BO3Si4O12(OH)	Triclinic
TIPTOPITE	K2(Na,Ca)2Li3Be6(PO4)6(OH)2.H2O	Hexagonal
TIRAGALLOITE	Mn++4As+++++Si3O12(OH)	Monoclinic
TIRODITE	Mn++2(Mg,Fe++)5Si8O22(OH)2	Monoclinic
TISINALITE	Na3H3(Mn++,Ca,Fe)TiSi6(O,OH)18.2H2O	Trigonal
TITANITE (SPHENE)	CaTiSiO5=CaTiOSiO4	Monoclinic
TITANIUM	Ti	Hexagonal
TITANOWODGINITE	(Mn++,Fe++)(Ti,Sn,Ta,Sc)(Ta,Nb)2O8	Monoclinic
TITANTARAMELLITE	Ba4(Ti,Fe+++,Fe++,Mg)4(B2Si8O27)O2Clx	Orthorhombic
TIVANITE	V+++TiO3(OH)	Monoclinic
TLALOCITE	(Cu,Zn)16(Te++++O3)(Te++++++O4)2Cl(OH)25.27H2O	Monoclinic
TLAPALLITE	H6(Ca,Pb)2(Cu,Zn)3(SO4)(Te++++O3)4(Te++++++O6)	Monoclinic
TOBELITE	(NH4,K)Al2(Si3Al)O10(OH)2	Monoclinic
TOBERMORITE	Ca5Si6O16(OH)2.4H2O	Orthorhombic
TOCHILINITE	6Fe0,9S.5(Mg,Fe++)(OH)2	Triclinic
TOCORNALITE	(Ag,Hg)I	
TODOROKITE	(Mn++,Ca,Mg)Mn++++3O7.H2O	Monoclinic
TOKKOITE	K2Ca4[Si7O18(OH)](F,OH)	Triclinic
TOLBACHITE	CuCl2	Monoclinic
TOLOVKITE	IrSbS	Isometric
TOMBARTHITE-	Y4(Si,H4)4O12-x(OH)4+2x	Monoclinic
TOMICHITE	AsTi3(V,Fe)4O13(OH)	Monoclinic
TONGBAITE	Cr3C2	Orthorhombic
TOOELEITE	Fe+++7,5-8(AsO4,SO4)6(OH)6.5H2O	Orthorhombic
TOPAZ	Al2SiO4(F,OH)2	Orthorhombic
TORBERNITE	Cu(UO2)2(PO4)2.8-12H2O	Tetragonal

Mineral Data

Mineral Name	Formula	Crystal System
TORNEBOHMITE-(Ce) (TOERNEBOHMITE-(Ce))	$(Ce,La)2Al(SiO4)2(OH)$	Monoclinic
TORNEBOHMITE-(La) (TOERNEBOHMITE-(La))	$(La,Ce)2Al(SiO4)2(OH)$	Monoclinic
TORREYITE	$(Mg,Mn)9Zn4(SO4)2(OH)22.8H2O$	Monoclinic
TOSUDITE (CHLORITE-SMECTITE)	$Na0,5(Al,Mg)6(Si,Al)8O18(OH)12.5H2O$	Orthorhombic
TOUNKITE	$(Na,Ca,K)8Al6Si6O24(SO4)2Cl.H2O$	Hexagonal
TOYOHAITE	$Ag2FeSn3S8$	Tetragonal
TRABZONITE	$Ca4Si3O10.2H2O$	Monoclinic
TRANQUILLITYITE	$Fe++8(Zr,Y)2Ti3Si3O24$	Hexagonal
TRASKITE	$Ba9Fe++2Ti2(SiO3)12(OH,Cl,F)6.6H2O$	Hexagonal
TREASURITE	$Ag7Pb6Bi15S32$	Monoclinic
TRECHMANNITE	$AgAsS2$	Trigonal
TREMBATHITE	$(Mg,Fe++)3B7O13Cl$	Trigonal
TREMOLITE	$Ca2(Mg,Fe++)5Si8O22(OH)2$	Monoclinic
TREVORITE	$NiFe+++2O4$	Isometric
TRIANGULITE	$Al3(UO2)4(PO4)4(OH)5.5H2O$	Triclinic
TRIDYMITE	$SiO2$	Monoclinic ps Hexagonal
TRIGONITE	$Pb3Mn(As+++O3)2(As+++O2OH)$	Monoclinic
TRIKALSILITE	$(K,Na)AlSiO4$	Hexagonal
TRIMERITE	$Ca,Mn2Be3(SiO4)3$	Monoclinic
TRIMOUNSITE-	$(Y,Dy,Er,Yb,Gd,Ho,Tb,Sm)2Ti2SiO9$	Monoclinic
TRIPHYLITE	$LiFe++PO4$	Orthorhombic
TRIPLITE	$(Mn,Fe++,Mg,Ca)2(PO4)(F,OH)$	Monoclinic
TRIPLOIDITE	$(Mn,Fe++)2(PO4)(OH)$	Monoclinic
TRIPPKEITE	$CuAs+++2O4$	Tetragonal
TRIPUHYITE	$Fe++Sb+++++2O6$	Tetragonal
TRISTRAMITE	$(Ca,U++++,Fe+++)(PO4,SO4).2H2O$	Hexagonal
TRITOMITE- (SPENCITE)	$(Y,Ca,La,Fe++)5(Si,B,Al)3(O,OH,F)13$	Trigonal
TRITOMITE-(Ce)	$(Ce,La,Y,Th)5(Si,B)3(O,OH,F)13$	Trigonal
TROGERITE (TROEGERITE)	$(UO2)3(AsO4)2.12H2O$	Tetragonal
TROGTALITE	$CoSe2$	Isometric
TROILITE	FeS	Hexagonal
TROLLEITE	$Al4(PO4)3(OH)3$	Monoclinic
TRONA	$Na3(CO3)(HCO3).2H2O$	Monoclinic
TRUSCOTTITE	$(Ca,Mn)14Si24O58(OH)8.2H2O$	Hexagonal
TRUSTEDTITE (TRUESTEDTITE)	$Ni3Se4$	Isometric
TSAREGORODTSEVITE	$N(CH3)4AlSi5O12$	Orthorhombic ps Cubic
TSCHERMAKITE	$Ca2(Mg,Fe++)3Al2(Si7Al)O22(OH)2$	Monoclinic
TSCHERMIGITE	$(NH4)Al(SO4)2.12H2O$	Isometric
TSCHERNICHITE	$Ca(Al2Si6O16).8H2O$	Tetragonal/Monoclinic
TSNIGRIITE	$Ag9SbTe3(S,Se)3$	Monoclinic
TSUMCORITE	$PbZnFe++(AsO4)2.H2O$	Monoclinic
TSUMEBITE	$Pb2Cu(PO4)(SO4)(OH)$	Monoclinic
TSUMOITE	$BiTe$	Trigonal
TUCEKITE	$Ni9Sb2S8$	Tetragonal

Mineral Data

Mineral Name	Formula	Crystal System
TUGARINOVITE	MoO2	Monoclinic
TUGTUPITE	Na4AlBeSi4O12Cl	Tetragonal
TUHUALITE	(Na,K)Fe++Fe+++Si6O15	Orthorhombic
TULAMEENITE	Pt(Cu,Fe)	Tetragonal
TULIOKITE	Na6BaTh(CO3)6.6H2O	Hexagonal
TUNDRITE-(Ce)	Na3(Ce,La)4(Ti,Nb)2(SiO4)2(CO3)3O4(OH).2H2O	Triclinic
TUNDRITE-(Nd)	Na3(Nd,La)4(Ti,Nb)2(SiO4)2(CO3)3O4(OH).2H2O	Triclinic
TUNELLITE	SrB6O9(OH)2.3H2O	Monoclinic
TUNGSTENITE-2H	WS2	Hexagonal
TUNGSTENITE-3R	WS2	Trigonal
TUNGSTITE	WO3.H2O	Orthorhombic
TUNGUSITE	Ca14Fe++9(Si4O10)6(OH)22	Triclinic
TUNISITE	NaCa2Al4(CO3)4(OH)8Cl	Tetragonal
TUPERSSUATSIAITE	NaFe+++3Si8O20.5H2O	Monoclinic
TURANITE	Cu5(VO4)2(OH)4	Orthorhombic
TURNEAUREITE	Ca5[(As,P)O4]3Cl	Hexagonal
TURQUOISE	CuAl6(PO4)4(OH)8.4H2O	Triclinic
TUSCANITE	K(Ca,Na)6(Si,Al)10O22(SO4,CO3,(OH)2).H2O	Monoclinic
TUSIONITE	MnSn++++(BO3)2	Trigonal
TUZLAITE	NaCaB5O8(0H)2.3H2O	Monoclinic
TVALCHRELIDZEITE	Hg12(Sb,As)8S15	Monoclinic
TVEDALITE	(Ca,Mn++,Fe++)4Be3Si6O17(OH)4.3H2O	Orthorhombic
TVEITITE-	Ca1-xYxF2+x	Monoclinic ps Cubic
TWINNITE	Pb(Sb,As)2S4	Orthorhombic
TYCHITE	Na6Mg2(CO3)4(SO4)	Isometric
TYRETSKITE-1Tc	Ca2B5O9(OH).H2O	Triclinic
TYROLITE	CaCu5(AsO4)2(CO3)(OH)4.6H2O	Orthorhombic
TYRRELLITE	(Cu,Co,Ni)3Se4	Isometric
TYUYAMUNITE	Ca(UO2)2V2O8.5-8H2O	Orthorhombic
UCHUCCHACUAITE	AgPb3MnSb5S12	Monoclinic ps Orthorhombic
UHLIGITE	Ca3(Ti,Al,Zr)9O20	Isometric
UKLONSKOVITE	NaMg(SO4)F.2H2O	Monoclinic
ULEXITE	NaCaB5O6(OH)6.5H2O	Triclinic
ULLMANNITE	NiSbS	Triclinic ps Cubic
ULRICHITE	CaCu(UO2)(PO4)2.4H2O	Monoclinic
ULVOSPINEL (ULVITE)	TiFe++2O4	Isometric
UMANGITE	Cu3Se2	Tetragonal
UMBITE	K2ZrSi3O9.H2O	Orthorhombic
UMBOZERITE	Na3Sr4ThSi8(O,OH)24	Amorphous
UMOHOITE	(UO2)MoO4.4H2O	Monoclinic/Orthorhombic
UNGARETTITE	(Na,K)Na2(Mn++,Mg)2Mn+++Si8O22O2	Monoclinic
UNGEMACHITE	K3Na8Fe+++(SO4)6(NO3)2.6H2O	Trigonal
UNGURSAITE =SODIANCALCIOTANTITE	NaCa5(Ta,Nb)24O65(OH)	Hexagonal
UPALITE	Al(UO2)3(PO4)2O(OH).7H2O	Monoclinic

Mineral Name	Formula	Crystal System
URALBORITE	$CaB_2O_2(OH)_4$	Monoclinic
URALOLITE	$Ca_2Be_4(PO_4)_3(OH).3.5H_2O$	Monoclinic
URAMPHITE	$(NH_4)(UO_2)(PO_4).3H_2O$	Orthorhombic
URANCALCARITE	$Ca(UO_2)_3(CO_3)(OH)_6.3H_2O$	Orthorhombic
URANINITE	UO_2	Isometric
URANMICROLITE (DJALMAITE)	$(U,Ca,Ce)_2(Ta,Nb)_2O_6(OH,F)$	Isometric
URANOCIRCITE	$Ba(UO_2)_2(PO_4)_2.12H_2O$	Tetragonal
URANOPHANE (URANOTILE)	$Ca(UO_2)_2SiO_3(OH)_2.5H_2O$	Monoclinic
URANOPHANEBETA	$Ca(UO_2)_2SiO_3(OH)_2.5H_2O$	Monoclinic
URANOPILITE	$(UO_2)_6(SO_4)(OH)_{10}.12H_2O$	Monoclinic
URANOPOLYCRASE	$(U,Y)(Ti,Nb,Ta)_2O_6$	Orthorhombic
URANOSILITE	$U^{++++++}Si_7O_{17}$	Orthorhombic
URANOSPATHITE	$HAl(UO_2)_4(PO_4)_4.40H_2O$	Tetragonal
URANOSPHAERITE	$Bi_2U_2O_9.3H_2O$	Monoclinic
URANOSPINITE	$Ca(UO_2)_2(AsO_4)_2.10H_2O$	Tetragonal
URANOTUNGSTITE	$(Ba,Pb,Fe^{++})(UO_2)_2WO_4(OH)_4.12H_2O$	Orthorhombic
URANPYROCHLORE	$(U,Ca,Ca)_2(Nb,Ta)_2O_6(OH,F)$	Isometric
UREA	$CO(NH_2)_2$	Tetragonal
URICITE	$C_5H_4N_4O_3$	Monoclinic
URSILITE	$(Mg,Ca)_4[(UO_2)_4(OH)_5/(Si_2O_5)_5,_5].13H_2O$	Orthorhombic
URVANTSEVITE	$Pd(Bi,Pb)_2$	Tetragonal
USHKOVITE	$MgFe^{+++}_2(PO_4)_2(OH)_2.8H_2O$	Triclinic
USOVITE	$Ba_2CaMgAl_2F_{14}$	Monoclinic
USSINGITE	$Na_2AlSi_3O_8(OH)$	Triclinic
USTARASITE	$Pb(Bi,Sb)_6S_{10}$	Orthorhombic
UVANITE	$U^{++++++}_2V^{+++++}_6O_{21}.15H_2O$	Orthorhombic
UVAROVITE	$Ca_3Cr_2(SiO_4)_3$	Isometric
UVITE	$(Ca,Na)(Mg,Fe^{++})_3Al_5Mg(BO_3)_3Si_6O_{18}(OH,F)_4$	Trigonal
UYTENBOGAARDTITE	Ag_3AuS_2	Tetragonal
UZONITE (USONITE)	As_4S_5	Monoclinic
VAESITE	NiS_2	Isometric
VALENTINITE	Sb_2O_3	Orthorhombic
VALLERIITE	$4(Fe,Cu)S.3(Mg,Al)(OH)_2$	Hexagonal
VANADINITE	$Pb_5(VO_4)_3Cl$	Hexagonal
VANADOMALAYAITE	$Ca(V^{++++},Ti^{++++})OSiO_4$	Monoclinic
VANALITE	$NaAl_8V_{10}O_{38}.30H_2O$	Monoclinic
VANDENBRANDEITE	$Cu(UO_2)(OH)_4$	Triclinic
VANDENDRIESSCHEITE	$PbU^{++++++}_7O_{22}.12H_2O$	Orthorhombic
VANMEERSSCHEITE	$U^{++++++}(UO_2)_3(PO_4)_2(OH)_6.4H_2O$	Orthorhombic
VANOXITE	$V^{++++}_4V^{+++++}_2O_{13}.8H_2O$	Trigonal
VANTASSELITE	$Al_4(PO_4)_3(OH)_3.9H_2O$	Orthorhombic
VANTHOFFITE	$Na_6Mg(SO_4)_4$	Monoclinic
VANURALITE	$(H_3O,Ba,Ca,K)_2(UO_2)_2V_2O_8.4H_2O$	Monoclinic
VANURANYLITE	$(H_3O,Ba,Ca,K)_{1,6}(UO_2)_2V_2O_8.4H_2O$	Orthorhombic
VARENNESITE	$Na_8(Mn,Fe^{+++},Ti)_2Si_{10}O_{25}(OH,Cl)_2.12H_2O$	Orthorhombic

Mineral Name	Formula	Crystal System
VARISCITE	AlPO4.2H2O	Orthorhombic
VARLAMOFFITE	(Sn,Fe)(O,OH)2	Isometric
VARULITE	NaCaMn(Mn,Fe++,Fe+++)2(PO4)3	Monoclinic
VASHEGYITE	Al11(PO4)9(OH)6.38H2OorAl6(PO4)5(OH)3.23H2O	Orthorhombic
VASILITE	(Pd,Cu)16(S,Te)7	Isometric
VATERITE	CaCO3	Hexagonal
VAUGHANITE	TlHgSb4S7	Triclinic
VAUQUELINITE	Pb2Cu(CrO4)(PO4)(OH)	Monoclinic
VAUXITE	Fe++Al2(PO4)2(OH).6H2O	Triclinic
VAYRYNENITE	MnBe(PO4)(OH,F)	Monoclinic
VEATCHITE	Sr2B11O16(OH)5.H2O	Monoclinic
VEATCHITE-A	Sr2B11O16(OH)5.H2O	Triclinic
VEATCHITE-P (P-VEATCHITE)	Sr2B11O16(OH)5.H2O	Monoclinic
VEENITE	Pb2(Sb,As)2S5	Orthorhombic
VELIKITE	Cu2HgSnS4	Tetragonal
VERMICULITE	(Mg,Fe++,Al)3(Al,Si)4O10(OH)2.4H2O	Monoclinic
VERNADITE	(gamma-MnO2),(Mn++++,Fe+++,Ca,Na)(O,OH)2.nH2O	Hexagonal
VERPLANCKITE	Ba2(Mn,Fe++,Ti)Si2O6(O,OH,Cl,F)2.3H2O	Hexagonal
VERSILIAITE	Fe++4Fe+++8Sb+++12S2	Orthorhombic
VERTUMNITE	Ca2Al[(OH)6AlSiO2-3(OH)4-3].2,5H2O	Monoclinic ps Hexagonal
VESIGNIEITE	Cu3Ba(VO4)2(OH)2	Monoclinic
VESUVIANITE (IDOCRASE)	Ca10Mg2Al4(SiO4)5(Si2O7)2(OH)4	Tetragonal
VESZELYITE	(Cu,Zn)3(PO4)(OH)3.2H2O	Monoclinic
VIAENEITE	11FeS2.(Fe,Pb)(S2O3)	Monoclinic
VICANITE-(Ce)	Na0,5(Ce,Ca,Th)15Fe+++As+++0,5As+++++B4Si6O40F7	Trigonal
VIGEZZITE	(Ca,Ce)(Nb,Ta,Ti)2O6	Orthorhombic
VIHORLATITE	Bi8+x(Se,Te,S)11-x	Trigonal
VIITANIEMIITE	Na(Ca,Mn++)Al(PO4)(F,OH)3	Monoclinic
VIKINGITE	Ag5Pb8Bi13S30	Monoclinic
VILLAMANINITE	(Cu,Ni,Co,Fe)S2	Isometric
VILLIAUMITE	NaF	Isometric
VILLYAELLENITE	(Mn,Ca,Zn)5(AsO4)2(AsO3OH).4H2O	Monoclinic
VIMSITE	CaB2O2(OH)4	Monoclinic
VINCENTITE	(Pd,Pt)3(As,Sb,Te)	
VINCIENNITE	Cu11Fe4Sn(As,Sb)S16	Tetragonal ps Cubic
VINOGRADOVITE	Na8Ti8O8(Si2O6)4[(Si3Al)O10]2[(H2O),(Na,K)2]	Monoclinic
VIOLARITE	Fe++Ni+++2S4	Isometric
VIRGILITE	LixAlxSi3-xO6	Hexagonal
VISEITE	Ca10Al24(SiO4)6(PO4)7O22F3.72H2O	Isometric
VISHNEVITE	(Na,Ca,K)6(Si,Al)12O24[(SO4),(CO3),Cl2]2-4.nH2O	Hexagonal
VISMIRNOVITE	ZnSn++++(OH)6	Isometric
VISTEPITE	Mn++5Sn++++B2(SiO4)5	Orthorhombic
VITUSITE-(Ce)	Na3(Ce,La,Nd)(PO4)2	Orthorhombic

Mineral Name	Formula	Crystal System
VIVIANITE	Fe++3(PO4)2.8H2O	Monoclinic
VLADIMIRITE	Ca5H2(AsO4)4.5H2O	Monoclinic
VLASOVITE	Na2ZrSi4O11	Monoclinic/Triclinic
VOCHTENITE	(Fe++,Mg)Fe+++[UO2/PO4]4(OH).12-13H2O	Monoclinic
VOGGITE	Na2Zr(PO4)(CO3)(OH).2H2O	Monoclinic
VOGLITE	Ca2Cu(UO2)(CO3)4.6H2O	Monoclinic
VOLBORTHITE	Cu++3V+++++2O7(OH)2.2H2O	Monoclinic
VOLFSONITE	Cu+10Cu++Fe++Fe+++2Sn++++3S16	Tetragonal
VOLKONSKOITE (VOLCHONSKOITE)	Ca0,3(Cr+++,Mg,Fe+++)2(Si,Al)4O10(OH)2.4H2O	Monoclinic
VOLKOVSKITE	KCa4[B5O8(OH)4][B(OH)3]Cl.4H2O	Triclinic
VOLTAITE	K2Fe++5Fe+++4(SO4)12.18H2O	Isometric
VOLYNSKITE	AgBiTe2	Orthorhombic
VONSENITE	Fe++2Fe+++BO5	Orthorhombic
VOZHMINITE	(Ni,Co)4(As,Sb)S2	Hexagonal
VRBAITE	Tl4Hg3Sb2As8S20	Orthorhombic
VUAGNATITE	CaAlSiO4(OH)	Orthorhombic
VULCANITE	CuTe	Orthorhombic
VUONNEMITE	Na11Nb2Ti+++Si4O12(PO4)2O5F2	Triclinic
VUORELAINENITE	(Mn++,Fe++)(V+++,Cr+++)2O4	Isometric
VYACHESLAVITE	U++++(PO4)(OH).2.5H2O	Orthorhombic
VYALSOVITE	FeS.Ca(OH)2.Al(OH)3	Orthorhombic
VYSOTSKITE	(Pd,Ni)S	Tetragonal
VYUNTSPAKHKITE-	Y4Al2AlSi5O18(OH)5	Monoclinic
WADALITE	(Ca,Mg)6(Al,Fe+++)4(AlSi2O16)Cl3	Isometric
WADEITE	(K,Na)2ZrSi3O9	Hexagonal
WADSLEYITE	(Mg,Fe++)2SiO4	Orthorhombic
WAGNERITE	(Mg,Fe++)2(PO4)F	Monoclinic
WAIRAKITE	CaAl2Si4O12.2H2O	Monoclinic
WAIRAUITE	CoFe	Isometric
WAKABAYASHILITE	(As,Sb)11S18	Monoclinic
WAKEFIELDITE-	YVO4	Tetragonal
WAKEFIELDITE-(Ce)	(Ce,La,Nd,Y,Pr,Sm)(V,As)O4	Tetragonal
WAKEFIELDITE-(Ce) (KUSUITE)	(Ce+++,Pb++,Pb++++)VO4	Tetragonal
WALENTAITE	H(Ca,Mn,Fe++)Fe+++3[(AsO4,PO4)]4.7H2O	Orthorhombic
WALLISITE	PbTl(Cu,Ag)As2S5	Triclinic
WALLKILLDELLITE	Ca4Mn++6As+++++4O16(OH)8.18H2O	Hexagonal
WALPURGITE	Bi4(UO2)(AsO4)2O4.2H2O	Triclinic
WALSTROMITE	BaCa2Si3O9	Triclinic
WALTHIERITE	Ba0,5Al3(SO4)2(OH)6	Trigonal
WARDITE	NaAl3(PO4)2(OH)4.2H2O	Tetragonal
WARDSMITHITE	Ca5MgB24O42.30H2O	Hexagonal
WARIKAHNITE	Zn3(AsO4)2.2H2O	Triclinic
WARWICKITE	(Mg,Ti,Fe+++,Al)2(BO3)O	Orthorhombic
WATANABEITE	Cu4(As,Sb)2S5	Orthorhombic ps Cubic
WATKINSONITE	PbCu2Bi4(Se,S)8	Monoclinic

Mineral Data

Mineral Name	Formula	Crystal System
WATTERSITE	Hg+4Hg++Cr++++++O6	Monoclinic
WATTEVILLITE	Na2Ca(SO4)2.4H2O	Orthorhombic
WAVELLITE	Al3(PO4)2(OH,F)3.5H2O	Orthorhombic
WAWAYANDAITE	Ca12Mn4B2Be18Si12O46(OH,Cl)30[Ortho.]	Monoclinic
WAYLANDITE	BiAl3(PO4)2(OH)6	Trigonal
WEBERITE	Na2MgAlF7	Orthorhombic
WEDDELLITE (CALCIUMOXALATE)	Ca(C2O4).2H2O	Tetragonal
WEEKSITE	K2(UO2)2Si6O15.4H2O	Orthorhombic
WEGSCHEIDERITE	Na5(CO3)(HCO3)3	Triclinic
WEIBULLITE	(Pb,Ag)6Bi8(S,Se)18	Orthorhombic
WEILITE	CaHAsO4	Triclinic
WEINEBENEITE	CaBe3(PO4)2(OH)2.5H2O	Monoclinic
WEISHANITE	(Au,Ag)3Hg2	Hexagonal
WEISSBERGITE	TlSbS2	Triclinic
WEISSITE	CuTe	Hexagonal Cubic
WELINITE	Mn++3(W++++++,Mg)0,7SiO4(O,OH)3	Hexagonal
WELLSITE	(Ba,Ca,K2)Al2Si6O16.6H2O	Monoclinic
WELOGANITE	Sr3Na2Zr(CO3)6.3H2O	Triclinic psTrigonal
WELSHITE	Ca2Sb+++++Mg4Fe+++Si4Be2O20	Triclinic
WENDWILSONITE	Ca2(Mg,Co)(AsO4)2.2H2O	Monoclinic
WENKITE	Ba4Ca6(Si,Al)20O39(OH)2(SO4)3.nH2O	Hexagonal
WERDINGITE	(Mg,Fe)2Al12(Al,Fe)2Si4(B,Al)4O37	Triclinic
WERMLANDITE	(Ca,Mg)Mg7(Al,Fe+++)2(SO4)2(OH)18.12H2O	Trigonal
WESTERVELDITE	(Fe,Ni,Co)As	Orthorhombic
WHEATLEYITE	Na2Cu(C2O4)2.2H2O	Triclinic
WHERRYITE	Pb7Cu2(SO4)4(SiO4)2(OH)2	Monoclinic
WHEWELLITE	Ca(C2O4).H2O	Monoclinic
WHITEITE-(CaFeMg)	Ca(Fe++,Mn++)Mg2Al2(PO4)4(OH)2.8H2O	Monoclinic
WHITEITE-(CaMnMg)	CaMnMg2Al2(PO4)4(OH)2.8H2O	Monoclinic
WHITEITE-(MnFeMg)	(Mn++,Ca)(Fe++,Mn++)Mg2Al2(PO4)4(OH)2.8H2O	Monoclinic
WHITLOCKITE	Ca9(Mg,Fe++)(PO4)6[PO3(OH)]	Trigonal
WHITMOREITE	Fe++Fe+++2(PO4)2(OH)2.4H2O	Monoclinic
WICKENBURGITE	Pb3CaAl2Si10O27.3H2O	Trigonal
WICKMANITE	Mn++Sn++++(OH)6	Isometric
WICKSITE	NaCa2(Fe++,Mn++)4MgFe+++(PO4)6.2H2O	Orthorhombic
WIDENMANNITE	Pb2(UO2)(CO3)3	Orthorhombic
WIDGIEMOOLTHALITE	(Ni,Mg)5(CO3)4(OH)2.4-5H2O	Monoclinic
WIGHTMANITE	Mg5(BO3)O(OH)5.2H2O	Monoclinic
WILCOXITE	MgAl(SO4)2F.18H2O	Triclinic
WILHELMVIERLINGITE	CaMn++Fe+++(PO4)2(OH).2H2O	Orthorhombic
WILKINSONITE	Na2Fe++4Fe+++2Si6O20	Triclinic
WILKMANITE	Ni3Se4	Monoclinic
WILLEMITE	Zn2SiO4	Trigonal
WILLEMSEITE	(Ni,Mg)3Si4O10(OH)2	Monoclinic
WILLHENDERSONITE	KCaAl3Si3O12.5H2O	Triclinic

Mineral Name	Formula	Crystal System
WILLYAMITE	(Co,Ni)SbS	Monoclinic/Triclinic
WINCHITE	NaCa(Mg,Fe++)4AlSi8O22(OH)2	Monoclinic
WINSTANLEYITE	TiTe++++3O8	Isometric
WISERITE	Mn4B2O5(OH,Cl)4	Tetragonal
WITHERITE	BaCO3	Orthorhombic
WITTICHENITE	Cu3BiS3	Orthorhombic
WITTITE	Pb3Bi4(S,Se)9	Monoclinic
WODGINITE	(Ta,Nb,Sn,Mn,Fe)16O32	Monoclinic
WOHLERITE (WOEHLERITE)	NaCa2(Zr,Nb)Si2O7(O,OH,F)2	Monoclinic
WOLFEITE	(Fe++,Mn++)2(PO4)(OH)	Monoclinic
WOLFRAMOIXIOLITE	(Fe,Mn,Nb)(Nb,W,Ta)O4	Monoclinic
WOLLASTONITE-1T	CaSiO3	Triclinic
WOLLASTONITE-2M	CaSiO3	Monoclinic
WOLLASTONITE-7T	CaSiO3	Triclinic
WOLSENDORFITE (WOELSENDORFITE)	(Pb,Ca)U2O7.2H2O	Orthorhombic
WONESITE	(Na,K)(Mg,Fe,Al)6(Si,Al)8O20(OH,F)4	Monoclinic
WOODHOUSEITE	CaAl3(PO4)(SO4)(OH)6	Trigonal
WOODRUFFITE	(Zn,Mn++)Mn++++3O7.1-2H2O	Monoclinic
WOODWARDITE	Cu4Al2(SO4)(OH)12.2-4H2O	Hexagonal
WROEWOLFEITE	Cu4(SO4)(OH)6.2H2O	Monoclinic
WULFENITE	PbMoO4	Tetragonal
WULFINGITE (WUELFINGITE)	Zn(OH)2	Orthorhombic
WUPATKIITE	(Co,Mg)Al2(SO4)4.22H2O	Monoclinic
WURTZITE	(Zn,Fe)S	Hexagonal
WUSTITE (WUESTITE)	FeO	Isometric
WYARTITE	CaU++++(UO2)6(CO3)2(OH)18.3-5H2O	Orthorhombic
WYCHEPROOFITE	NaAlZr(PO4)2(OH)2.H2O	Triclinic
WYLLIEITE	(Na,Ca,Mn++)(Mn++,Fe++)(Fe++,Fe+++,Mg)Al(PO4)3	Monoclinic
XANTHIOSITE	Ni3(AsO4)2	Monoclinic
XANTHOCONITE	Ag3AsS3	Monoclinic
XANTHOXENITE	Ca4Fe+++2(PO4)4(OH)2.3H2O	Triclinic
XENOTIME-	YPO4	Tetragonal
XIANGJIANGITE	(Fe+++,Al)(UO2)4(PO4)2(SO4)2(OH).22H2O	Tetragonal
XIFENGITE	Fe5Si3	Hexagonal
XILINGOLITE	Pb3Bi2S6	Monoclinic
XIMENGITE	BiPO4	Hexagonal
XINGZHONGITE	(Pb,Cu,Fe)(Ir.Pt,Rh)2S4	Isometric
XITIESHANITE	Fe+++(SO4)(OH).7H2O	Monoclinic
XOCOMECATLITE	Cu3Te++++++O4(OH)4	Orthorhombic
XONOTLITE	Ca6Si6O17(OH)2	Monoclinic/Triclinic
YAFSOANITE	Ca3Te2Zn3O12	Isometric
YAGIITE	(Na,K)3Mg4(Al,Mg)6(Si,Al)24O60	Hexagonal
YAKHONTOVITE	(Ca,Na,K)0,3(CuFe++Mg)2Si4O10(OH)2.3H2O	Monoclinic
YAMATOITE	(Mn++,Ca)3(V+++,Al)2(SiO4)3	Isometric

Mineral Data

Mineral Name	Formula	Crystal System
YANOMAMITE	In(AsO4).2H2O	Orthorhombic
YAROSLAVITE	Ca3Al2F10(OH)2.H2O	Orthorhombic
YARROWITE	Cu9S8	Hexagonal
YAVAPAIITE	KFe+++(SO4)2	Monoclinic
YEATMANITE	Mn++9Zn6Sb+++++2Si4O28	Triclinic
YECORAITE	Bi5Fe+++3(Te++++O3)(Te++++++O4)2O9.9H2O	
YEDLINITE	Pb6CrCl6(O,OH)8	Trigonal
YEELIMITE	Ca4Al6O12(SO4)	Isometric
YFTISITE-	(Y,Dy,Er)4(Ti,Sn)O(SiO4)2(F,OH)6	Orthorhombic
YIMENGITE	K(Cr,Ti,Fe,Mg)12O19	Hexagonal
YINGJIANGITE	(K2,Ca)(UO2)7(PO4)4(OH)6.6H2O	Orthorhombic
YIXUNITE	PtIn	Isometric
YODERITE	(Mg,Al,Fe+++)8Si4(O,OH)20	Monoclinic
YOFORTIERITE	(Mn,Mg)5Si8O20(OH)2.8-9H2O	Monoclinic
YOSHIMURAITE	(Ba,Sr)2TiMn2(SiO4)2(PO4,SO4)(OH,Cl)	Triclinic
YOSHIOKAITE	(Ca8-(x/2)_(x/2)Al16-xSixO32)	Hexagonal
YTTRIALITE-	(Y,Th)2Si2O7	Hexagonal
YTTROBETAFITE-	(Y,U,Ce)2(Ti,Nb,Ta)2O6(OH)	Isometric
YTTROCOLUMBITE-	(Y,U,Fe++)(Nb,Ta)O4	Orthorhombic
YTTROCRASITE-	(Y,Th,Ca,U)(Ti,Fe+++)2(O,OH)6	Orthorhombic
YTTROPYROCHLORE-	(Y,Na,Ca,U)1-2(Nb,Ta,Ti)2(O,OH)7	Isometric
YTTROTANTALITE-	(Y,U,Fe++)(Ta,Nb)O4	Orthorhombic
YTTROTUNGSTITE-	YW2O6(OH)3	Monoclinic
YUANFULIITE	Mg(Fe+++,Fe++,Al,Ti,Mg)(BO3)O	Orthorhombic
YUGAWARALITE	CaAl2Si6O16.4H2O	Monoclinic
YUKONITE	Ca2Fe+++3(AsO4)4(OH).12H2O	Amorphous
YUKSPORITE	(K,Ba)NaCa2(Si,Ti)4O11(F,OH).H2O	Orthorhombic
YUSHKINITE	V1-xS.n(Mg,Al)(OH)2	Hexagonal
YVONITE	Cu(AsO3.OH).2H2O	Triclinic
ZABUYELITE	Li2CO3	Monoclinic
ZAHERITE	Al12(SO4)5(OH)26.20H20	Triclinic
ZAIRITE	Bi(Fe+++,Al)3(PO4)2(OH)6	Trigonal
ZAKHAROVITE	Na4Mn++5Si10O24(OH)6.6H2O	Trigonal
ZANAZZIITE	(Ca,Mn)2(Mg,Fe)(Mg,Fe++,Mn,Fe+++)4Be4(PO4)6(OH)4.6H2O	Monoclinic
ZAPATALITE	Cu3Al4(PO4)3(OH)9.4H2O	Tetragonal
ZARATITE	Ni3(CO3)(OH)4.4H2O	Isometric
ZAVARITSKITE	BiOF	Tetragonal
ZDENEKITE	Na(Pb,Ca)Cu5(AsO4)4Cl.5H2O	Tetragonal
ZEKTZERITE	NaLiZrSi6O15	Orthorhombic
ZELLERITE	Ca(UO2)(CO3)2.5H2O	Orthorhombic
ZEMANNITE	(Zn,Fe++)2(Te++++O3)3NaxH2-x.nH2O	Hexagonal
ZEMKORITE	(Na,K)2Ca(CO3)2	Hexagonal
ZENZENITE	Pb3(Fe+++,Mn+++)4Mn++++3O15	Hexagonal
ZEOPHYLLITE	Ca4Si3O8(OH,F)4.2H2O	Triclinic ps Hexagonal
ZEUNERITE	Cu(UO2)2(AsO4)2.10-16H2O	Tetragonal

Mineral Name	Formula	Crystal System
ZHANGHENGITE	(Cu,Zn,Fe,Al,Cr)	Isometric
ZHARCHIKHITE	AlF(OH)2	Monoclinic
ZHEMCHUZHNIKOVITE	NaMg(Al,Fe+++)(C2O4)3.8H2O	Trigonal
ZHONGHUACERITE-(Ce)	Ba2Ce(CO3)3F	Trigonal
ZIESITE	Cu2V+++++2O7	Monoclinic
ZIMBABWEITE	(Na,K)2PbAs+++4(Ta,Nb,Ti)4O18	Orthorhombic
ZINALSITE	Zn2AlSi2O5(OH)4.2H2O	Monoclinic
ZINC	Zn	Hexagonal
ZINCALUMINITE	Zn6Al6(SO4)2(OH)26.5H2O	Hexagonal
ZINCITE	(Zn,Mn)O	Hexagonal
ZINCLAVENDULAN	NaCa(Zn,Cu)5(AsO4)4Cl.4-5H2O	Orthorhombic
ZINCMELANTERITE	(Zn,Cu,Fe++)SO4.7H2O	Monoclinic
ZINCOBOTRYOGEN	(Zn,Mg,Mn)Fe+++(SO4)2(OH).7H2O	Monoclinic
ZINCOCHROMITE	ZnCr2O4	Isometric
ZINCOCOPIAPITE	ZnFe+++4(SO4)6(OH)2.18H2O	Triclinic
ZINCROSASITE	(Zn,Cu)2(CO3)(OH)2	Monoclinic
ZINCROSELITE	Ca2Zn(AsO4)2.2H2O	Monoclinic
ZINCSILITE	Zn3Si4O10(OH)2.4H2O	Monoclinic
ZINCVOLTAITE	K2Zn5Fe+++3Al(SO4)12.18H2O	Isometric
ZINCZIPPEITE	Zn++0,5(UO2)2(SO4)(OH)3.H2O	Monoclinic
ZINKENITE	Pb9Sb22S42	Hexagonal
ZINKOSITE	ZnSO4	Orthorhombic
ZINNWALDITE	KLiFe++Al(AlSi3)O10(F,OH)2	Monoclinic
ZIPPEITE	K(UO2)2(SO4)(OH)3.H2O	Monoclinic
ZIRCON	ZrSiO4	Tetragonal
ZIRCONOLITE-2M	CaZrTi2O7	Monoclinic
ZIRCONOLITE-3O	(Ca,Fe,Y,Th)2Fe(Ti,Nb)3Zr2O14	Orthorhombic
ZIRCONOLITE-3T	CaZrTi2O7	Tetragonal
ZIRCOPHYLLITE	(K,Na,Ca)3(Mn,Fe++)7(Zr,Nb)2Si8O27(OH,F)4	Triclinic
ZIRCOSULFATE	Zr(SO4)2.4H2O	Orthorhombic
ZIRKELITE	(Ca,Th,Ce)Zr(Ti,Nb)2O7	Monoclinic ps Cubic
ZIRKLERITE	(Fe++,Mg)9Al4Cl18(OH)12.14H2O	Trigonal
ZIRSINALITE	Na6(Ca,Mn,Fe++)ZrSi6O18	Trigonal
ZNUCALITE	CaZn11(UO2)(CO3)3(OH)20.4H2O	Triclinic
ZODACITE	Ca4Mn++Fe+++4(PO4)6(OH)4.12H2O	Monoclinic
ZOISITE	Ca2Al3(SiO4)3(OH)=Ca2AlAl2(SiO4)(Si2O7)O(OH)	Orthorhombic
ZORITE	Na6(Ti,Nb)2(Si6O17)2(O,OH).11H2O	Orthorhombic
ZOUBEKITE	AgPb4Sb4S10	Orthorhombic
ZUNYITE	Al13Si5O20(OH,F)18Cl	Isometric
ZUSSMANITE	K(Fe++,Mg,Mn)13(Si,Al)18O42(OH)14	Trigonal
ZVYAGINTSEVITE	(Pd,Pt,Au)3(Pb,Sn)	Isometric
ZWIESELITE	Fe++,Mn)2(PO4)F	Monoclinic
ZYKAITE	Fe+++4(AsO4)3(SO4)(OH).15H2O	Orthorhombic

Mineral Data by Chemical Formula

Alphabetical listing of mineral name, formula and crystal system listed by chemical formula.

Mineral Name	Formula	Crystal System
KURILITE	(Ag,Au)2(Te,Se,S)	Isometric
MUTHMANNITE	(Ag,Au)Te	Orthorhombic
PAVONITE	(Ag,Cu)(Bi,Pb)3S5	Monoclinic
MAKOVICKYITE	(Ag,Cu)1,5(Bi,Pb)5,5S9	Monoclinic
ARGENTOTENNANTITE	(Ag,Cu)10(Zn,Fe)2(As,Sb)4S13	Isometric
ARSENPOLYBASITE	(Ag,Cu)16(As,Sb)2S11	Monoclinic
ANTIMONPEARCEITE	(Ag,Cu)16(Sb,As)2S11	Monoclinic
POLYBASITE	(Ag,Cu)16Sb2S11	Monoclinic ps Hexagonal
MCKINSTRYITE	(Ag,Cu)2S	Orthorhombic
BENJAMINITE	(Ag,Cu)3(Bi,Pb)7S12	Monoclinic
PENZHINITE	(Ag,Cu)4Au(S,Se)4	Hexagonal
MIERSITE	(Ag,Cu)I	Isometric
FREIBERGITE	(Ag,Cu,Fe)12(Sb,As)4S13	Isometric
GEFFROYITE	(Ag,Cu,Fe)9(Se,S)8	Isometric
TOCORNALITE	(Ag,Hg)I	
RAYITE	(Ag,Tl)2Pb8Sb8S21	Monoclinic
MILLOSEVICHITE	(Al,Fe+++)2(SO4)3	Trigonal
LISKEARDITE	(Al,Fe+++)3(AsO4)(OH)6.5H2O	Orthorhombic
GUTSEVICHITE	(Al,Fe+++)3(PO4,VO4)2(OH)3.8H2O	
SASAITE	(Al,Fe+++)6(PO4,SO4)5(OH)3.36H2O	Orthorhombic
GOUDEYITE	(Al,Y)Cu6(AsO4)3(OH)6.3H2O	Hexagonal
WAKABAYASHILITE	(As,Sb)11S18	Monoclinic
HEMLOITE	(As,Sb)2(Ti,V,Fe,Al)12O23OH	Triclinic
HUNCHUNITE	(Au,Ag)2Pb	Isometric
SYLVANITE	(Au,Ag)2Te4	Monoclinic
WEISHANITE	(Au,Ag)3Hg2	Hexagonal
GOLDAMALGAM	(Au,Ag)Hg	Isometric
MONTBRAYITE	(Au,Sb)2Te3	Triclinic
BOGDANOVITE	(Au,Te,Pb)3(Cu,Fe)	Isometric
BERGENITE	(Ba,Ca)2(UO2)3(PO4)2(OH)4,5.5H2O	Monoclinic
JINSHAJIANGITE	(Ba,Ca)4(Na,K)5(Fe,Mn)15(Ti,Fe,Nb,Zr)8Si15O64(F,OH)6	Monoclinic
WELLSITE	(Ba,Ca,K2)Al2Si6O16.6H2O	Monoclinic
MORELANDITE	(Ba,Ca,Pb)5(AsO4,PO4)3Cl	Hexagonal
ROMANECHITE (PSILOMELANE)	(Ba,H2O)(Mn++++,Mn+++)5O10	Monoclinic

Mineral Name	Formula	Crystal System
ANANDITE	(Ba,K)(Fe++,Mg)3(Si,Al,Fe)4O10(O,OH)2	Monoclinic
KINOSHITALITE	(Ba,K)(Mg,Mn,Al)3Si2Al2O10(OH)2	Monoclinic
INNELITE	(Ba,K)4(Na,Ca)3Ti3(Si2O7)(SO4)O4	Triclinic
BATISITE	(Ba,K,Na)3Ti2Si4O14	Orthorhombic
CHERNYKHITE	(Ba,Na)V+++,Al)2(Si,Al)4O10(OH)2	Monoclinic
HARMOTOME	(Ba,Na,K)1-2(Si,Al)8O16.6H2O	Monoclinic
FRANCEVILLITE	(Ba,Pb)(UO2)2V2O8.5H2O	Orthorhombic
OWENSITE	(Ba,Pb)6(Cu,Fe,Ni)25S27	Isometric
HYALOTEKITE	(Ba,Pb,Ca,K)6(B,Si,Al)2(Si,Be)10O28(F,Cl)	Triclinic ps Monoclinic
URANOTUNGSTITE	(Ba,Pb,Fe++)(UO2)2WO4(OH)4.12H2O	Orthorhombic
EWALDITE	(Ba,Sr)(Ca,Na,Y,Ce)(CO3)2.10-12H2O	Hexagonal
BJAREBYITE	(Ba,Sr)(Mn++,Fe++,Mg)2Al2(PO4)3(OH)3	Monoclinic
BARIOPYROCHLORE (PANDAITE)	(Ba,Sr)2(Nb,Ti)2(O,OH)7	Isometric
TAIKANITE	(Ba,Sr)2Mn+++2Si4O12	Monoclinic
YOSHIMURAITE	(Ba,Sr)2TiMn2(SiO4)2(PO4,SO4)(OH,Cl)	Triclinic
BARIOORTHOJOAQUINITE	(Ba,Sr)4Fe++2Ti2Si8O26.H2O	Orthorhombic
BENSTONITE	(Ba,Sr)6(Ca,Mn)6Mg(CO3)13	Trigonal
BARIUMBREWSTERITE	(Ba,Sr)Al2Si6O16.5H2O	Monoclinic
LINDSLEYITE	(BaSr)(Ti,Cr,Fe,Mg,Zr)21O38	Trigonal
BISMUTOMICROLITE (WESTGRENITE)	(Bi,Ca)(Ta,Nb)2O6(OH)	Isometric
PETITJEANITE	(Bi,Pb)3(PO4,VO4,AsO4)2(O,OH)2	Triclinic
PLATYNITE	(Bi,Pb)3(Se,S)4	Trigonal
RUCKLIDGEITE	(Bi,Pb)3Te4	Trigonal
CHEKHOVICHITE	(Bi,Pb,Fe)2Te4O11	Monoclinic
HOELITE (ANTHRAQUINONE)	(C6H4)2(CO)2	Orthorhombic
KRATOCHVILITE	(C6H4)2CH2	Orthorhombic
SAMUELSONITE	(Ca,Ba)Ca8(Fe++,Mn)4Al2(PO4)10(OH)2	Monoclinic
CALCIOURANOITE	(Ca,Ba,Pb)U2O7.5H2O	Amorphous
HIBONITE	(Ca,Ce)(Al,Ti,Mg)12O19	Hexagonal
KHRISTOVITE-(Ce)	(Ca,Ce)(Ce,La,Nd,Dy)Mn++2AlSiO4Si2O7(OH,F,O)2	Monoclinic
VIGEZZITE	(Ca,Ce)(Nb,Ta,Ti)2O6	Orthorhombic
LOVERINGITE	(Ca,Ce)(Ti,Fe+++,Cr,Mg)21O38	Trigonal
SEMENOVITE-(Ce)	(Ca,Ce,La,Na)10-12(Fe++,Mn)(Si,Be)20(O,OH,F)48	Orthorhombic
FERSMITE	(Ca,Ce,Na)(Nb,Ta,Ti)2(O,OH,F)6	Orthorhombic
COERULEOLACTITE	(Ca,Cu)Al6(PO4)4(OH)8.4-5H2O	Triclinic
RICHELLITE	(Ca,Fe++)(Fe+++,Al)(PO4)2(OH,F)2	Tetragonal
LAZARENKOITE	(Ca,Fe++)Fe+++As+++3O7.3H2O	Orthorhombic
ROMEITE	(Ca,Fe++,Mn,Na)2(Sb,Ti)2O6(O,OH,F)	Isometric
LEWISITE	(Ca,Fe++,Na)2(Sb,Ti)2O7	Isometric
ZIRCONOLITE-3O	(Ca,Fe,Y,Th)2Fe(Ti,Nb)3Zr2O14	Orthorhombic
LATIUMITE	(Ca,K)8(Al,Mg,Fe)(Si,Al)10O25(SO4)	Monoclinic
STRACZEKITE	(Ca,K,Ba,Na)(V++++,V+++++)8O20.3H2O	Monoclinic

Mineral Name	Formula	Crystal System
PAHASAPAITE	(Ca,Li,K,Na)27Li16Be48(PO4)48.76H2O	Isometric
CHELYABINSKITE	(Ca,Mg)3Si(OH)6(SO4,CO3)2.9H2O	Orthorhombic
WADALITE	(Ca,Mg)6(Al,Fe+++)4(AlSi2O16)Cl3	Isometric
RORISITE	(Ca,Mg)FCl	Tetragonal
WERMLANDITE	(Ca,Mg)Mg7(Al,Fe+++)2(SO4)2(OH)18.12H2O	Trigonal
HYDROUGRANDITE	(Ca,Mg,Fe++)3(Fe+++,Al)2(SiO4)3-x(OH)4x	Isometric
TRUSCOTTITE	(Ca,Mn)14Si24O58(OH)8.2H2O	Hexagonal
ZANAZZIITE	(Ca,Mn)2(Mg,Fe)(Mg,Fe++,Mn,Fe+++)4Be4(PO4)6(OH)4.6H2O	Monoclinic
OLDHAMITE	(Ca,Mn)S	Isometric
THADEUITE	(Ca,Mn++)(Mg,Fe++,Mn+++)3(PO4)2(OH,F)2	Orthorhombic
STRONTIOPIEMONTITE	(Ca,Mn++)(Sr,Ca)Mn+++(Al,Fe+++)2(SiO4)(Si2O7)O(OH)	Monoclinic
POLDERVAARTITE	(Ca,Mn++)2SiO3(OH)2	Orthorhombic
KINGSMOUNTITE	(Ca,Mn++)4(Fe++,Mn++)Al4(PO4)6(OH)4.12H2O	Monoclinic
RANCIEITE	(Ca,Mn++)Mn++++4O9.3H2O	Hexagonal
TVEDALITE	(Ca,Mn++,Fe++)4Be3Si6O17(OH)4.3H2O	Orthorhombic
TINZENITE	(Ca,Mn,Fe)3Al2BO3Si4O12(OH)	Triclinic
KOMAROVITE	(Ca,Mn,Na)2Nb4Si4O12O2(OH,F)4.7H2O	Orthorhombic
FERUVITE	(Ca,Na)(Fe,Mg,Ti)3(Al,Mg,Fe)6(BO3)3Si6O18(OH)4	Trigonal
UVITE	(Ca,Na)(Mg,Fe++)3Al5Mg(BO3)3Si6O18(OH,F)4	Trigonal
OMPHACITE	(Ca,Na)(Mg,Fe++,Fe+++,Al)Si2O6	Monoclinic
AUGITE	(Ca,Na)(Mg,Fe,Al,Ti)(Si,Al)2O6	Monoclinic
LATRAPPITE	(Ca,Na)(Nb,Ti,Fe)O3	Orthorhombic
BYTOWNITE	(Ca,Na)(Si,Al)4O8	Triclinic
LABRADORITE	(Ca,Na)(Si,Al)4O8	Triclinic
CARBOCERNAITE	(Ca,Na)(Sr,Ce,Ba)(CO3)2	Orthorhombic
SWINEFORDITE	(Ca,Na)0,3(Al,Li,Mg)2(Si,Al)4O10(OH,F)2.2H2O	Monoclinic
FERSMANITE	(Ca,Na)16(Ti,Nb)8Si4O12O18F6	Monoclinic
MELILITE	(Ca,Na)2(Al,Mg,Fe++)(Si,Al)2O7	Tetragonal
JEFFREYITE	(Ca,Na)2(Be,Al)Si2(O,OH)7	Orthorhombic ps Tetragonal
HOGTUVAITE (HOEGTUVAITE)	(Ca,Na)2(Fe++,Fe+++,Ti,Mg,Mn)6(Si,Be,Al)6O20	Triclinic
MELIPHANITE (MELINOPHANE)	(Ca,Na)2Be(Si,Al)2(O,OH,F)7	Tetragonal
MENGXIANMINITE	(Ca,Na)3(Fe++,Mn++)2Mg2(Sn++++,Zn)5Al8O29	Orthorhombic
BERZELIITE	(Ca,Na)3(Mg,Mn)2(AsO4)3	Isometric
MANGANBERZELIITE	(Ca,Na)3(Mn,Mg)2(AsO4)3	Isometric
GOTZENITE (GOETZENITE)	(Ca,Na)3(Ti,Al)Si2O7(F,OH)2	Triclinic
ROSENBUSCHITE	(Ca,Na)3(Zr,Ti)Si2O8F	Triclinic
HIORTDAHLITE	(Ca,Na)3(Zr,Ti,Y)Si2O7(O,OH,F)2	Triclinic
PALENZONAITE	(Ca,Na)3Mn++(V+++++,As+++++,Si)3O12	Isometric
MACHATSCHKIITE	(Ca,Na)6(As+++++O4)(As++++O3OH)3(PO4,SO4).15H2O	Trigonal
METACALCIOURANOITE	(Ca,Na,Ba)U2O7.2H2O	
YAKHONTOVITE	(Ca,Na,K)0,3(CuFe++Mg)2Si4O10(OH)2.3H2O	Monoclinic

Mineral Name	Formula	Crystal System
LIOTTITE	(Ca,Na,K)8(Si,Al)12O24[(SO4),(CO3),Cl,OH]4.H2O	Hexagonal
HOCHELAGAITE	(Ca,Na,Sr)Nb4O11.8H2O	Monoclinic
BETAFITE	(Ca,Na,U)2(Ti,Nb,Ta)2O6(OH)	Isometric
BRAITSCHITE-(Ce)	(Ca,Na2)7(Ce,La)2B22O43.7H2O	Hexagonal
MOUNTAINITE	(Ca,Na2,K2)2Si4O10.3H2O	Monoclinic
DACHIARDITE	(Ca,Na2,K2)5Al10Si38O96.25H2O	Monoclinic
RHODESITE	(Ca,Na2,K2)8Si16O40.11H2O	Orthorhombic
MORDENITE	(Ca,Na2,K2)Al2Si10O24.7H2O	Orthorhombic
LEVYNE (LEVYNITE)	(Ca,Na2,K2)Al2Si4O12.6H2O	Trigonal
BEYERITE	(Ca,Pb)Bi2(CO3)2O2	Tetragonal
FERMORITE	(Ca,Sr)5(AsO4,PO4)3(OH)	Hexagonal
ARSENOCRANDALLITE	(Ca,Sr)Al3[(As,P)O4]2(OH)5.H2O	Trigonal
CALCIOANCYLITE-(Ce)	(Ca,Sr)Ce(CO3)2(OH).H2O	Orthorhombic
ATTAKOLITE (ATTACOLITE)	(Ca,Sr)Mn++(Al,Fe+++)4[(Si,P)O4]H(PO4)3(OH)4	Monoclinic
BROCKITE	(Ca,Th,Ce)(PO4).H2O	Hexagonal
ZIRKELITE	(Ca,Th,Ce)Zr(Ti,Nb)2O7	Monoclinic ps Cubic
THOROSTEENSTRUPINE	(Ca,Th,Mn)3Si4O11F.6H2O	Amorphous
TRISTRAMITE	(Ca,U++++,Fe+++)(PO4,SO4).2H2O	Hexagonal
HELLANDITE-	(Ca,Y)6(Al,Fe+++)Si4B4O20(OH)4	Monoclinic
SCHETELIGITE	(Ca,Y,Sb,Mn)2(Ti,Ta,Nb,W)2O6(O,OH)	Orthorhombic
STEVENSITE	(Ca/2)0,3Mg3Si4O10(OH)2	Monoclinic
SAPONITE	(Ca/2,Na)0,3(Mg,Fe++)3(Si,Al)4O10(OH)2.4H2O	Monoclinic
YOSHIOKAITE	(Ca8-(x/2)_(x/2)Al16-xSixO32)	Hexagonal
CERIANITE-(Ce)	(Ce++++,Th)O2	Isometric
WAKEFIELDITE-(Ce) (KUSUITE)	(Ce+++,Pb++,Pb++++)VO4	Tetragonal
MELANOCERITE-(Ce)	(Ce,Ca)5(Si,B)3O12(OH,F).nH2O	Hexagonal
LESSINGITE-(Ce)	(Ce,Ca)5(SiO4)3(F,OH)	Monoclinic
BRITHOLITE-(Ce)	(Ce,Ca)5(SiO4,PO4)3(OH,F)	Hexagonal
AGARDITE-(Ce)	(Ce,Ca)Cu6(AsO4)3(OH)2.3H2O	Hexagonal
AESCHYNITE-(Ce)	(Ce,Ca,Fe,Th)(Ti,Nb)2(O,OH)6	Orthorhombic
PERRIERITE	(Ce,Ca,La,Nd,Th)4(Fe++,Mg)2(Ti,Al,Zr,Fe+++)2Ti2(Si2O7)2O8	Monoclinic
NIOBOAESCHYNITE-(Ce)	(Ce,Ca,Th)(Nb,Ti)2(O,OH)6	Orthorhombic
CHERALITE-(Ce)	(Ce,Ca,Th,U)(P,Si)O4	Monoclinic
ALLANITE-(Ce) (ORTHITE)	(Ce,Ca,Y)2(Al,Fe+++)3(SiO4)3(OH)	Monoclinic
CERIOPYROCHLORE-(Ce) (MARIGNACITE)	(Ce,Ca,Y)2(Nb,Ta)2O6(OH,F)	Isometric
HYDROXYLBASTNAESITE-(Ce)	(Ce,La)(CO3)(OH,F)	Hexagonal
BASTNASITE-(Ce) (BASTNAESITE-(Ce))	(Ce,La)(CO3)F	Hexagonal
NORDITE-(Ce)	(Ce,La)(Sr,Ca)Na2(Na,Mn)(Zn,Mg)Si6O17	Orthorhombic
DAVIDITE-(Ce)	(Ce,La)(Y,U,Fe++)(Ti,Fe+++)20(O,OH)38	Trigonal
CALKINSITE-(Ce)	(Ce,La)2(CO3)3.4H2O	Orthorhombic
LANTHANITE-(Ce)	(Ce,La)2(CO3)3.8H2O	Orthorhombic

Mineral Name	Formula	Crystal System
TORNEBOHMITE-(Ce) (TOERNEBOHMITE-(Ce))	(Ce,La)2Al(SiO4)2(OH)	Monoclinic
CHEVKINITE-(Ce)	(Ce,La)4(Ti,Fe)5Si4O12O10	Monoclinic
PEPROSSIITE-(Ce)	(Ce,La)Al2(BO3)3	Hexagonal
ARSENOFLORENCITE-(Ce)	(Ce,La)Al3(AsO4)2(OH)6	Trigonal
FLUOCERITE-(Ce) (TYSONITE)	(Ce,La)F3	Hexagonal
RHABDOPHANE-(Ce)	(Ce,La)PO4.H2O	Hexagonal
LUCASITE-(Ce)	(Ce,La)Ti2(O,OH)6	Monoclinic
STILLWELLITE-(Ce)	(Ce,La,Ca)BSiO5	Trigonal
FERGUSONITE-BETA-(Ce)	(Ce,La,Nd)NbO4	Monoclinic
MONAZITE-(Ce)	(Ce,La,Nd,Th)PO4	Monoclinic
GADOLINITE-(Ce)	(Ce,La,Nd,Y)2Fe++Be2Si2O10	Monoclinic
WAKEFIELDITE-(Ce)	(Ce,La,Nd,Y,Pr,Sm)(V,As)O4	Tetragonal
KARNASURTITE-(Ce)	(Ce,La,Th)(Ti,Nb)(Al,Fe+++)(Si,P)2O7(OH)4.3H2O	Hexagonal
TRITOMITE-(Ce)	(Ce,La,Y,Th)5(Si,B)3(O,OH,F)13	Trigonal
LOPARITE-(Ce)	(Ce,Na,Ca)2(Ti,Nb)2O6	Isometric
CERVANDONITE-(Ce)	(Ce,Nd,La)(Fe+++,Fe++,Ti++++,Al)3SiAs(Si,As)O13	Monoclinic
FERGUSONITE-(Ce)	(Ce,Nd,La)NbO4.0,3H2O	Monoclinic
CHERNOVITE-(Ce)	(Ce,Y)(AsO4)	Tetragonal
HINGGANITE-(Ce)	(Ce,Y)2(_,Fe++)Be2Si2O8(OH,O)2	Monoclinic
EVENKITE	(CH3)2(CH2)22	Monoclinic
MODDERITE	(Co,Fe)As	Orthorhombic
ALLOCLASITE	(Co,Fe)AsS	Monoclinic
GLAUCODOT	(Co,Fe)AsS	Orthorhombic ps Cubic
CLINOSAFFLORITE	(Co,Fe,Ni)As2	Monoclinic
OURSINITE	(Co,Mg)(UO2)2Si2O7.6H2O	Orthorhombic
WUPATKIITE	(Co,Mg)Al2(SO4)4.22H2O	Monoclinic
APLOWITE	(Co,Mn,Ni)SO4.4H2O	Monoclinic
ASBOLAN	(Co,Ni)1-y(Mn++++O2)2-x(OH)2-2y+2x.nH2O	Hexagonal
LANGISITE	(Co,Ni)As	Hexagonal
SKUTTERUDITE	(Co,Ni)As3-x	Isometric
WILLYAMITE	(Co,Ni)SbS	Monoclinic/Triclinic
COCHROMITE	(Co,Ni,Fe++)(Cr,Al)2O4	Isometric
SMOLIANINOVITE	(Co,Ni,Mg,Ca)3(Fe+++,Al)2(AsO4)4.11H2O	Orthorhombic
MOORHOUSEITE	(Co,Ni,Mn)SO4.6H2O	Monoclinic
COBALTKORITNIGITE	(Co,Zn)(As+++++O3)(OH).H2O	Triclinic
THERESMAGNANITE	(Co,Zn,Ni)6(SO4)(OH,Cl)10.8H2O	Hexagonal
RILANDITE	(Cr,Al)6SiO11.5H2O	
OLKHONSKITE	(Cr,V+++)2Ti3O9	Monoclinic
MARGARITASITE	(Cs,K,H3O)2(UO2)2V2O8.H2O	Monoclinic
CESIUMKUPLETSKITE	(Cs,K,Na)3(Mn,Fe++)7(Ti,Nb)2Si8O24(O,OH,F)7	Triclinic
CESPLUMTANTITE	(Cs,Na)2(Pb,Sb+++)3Ta8O24	Tetragonal
POLLUCITE	(Cs,Na)2Al2Si4O12.H2O	Isometric
CESSTIBTANTITE	(Cs,Na)SbTa4O12	Isometric

Mineral Name	Formula	Crystal System
GALKHAITE	(Cs,Tl)(Hg,Cu,Zn)6(As,Sb)4S12	Isometric
ABSWURMBACHITE	(Cu++,Mn++)Mn+++6(SiO4)O8	Tetragonal
DANIELSITE	(Cu,Ag)14HgS8	Orthorhombic
LAROSITE	(Cu,Ag)21(Pb,Bi)2S13	Orthorhombic
NOVAKITE	(Cu,Ag)21As10	Monoclinic ps Tetragonal
FURUTOBEITE	(Cu,Ag)6PbS4	Monoclinic
CHRYSOCOLLA	(Cu,Al)2H2Si2O5(OH)4.nH2O	Monoclinic
KOLWEZITE	(Cu,Co)2(CO3)(OH)2	Triclinic
LINDACKERITE	(Cu,Co)5(AsO4)2(AsO3OH)2.10H2O	Triclinic
TYRRELLITE	(Cu,Co,Ni)3Se4	Isometric
TENNANTITE	(Cu,Fe)12As4S13	Isometric
TETRAHEDRITE	(Cu,Fe)12Sb4S13	Isometric
CHAMEANITE	(Cu,Fe)4As(Se,S)4	Isometric
NUKUNDAMITE	(Cu,Fe)4S4	Hexagonal
GORTDRUMITE	(Cu,Fe)6Hg2S5	Orthorhombic
POITEVINITE	(Cu,Fe++,Zn)SO4.H2O	Monoclinic
PETRUKITE	(Cu,Fe,Zn)2(Sn,In)S4	Orthorhombic
HAKITE	(Cu,Hg,Ag)12Sb4(Se,S)13	Isometric
CUPROSPINEL	(Cu,Mg)Fe+++2O4	Isometric
GLAUKOSPHAERITE	(Cu,Ni)2(CO3)(OH)2	Monoclinic
VILLAMANINITE	(Cu,Ni,Co,Fe)S2	Isometric
RENIERITE	(Cu,Zn)11(Ge,As)2Fe4S16	Tetragonal ps Cubic
RAMSBECKITE	(Cu,Zn)15(SO4)4(OH)22.6H2O	Monoclinic
TLALOCITE	(Cu,Zn)16(Te++++O3)(Te++++++O4)2Cl(OH)25.27H2O	Monoclinic
ROSASITE	(Cu,Zn)2(CO3)(OH)2	Monoclinic
PARATACAMITE	(Cu,Zn)2(OH)3Cl	Trigonal
SABELLIITE	(Cu,Zn)2Zn[(As,Sb)O4](OH)3	Trigonal
CLARAITE	(Cu,Zn)3(CO3)(OH)4.4H2O	Tetragonal ps Hexagonal
VESZELYITE	(Cu,Zn)3(PO4)(OH)3.2H2O	Monoclinic
BAYLDONITE	(Cu,Zn)3Pb(AsO4)2(OH)2	Monoclinic
KTENASITE	(Cu,Zn)5(SO4)2(OH)6.6H2O	Monoclinic
KIPUSHITE	(Cu,Zn)5Zn(PO4)2(OH)6.H2O	Monoclinic
PHILIPSBURGITE	(Cu,Zn)6(AsO4,PO4)2(OH)6.H2O	Monoclinic
SCHULENBERGITE	(Cu,Zn)7(SO4,CO3)2(OH)10.3H2O	Trigonal
CUPALITE	(Cu,Zn)Al	Orthorhombic
KHATYRKITE	(Cu,Zn)Al2	Tetragonal
GIRAUDITE	(Cu,Zn,Ag)12(As,Sb)4(Se,S)13	Isometric
MUSHISTONITE	(Cu,Zn,Fe)Sn++++(OH)6	Isometric
ZHANGHENGITE	(Cu,Zn,Fe,Al,Cr)	Isometric
SAKURAIITE	(Cu,Zn,Fe,In,Sn)S	Cubic or Tetragonal
AGARDITE-(Dy)	(Dy,La,Ca)Cu6(AsO4)3(OH)6.3H2O	Hexagonal
CHURCHITE-(Dy)	(Dy,Sm,Gd,Nd)(PO4).2H2O	Monoclinic
XIANGJIANGITE	(Fe+++,Al)(UO2)4(PO4)2(SO4)2(OH).22H2O	Tetragonal
MIKASAITE	(Fe+++,Al)2(SO4)3	Trigonal
MACAULAYITE	(Fe+++,Al)24Si4O43(OH)2	Monoclinic

Mineral Data

Mineral Name	Formula	Crystal System
CACOXENITE	(Fe+++,Al)25(PO4)17O6(OH)12.75H2O	Hexagonal
RUSAKOVITE	(Fe+++,Al)5(VO4,PO4)2(OH)9.3H2O	
PSEUDOBROOKITE	(Fe+++,Fe++)2(Fe++,Ti)O5	Orthorhombic
ODINITE	(Fe+++,Mg,Al,Fe++,Ti,Mn)2.4(Si1,8Al0,2)O5(OH)4	Monoclinic/Trigonal
JEANBANDYITE	(Fe+++,Mn++)Sn++++(OH)6	Tetragonal ps Cubic
OXIBERAUNITE	(Fe+++,Mn+++)Fe+++5(PO4)4O(OH)4.6H2O	Monoclinic
SQUAWCREEKITE	(Fe+++,Sb+++++,Sn++++,Ti)O2	Tetragonal
GREENALITE	(Fe++,Fe+++)2-3Si2O5(OH)4	Monoclinic
BERTHIERINE	(Fe++,Fe+++,Mg)2-3(Si,Al)2O5(OH)4	Monoclinic
ERLIANITE	(Fe++,Fe+++,Mg)24(Fe+++V)6Si36O90(OH,O)48	Orthorhombic
FETIASITE	(Fe++,Fe+++,Ti)3O2(As2O5)	Monoclinic
DONATHITE	(Fe++,Mg)(Cr,Fe+++)2O4	Tetragonal
AMAKINITE	(Fe++,Mg)(OH)2	Trigonal
HULSITE	(Fe++,Mg)2(Fe+++,Sn)BO5	Monoclinic
SEKANINAITE	(Fe++,Mg)2Al4Si5O18	Orthorhombic
CLINOFERROSILITE	(Fe++,Mg)2Si2O6	Monoclinic
FERROSILITE (ORTHOFERROSILITE)	(Fe++,Mg)2Si2O6	Orthorhombic
MINNESOTAITE	(Fe++,Mg)3Si4O10(OH)2	Monoclinic
FERROGEDRITE	(Fe++,Mg)5Al2(Si6Al2)O22(OH)2	Orthorhombic
FERROANTHOPHYLLITE	(Fe++,Mg)7Si8O22(OH)2	Orthorhombic
GRUNERITE	(Fe++,Mg)7Si8O22(OH)2	Monoclinic
ZIRKLERITE	(Fe++,Mg)9Al4Cl18(OH)12.14H2O	Trigonal
SCORZALITE	(Fe++,Mg)Al2(PO4)2(OH)2	Monoclinic
FERROCARPHOLITE	(Fe++,Mg)Al2Si2O6(OH)4	Orthorhombic
VOCHTENITE	(Fe++,Mg)Fe+++[UO2/PO4]4(OH).12-13H2O	Monoclinic
SATTERLYITE	(Fe++,Mg,Fe+++)2(PO4)(OH)	Hexagonal
CHAMOSITE	(Fe++,Mg,Fe+++)5Al(Si3Al)O10(OH,O)8	Monoclinic
ORTHOCHAMOSITE	(Fe++,Mg,Fe+++)5Al(Si3Al)O10(OH,O)8	Orthorhombic
CHLORITOID	(Fe++,Mg,Mn)2Al4Si2O10(OH)4	Monoclinic/Triclinic
LUDLAMITE	(Fe++,Mg,Mn)3(PO4)2.4H2O	Monoclinic
CONGOLITE	(Fe++,Mg,Mn)3B7O13Cl	Trigonal
ERICAITE	(Fe++,Mg,Mn)3B7O13Cl	Orthorhombic
EKMANITE	(Fe++,Mg,Mn,Fe+++)3(Si,Al)4O10(OH)2.2H2O	Orthorhombic
REDINGTONITE	(Fe++,Mg,Ni)(Cr,Al)2(SO4)4.22H2O	Monoclinic
KINICHILITE	(Fe++,Mg,Zn)2(Te++++O3)3(NaxH2-x).3H2O	Hexagonal
STAUROLITE (STAUROTIDE)	(Fe++,Mg,Zn)2Al9(Si,Al)4O22(OH)2	Monoclinic ps Orthorhombic
QITIANLINGITE	(Fe++,Mn)2(Nb,Ta)2W++++++O10	Orthorhombic
PHOSPHOFERRITE	(Fe++,Mn)3(PO4)2.3H2O	Orthorhombic
DEERITE	(Fe++,Mn)6(Fe+++,Al)3Si6O20(OH)5	Monoclinic ps Orthorhombic
FERROPYROSMALITE	(Fe++,Mn)8Si6O15(Cl,OH)10	Hexagonal
LIPSCOMBITE	(Fe++,Mn)Fe+++2(PO4)2(OH)2	Tetragonal
ROCKBRIDGEITE	(Fe++,Mn)Fe+++4(PO4)3(OH)5	Orthorhombic
FERROWODGINITE	(Fe++,Mn++)(Sn,Ti,Fe+++,Ta)(Ta,Nb)2O8	Monoclinic
FERROTAPIOLITE	(Fe++,Mn++)(Ta,Nb)2O6	Tetragonal

Mineral Name	Formula	Crystal System
WOLFEITE	(Fe++,Mn++)2(PO4)(OH)	Monoclinic
GRAFTONITE	(Fe++,Mn,Ca)3(PO4)2	Monoclinic
PYROXFERROITE	(Fe++,Mn,Ca)SiO3	Triclinic
SARCOPSIDE	(Fe++,Mn,Mg)3(PO4)2	Monoclinic
LAWRENCITE	(Fe++,Ni)Cl2	Trigonal
AHEYLITE	(Fe++,Zn)Al6(PO4)4(OH)8.4H2O	Triclinic
PEHRMANITE	(Fe++,Zn,Mg)2Al6BeO12	Trigonal
HEIDEITE	(Fe,Cr)1+x(Ti,Fe)2S4	Monoclinic
HIBBINGITE	(Fe,Mg)2(OH)3Cl	Orthorhombic
WOLFRAMOIXIOLITE	(Fe,Mn,Nb)(Nb,W,Ta)O4	Monoclinic
KAMACITE	(Fe,Ni)	Isometric
SEINAJOKITE	(Fe,Ni)(Sb,As)2	Orthorhombic
BARRINGERITE	(Fe,Ni)2P	Hexagonal
SCHREIBERSITE (RHABDITE)	(Fe,Ni)3P	Tetragonal
SUESSITE	(Fe,Ni)3Si	Isometric
SMYTHITE	(Fe,Ni)9S11or(Fe,Ni)13S16	Trigonal
MACKINAWITE	(Fe,Ni)9S8	Tetragonal
PENTLANDITE	(Fe,Ni)9S8	Isometric
COHENITE	(Fe,Ni,Co)3C	Orthorhombic
WESTERVELDITE	(Fe,Ni,Co)As	Orthorhombic
POLKOVICITE	(Fe,Pb)3(Ge,Fe)1-xS4	Isometric
HAXONITE	(FeNi)23C6	Isometric
NSUTITE	(g-MnO2)Mn++xMn++++1-xO2-2x(OH)2x(xissmall)	Hexagonal
VERNADITE	(gamma-MnO2),(Mn++++,Fe+++,Ca,Na)(O,OH)2.nH2O	Hexagonal
BRUNOGEIERITE	(Ge++,Fe++)Fe+++2O4	Isometric
BETPAKDALITE	(H,K)6Ca4Fe+++6As+++++4Mo++++++16O74.24-40H2O	Monoclinic
KALIPYROCHLORE	(H2O,K)2(Nb,Ti)2O4(OH)2	Isometric
SODIUMBOLTWOODITE	(H3O)(Na,K)(UO2)SiO4.H2O	Orthorhombic
CHERNIKOVITE (HYDROGENAUTUNITE)	(H3O)2(UO2)2(PO4)2.6H2O	Tetragonal
CUPROSKLODOWSKITE	(H3O)2Cu(UO2)2(SiO4)2.2H2O	Triclinic
SKLODOWSKITE	(H3O)2Mg(UO2)2(SiO4)2.2H2O	Monoclinic
PSEUDOAUTUNITE	(H3O)4Ca2(UO2)2(PO4)4.5H2O	Tetragonal
HYDRONIUMJAROSITE	(H3O)Fe+++3(SO4)2(OH)6	Trigonal
VANURANYLITE	(H3O,Ba,Ca,K)1,6(UO2)2V2O8.4H2O	Orthorhombic
VANURALITE	(H3O,Ba,Ca,K)2(UO2)2V2O8.4H2O	Monoclinic
SCHLOSSMACHERITE	(H3O,Ca)Al3(AsO4,SO4)2(OH)6	Trigonal
HYDROASTROPHYLLITE	(H3O,K,Ca)3(Fe++,Mn)5-6Ti2Si8(O,OH)31	Triclinic
IRIDISITE	(Ir,Cu,Rh,Ni,Pt)S2	ps Cubic
IRIDIUM	(Ir,Os,Ru)	Isometric
RUTHENIRIDOSMINE	(Ir,Os,Ru)	Hexagonal
KASHINITE	(Ir,Rh)2S3	Orthorhombic
IRIDARSENITE	(Ir,Ru)As2	Monoclinic

Mineral Name	Formula	Crystal System
IRARSITE	(Ir,Ru,Rh,Pt)AsS	Isometric
DICKINSONITE	(K,Ba)(Na,Ca)5(Mn++,Fe++,Mg)14Al(PO4)12(OH,F)2	Monoclinic
PRIDERITE	(K,Ba)(Ti,Fe+++)8O16	Tetragonal
JEPPEITE	(K,Ba)2(Ti,Fe+++)6O13	Monoclinic
LOURENSWALSITE	(K,Ba)2(Ti,Mg,Ca,Fe)4(Si,Al,Fe)6O14(OH)12	Hexagonal
HYALOPHANE	(K,Ba)Al(Si,Al)3O8	Monoclinic
YUKSPORITE	(K,Ba)NaCa2(Si,Ti)4O11(F,OH).H2O	Orthorhombic
BELLBERGITE	(K,Ba,Sr)(Ca,Sr,Na)4Al9Si9O36.15H2O	Hexagonal
FERRITUNGSTITE	(K,Ca,Na)(W,Fe+++)2(O,OH)6.H2O	Isometric
MERLINOITE	(K,Ca,Na,Ba)7(Si25Al9)O64.23H2O	Orthorhombic
MATHIASITE	(K,Ca,Sr)(Ti,Cr,Fe,Mg)21O38	Trigonal
RHODIZITE	(K,Cs)Al4Be4(B,Be)12O28	Isometric
AVOGADRITE	(K,Cs)BF4	Orthorhombic
BENYACARITE	(K,H2O)(Mn++,Fe++)2(Fe+++,Ti,Al)2Ti(PO4)4(O,F)2.14H2O	Orthorhombic
ILLITE	(K,H3O)(Al,Mg,Fe)2(Si,Al)4O10[(OH)2,H2O]	Monoclinic
BARIUMBANNISTERITE	(K,H3O)(Ba,Ca)(Mn++,Fe++,Mg)21(Si,Al)32O80(O,OH)16.4-12H2O	Monoclinic
POTASSIUMFLUORRICHTERITE	(K,Na)(Ca,Na)2(Mg,Fe)5Si8O22(F,OH)2	Monoclinic
POTASSIUMRICHTERITE	(K,Na)(Ca,Na)2(Mg,Fe)5Si8O22(OH,F)2	Monoclinic
GLAUCONITE	(K,Na)(Fe+++,Al,Mg)2(Si,Al)4O10(OH)2	Monoclinic
OSUMILITE	(K,Na)(Fe++,Mg)2(Al,Fe+++)3(Si,Al)12O30	Hexagonal
OSUMILITE-(Mg)	(K,Na)(Mg,Fe++)2(Al,Fe+++)3(Si,Al)12O30	Hexagonal
KORNITE	(K,Na)(Na,Li)2(Mg,Mn+++,Fe+++,Li)5Si8O22(OH,F)2	Monoclinic
SANIDINE	(K,Na)(Si,Al)4O8	Monoclinic
FRANKLINPHILITE	(K,Na)<1(Mn++,Mg,Zn,Fe+++)8(Si,Al,Fe+++)12(O,OH)36.2-3H2O	Triclinic ps Hexagonal
MERRIHUEITE	(K,Na)2(Fe++,Mg)5Si12O30	Hexagonal
SOGDIANITE	(K,Na)2(Li,Fe+++,Al)3ZrSi12O30	Hexagonal
GANOPHYLLITE	(K,Na)2(Mn,Al,Mg)8(Si,Al)12O29(OH)7.8-9H2O	Monoclinic
MINEHILLITE	(K,Na)2-3Ca28(Zn4Al4Si40)O112(OH)16	Hexagonal
PALMIERITE	(K,Na)2Pb(SO4)2	Trigonal
WADEITE	(K,Na)2ZrSi3O9	Hexagonal
NIOBOPHYLLITE	(K,Na)3(Fe++,Mn)6(Nb,Ti)2Si8(O,OH,F)31	Triclinic
ASTROPHYLLITE	(K,Na)3(Fe++,Mn)7Ti2Si8O24(O,OH)7	Triclinic
KUPLETSKITE	(K,Na)3(Mn,Fe++)7(Ti,Nb)2Si8O24(O,OH)7	Triclinic
CETINEITE	(K,Na)3+x(Sb2O3)3(Sb2S3)(OH)x.(2,8-x)H2O	Hexagonal
APHTHITALITE (GLASERITE)	(K,Na)3Na(SO4)2	Trigonal
FEDORITE	(K,Na)5(Ca,Na)14Si32O76(OH,F)4.2H2O	Triclinic
SOSEDKOITE	(K,Na)5Al2(Ta,Nb)22O60	Orthorhombic
ALTISITE	(K,Na)6Na3Ti2Al2Si8O26Cl3	Monoclinic
LABUNTSOVITE	(K,Na)8(Ti++++,Nb,Fe)9(Si4O12)4O8.16H2O	Monoclinic
MONTESOMMAITE	(K,Na)9Al9Si23O64.10H2O	Orthorhombic ps Tetragonal
PANUNZITE	(K,Na)AlSiO4	Hexagonal

Mineral Name	Formula	Crystal System
TRIKALSILITE	(K,Na)AlSiO4	Hexagonal
SADANAGAITE	(K,Na)Ca2(Fe++,Mg,Al,Fe+++,Ti)5(Si,Al)8O22(OH)2	Monoclinic
MAGNESIOSADANAGAITE	(K,Na)Ca2(Mg,Fe++,Al,Fe+++,Ti)5(Si,Al)8O22(OH)2	Monoclinic
DENISOVITE	(K,Na)Ca2Si3O8(F,OH)	Monoclinic
FLUORAPOPHYLLITE	(K,Na)Ca4Si8O20(F,OH).8H2O	Orthorhombic
AJOITE	(K,Na)Cu7AlSi9O24(OH)6.3H2O	Triclinic
SHCHERBAKOVITE	(K,Na,Ba)3(Ti,Nb)2Si4O14	Orthorhombic
PARSETTENSITE	(K,Na,Ca)(Mn,Al)7Si8O20(OH)8.2H2O	Monoclinic ps Hexagonal
PHILLIPSITE	(K,Na,Ca)1-2(Si,Al)8O16.6H2O	Monoclinic
ZIRCOPHYLLITE	(K,Na,Ca)3(Mn,Fe++)7(Zr,Nb)2Si8O27(OH,F)4	Triclinic
FENAKSITE	(K,Na,Ca)4(Fe++,Fe+++,Mn)2Si8O20(OH,F)	Triclinic
ARCHERITE	(K,NH4,)H2PO4	Tetragonal
YINGJIANGITE	(K2,Ca)(UO2)7(PO4)4(OH)6.6H2O	Orthorhombic
OFFRETITE	(K2,Ca)5Al10Si26O72.30H2O	Hexagonal
ERIONITE	(K2,Ca,Na2)2Al4Si14O36.15H2O	Hexagonal
PAULINGITE	(K2,Ca,Na2,Ba)5Al10Si35O90.45H2O	Isometric
AGRINIERITE	(K2,Ca,Sr)U3O10.4H2O	Orthorhombic
AGARDITE-(La)	(La,Ca)Cu6(AsO4)3(OH)6.3H2O	Hexagonal
HYDROXYLBASTNAESITE-(La)	(La,Ce)(CO3)(OH,F)	Hexagonal
BASTNASITE-(La) (BASTNAESITE-(La))	(La,Ce)(CO3)F	Hexagonal
NORDITE-(La)	(La,Ce)(Sr,Ca)Na2(Na,Mn)(Zn,Mg)Si6O17	Orthorhombic
DAVIDITE-(La)	(La,Ce)(Y,U,Fe++)(Ti,Fe+++)20(O,OH)38	Trigonal
LANTHANITE-(La)	(La,Ce)2(CO3)3.8H2O	Orthorhombic
TORNEBOHMITE-(La) (TOERNEBOHMITE-(La))	(La,Ce)2Al(SiO4)2(OH)	Monoclinic
FLORENCITE-(La)	(La,Ce)Al3(PO4)2(OH)6	Trigonal
FLUOCERITE-(La)	(La,Ce)F3	Hexagonal
RHABDOPHANE-(La)	(La,Ce)PO4.H2O	Hexagonal
MONAZITE-(La)	(La,Ce,Nd)PO4	Monoclinic
ARSENOFLORENCITE-(La)	(La,Sr)Al3(AsO4,SO4,PO4)2(OH)6	Trigonal
LITHIOWODGINITE	(Li,Mn,Fe)(Ta,Nb,Sn)3O8	Monoclinic
NAMBULITE	(Li,Na)(Mn,Ca)4Si5O14(OH)	Triclinic
BERTOSSAITE	(Li,Na)2CaAl4(PO4)4(OH,F)4	Orthorhombic
AMBLYGONITE	(Li,Na)Al(PO4)(F,OH)	Triclinic
PALYGORSKITE	(Mg,Al)2Si4O10(OH).4H2O	Monoclinic
KARLITE	(Mg,Al)6(BO3)3(OH,Cl)4	Orthorhombic
SAPPHIRINE-1Tc	(Mg,Al)8(Al,Si)6O20	Triclinic
SAPPHIRINE-2M	(Mg,Al)8(Al,Si)6O20	Monoclinic
YODERITE	(Mg,Al,Fe+++)8Si4(O,OH)20	Monoclinic
URSILITE	(Mg,Ca)4[(UO2)4(OH)5/(Si2O5)5,5].13H2O	Orthorhombic
MONGSHANITE	(Mg,Cr,Fe++)2(Ti,Zr,Cr,FE+++)5O12	Hexagonal
MCGUINNESSITE	(Mg,Cu)2(CO3)(OH)2	Monoclinic
MAGNESIOAUBERTITE	(Mg,Cu)Al(SO4)2Cl.14H2O	Triclinic

Mineral Data

Mineral Name	Formula	Crystal System
WERDINGITE	(Mg,Fe)2Al12(Al,Fe)2Si4(B,Al)4O37	Triclinic
CHESTERITE	(Mg,Fe++)17Si20O54(OH)6	Orthorhombic
HOGBOMITE (HOEGBOMITE)	(Mg,Fe++)2(Al,Ti)5O10	Hexagonal/Trigonal
MAGNESIOHULSITE	(Mg,Fe++)2(Fe+++,Sn++++,Mg)BO5	Monoclinic
AZOPROITE	(Mg,Fe++)2(Fe+++,Ti,Mg)BO5	Orthorhombic
WAGNERITE	(Mg,Fe++)2(PO4)F	Monoclinic
QANDILITE	(Mg,Fe++)2(Ti,Fe+++,Al)O4	Isometric
RINGWOODITE	(Mg,Fe++)2SiO4	Isometric
WADSLEYITE	(Mg,Fe++)2SiO4	Orthorhombic
SOUZALITE	(Mg,Fe++)3(Al,Fe+++)4(PO4)4(OH)6.2H2O	Monoclinic
BARICITE	(Mg,Fe++)3(PO4)2.8H2O	Monoclinic
SURINAMITE	(Mg,Fe++)3Al4BeSi3O16	Monoclinic
TREMBATHITE	(Mg,Fe++)3B7O13Cl	Trigonal
ANTIGORITE	(Mg,Fe++)3Si2O5(OH)4	Monoclinic
CHLORMAGALUMINITE	(Mg,Fe++)4Al2(OH)12(Cl2,CO3).2H2O	Hexagonal
CHONDRODITE	(Mg,Fe++)5(SiO4)2(F,OH)2	Monoclinic
CLINOCHLORE	(Mg,Fe++)5Al(Si3Al)O10(OH)8	Monoclinic
GEDRITE	(Mg,Fe++)5Al2(Si6Al2)O22(OH)2	Orthorhombic
MAGNESIOGEDRITE	(Mg,Fe++)5Al2Si6Al2O22(OH)2	Orthorhombic
CLINOJIMTHOMPSONITE	(Mg,Fe++)5Si6O16(OH)2	Monoclinic
JIMTHOMPSONITE	(Mg,Fe++)5Si6O16(OH)2	Orthorhombic
HUMITE	(Mg,Fe++)7(SiO4)3(F,OH)2	Orthorhombic
ANTHOPHYLLITE	(Mg,Fe++)7Si8O22(OH)2	Orthorhombic
CUMMINGTONITE	(Mg,Fe++)7Si8O22(OH)2	Monoclinic
MAGNESIOANTHOPHYLLITE	(Mg,Fe++)7Si8O22(OH)2	Orthorhombic
MAGNESIOCUMMINGTONITE	(Mg,Fe++)7Si8O22(OH)2	Monoclinic
CLINOHUMITE	(Mg,Fe++)9(SiO4)4(F,OH)2	Monoclinic
GRANDIDIERITE	(Mg,Fe++)Al3(BO4)(SiO4)O	Orthorhombic
SAHAMALITE-(Ce)	(Mg,Fe++)Ce2(CO3)4	Monoclinic
ARMALCOLITE	(Mg,Fe++)Ti2O5	Orthorhombic
GARYANSELLITE	(Mg,Fe+++)3(PO4)2(OH,O).1,5H2O	Orthorhombic
BALANGEROITE	(Mg,Fe+++,Fe++,Mn++)42Si16O54(OH)40	Monoclinic
VERMICULITE	(Mg,Fe++,Al)3(Al,Si)4O10(OH)2.4H2O	Monoclinic
PIGEONITE	(Mg,Fe++,Ca)(Mg,Fe++)Si2O6	Monoclinic
MAGNOCOLUMBITE	(Mg,Fe++,Mn)(Nb,Ta)2O6	Orthorhombic
MAGNIOTRIPLITE	(Mg,Fe++,Mn)2(PO4)F	Monoclinic
NININGERITE	(Mg,Fe++,Mn)S	Isometric
CARLOSTURANITE	(Mg,Fe++,Ti)21(Si,Al)12O28(OH)34	Monoclinic
MUSGRAVITE	(Mg,Fe++,Zn)2Al6BeO12	Trigonal
CORRENSITE	(Mg,Fe,Al)9(Si,Al)8O20(OH)10.nH2O	Orthorhombic
SVYAZHINITE	(Mg,Mn)(Al,Fe+++)(SO4)2F.14H2O	Triclinic
GERSTMANNITE	(Mg,Mn)2ZnSiO4(OH)2	Orthorhombic
MAGNESIUMCHLOROPHOE NICITE	(Mg,Mn)3Zn2(AsO4)(OH,O)6	Monoclinic
RIMKOROLGITE	(Mg,Mn)5(Ba,Sr,Ca)(PO4)4.8H2O	Orthorhombic
TORREYITE	(Mg,Mn)9Zn4(SO4)2(OH)22.8H2O	Monoclinic

Mineral Name	Formula	Crystal System
TAKEUCHIITE	(Mg,Mn++)2(Mn+++,Fe+++)BO5	Orthorhombic
PINAKIOLITE	(Mg,Mn++)2(Mn+++,Sb+++)BO5	Monoclinic
HAUCKITE	(Mg,Mn++)24Zn18Fe+++3(SO4)4(CO3)2(OH)81	Hexagonal
ORTHOPINAKIOLITE	(Mg,Mn++)2Mn+++BO5	Orthorhombic
JIANSHUIITE	(Mg,Mn++)Mn++++3O7.3H2O	Triclinic
MOUNTKEITHITE	(Mg,Ni)11(Fe+++,Cr,Al)3(OH)24(SO4,CO3)3.5.11H2O	Hexagonal
KARPINSKITE	(Mg,Ni)2Si2O5(OH)2	Monoclinic
CEROLITE (KEROLITE)	(Mg,Ni)3Si4O10(OH)2.H2O	Monoclinic
MAUFITE	(Mg,Ni)Al4Si3O13.4H2O	
MAGNESIODUMORTIERITE	(Mg,Ti++++)<1(Al,Mg)2Al4Si3O18-y(OH)yB(y=2-3)	Orthorhombic
WARWICKITE	(Mg,Ti,Fe+++,Al)2(BO3)O	Orthorhombic
CHUDOBAITE	(Mg,Zn)5H2(AsO4)4.10H2O	Triclinic
PENGZHIZHONGITE-24R	(Mg,Zn,Fe+++,Al)4(Sn,Fe+++)2Al10O22(OH)2	Trigonal
PENGZHIZHONGITE-6H	(Mg,Zn,Fe+++,Al)4(Sn,Fe+++)2Al10O22(OH)2	Trigonal
MOOREITE	(Mg,Zn,Mn)15(SO4)2(OH)26.8H2O	Monoclinic
KANONAITE	(Mn+++,Al)AlSiO5	Orthorhombic
BIXBYITE	(Mn+++,Fe+++)2O3	Isometric
WHITEITE-(MnFeMg)	(Mn++,Ca)(Fe++,Mn++)Mg2Al2(PO4)4(OH)2.8H2O	Monoclinic
ANDROSITE-(La)	(Mn++,Ca)(La,Ce,Nd)Mn++(Al,Mn+++)(SiO4)(Si2O7)O(OH)	Monoclinic
CALDERITE	(Mn++,Ca)3(Fe+++,Al)2(SiO4)3	Isometric
YAMATOITE	(Mn++,Ca)3(V+++,Al)2(SiO4)3	Isometric
KALUGINITE	(Mn++,Ca)MgFe+++(PO4)2(OH).4H2O	Orthorhombic
JAHNSITE-(MnMnMn)	(Mn++,Ca)Mn++(Mn++,Fe++)2Fe+++2(PO4)4(OH)2.8H2O	Monoclinic
TAKANELITE	(Mn++,Ca)Mn++++4O8.H2O	Hexagonal
TODOROKITE	(Mn++,Ca,Mg)Mn++++3O7.H2O	Monoclinic
MANGANOTAPIOLITE	(Mn++,Fe++)(Ta,Nb)2O6	Tetragonal
TITANOWODGINITE	(Mn++,Fe++)(Ti,Sn,Ta,Sc)(Ta,Nb)2O8	Monoclinic
VUORELAINENITE	(Mn++,Fe++)(V+++,Cr+++)2O4	Isometric
SCHALLERITE	(Mn++,Fe++)16Si12As+++3O36(OH)17	Trigonal
SWITZERITE	(Mn++,Fe++)3(PO4)2.7H2O	Monoclinic
MANGANOSTIBITE	(Mn++,Fe++)7(SbO4)(AsO4,SiO4)O4	Orthorhombic
AKATOREITE	(Mn++,Fe++)Al2Si8O24(OH)8	Triclinic
BEUSITE	(Mn++,Fe++,Ca,Mg)3(PO4)2	Monoclinic
JACOBSITE	(Mn++,Fe++,Mg)(Fe+++,Mn+++)2O4	Isometric
MANGANGORDONITE	(Mn++,Fe++,Mg)Al2(PO4)2(OH)2.8H2O	Triclinic
RHODONITE	(Mn++,Fe++,Mg,Ca)SiO3	Triclinic
BLATTERITE	(Mn++,Mg)2(Mn+++,Sb+++,Fe+++)(BO3)O2	Orthorhombic
KRAISSLITE	(Mn++,Mg)24Zn3Fe+++(As+++O3)2(As+++++O4)3(SiO4)6(OH)18	Hexagonal
KANOITE	(Mn++,Mg)2Si2O6	Monoclinic
NCHWANINGITE	(Mn++,Mg)2Si2O6(OH)4.2H2O	Orthorhombic
RIBBEITE	(Mn++,Mg)5(SiO4)2(OH)2	Orthorhombic

Mineral Name	Formula	Crystal System
CARYOPILITE	$(Mn^{++},Mg)6Si4O10(OH)8$	Monoclinic
CHVALETICEITE	$(Mn^{++},Mg)SO4.6H2O$	Monoclinic
KELLYITE	$(Mn^{++},Mg,Al)3(Si,Al)2O5(OH)4$	Hexagonal
KORAGOITE	$(Mn^{++},Mn^{+++})3(Nb,Mn^{++})2(Nb,Ta)3W2O20$	Monoclinic
ERNSTITE	$(Mn^{++}1-xFe^{+++}x)Al(PO4)(OH)2-xOx$	Monoclinic
AURORITE	$(Mn,Ag,Ca)Mn^{++++}3O7.3H2O$	Triclinic
LUNOKITE	$(Mn,Ca)(Mg,Fe^{++},Mn)Al(PO4)2(OH).4H2O$	Orthorhombic
MANGANOSEGELERITE	$(Mn,Ca)(Mn,Fe^{++},Mg)Fe^{+++}(PO4)2(OH).4H2O$	Orthorhombic
GERASIMOVSKITE	$(Mn,Ca)(Nb,Ti)5O12.9H2O$	Amorphous
MANGANBELYANKINITE	$(Mn,Ca)(Ti,Nb)5O12.9H2O$	Amorphous
BUSTAMITE	$(Mn,Ca)3Si3O9$	Trigonal
MEDAITE	$(Mn,Ca)6(V^{+++++},As)Si5O18(OH)$	Monoclinic
LANGBANITE	$(Mn,Ca,Fe,Mg)^{++}4(Mn,Fe)9Sb^{+++++}[O16(SiO4)2]$	Monoclinic
ARDENNITE	$(Mn,Ca,Mg)4(Al,Mn,Fe,Mg)6(As,V,P,Si)(O,OH)4(SiO4)2Si3O10(OH,O)6$	Orthorhombic
VILLYAELLENITE	$(Mn,Ca,Zn)5(AsO4)2(AsO3OH).4H2O$	Monoclinic
BROKENHILLITE	$(Mn,Fe)32[Si24O60]OH29Cl11$	Hexagonal
MANGANOCHROMITE	$(Mn,Fe^{++})(Cr,V)2O4$	Isometric
MANGANOCOLUMBITE	$(Mn,Fe^{++})(Nb,Ta)2O6$	Orthorhombic
NELENITE (FERROSCHALLERITE)	$(Mn,Fe^{++})16Si12As^{+++}3O36(OH)17$	Monoclinic
TRIPLOIDITE	$(Mn,Fe^{++})2(PO4)(OH)$	Monoclinic
MCGILLITE	$(Mn,Fe^{++})8Si6O15(OH)8Cl2$	Monoclinic psTrigonal
MANGANPYROSMALITE	$(Mn,Fe^{++})8Si6O15(OH,Cl)10$	Hexagonal
EARLSHANNONITE	$(Mn,Fe^{++})Fe^{+++}2(PO4)2(OH)2.4H2O$	Monoclinic
NEOTOCITE	$(Mn,Fe^{++})SiO3.H2O$	Amorphous
GALAXITE	$(Mn,Fe^{++},Mg)(Al,Fe^{+++})2O4$	Isometric
OTTRELITE	$(Mn,Fe^{++},Mg)2Al4Si2O10(OH)4$	Monoclinic/Triclinic
TRIPLITE	$(Mn,Fe^{++},Mg,Ca)2(PO4)(F,OH)$	Monoclinic
SANTAFEITE	$(Mn,Fe,Al,Mg)8(Mn,Mn)8(Ca,Sr,Na)12(VO4,AsO4)16(OH)20.8H2O$	Orthorhombic
KATOPTRITE	$(Mn,Mg)13(Al,Fe^{+++})4Sb^{+++++}2Si2O28$	Monoclinic
RETZIAN-(La)	$(Mn,Mg)2(La,Ce,Nd)(AsO4)(OH)4$	Orthorhombic
FILIPSTADITE	$(Mn,Mg)2Sb^{+++++}Fe^{+++}O8$	Orthorhombic
MANGANESEHORNESITE	$(Mn,Mg)3(AsO4)2.8H2O$	Monoclinic
CHLOROPHOENICITE	$(Mn,Mg)3Zn2(AsO4)(OH,O)6$	Monoclinic
GONYERITE	$(Mn,Mg)5Fe^{+++}(Si3Fe^{+++})O10(OH)8$	Orthorhombic
PARWELITE	$(Mn,Mg)5Sb(As,Si)2O12$	Monoclinic
YOFORTIERITE	$(Mn,Mg)5Si8O20(OH)2.8-9H2O$	Monoclinic
HOLDENITE	$(Mn,Mg)6Zn3(AsO4)2(SiO4)(OH)8$	Orthorhombic
MANGANHUMITE	$(Mn,Mg)7(SiO4)3(OH)2$	Orthorhombic
LANDESITE	$(Mn,Mg)9Fe^{+++}3(PO4)8(OH)3.9H2O$	Orthorhombic
LAWSONBAUERITE	$(Mn,Mg)9Zn4(SO4)2(OH)22.8H2O$	Monoclinic
FAHEYITE	$(Mn,Mg)Fe^{+++}2Be2(PO4)4.6H2O$	Hexagonal
DONPEACORITE	$(Mn,Mg)MgSi2O6$	Orthorhombic
HEMATOLITE	$(Mn,Mg,Al)15(AsO3)(AsO4)2(OH)23$	Trigonal

Mineral Name	Formula	Crystal System
SYNADELPHITE	$(Mn,Mg,Ca,Pb)9(As+++O3)(As++++O4)2(OH)9.2H2O$	Triclinic ps Orthorhombic
MCGOVERNITE	$(Mn,Mg,Zn)22(AsO3)(AsO4)3(SiO4)3(OH)21$	Trigonal
GAGEITE-1A	$(Mn,Mg,Zn)42Si16O54(OH)40$	Triclinic
GAGEITE-2M	$(Mn,Mg,Zn)42Si16O54(OH)40$	Monoclinic
MANGANESESHADLUNITE	$(Mn,Pb,Cd)(Cu,Fe)8S8$	Isometric
SPIROFFITE	$(Mn,Zn)2Te3O8$	Monoclinic
LOSEYITE	$(Mn,Zn)7(CO3)2(OH)10$	Monoclinic
DENNINGITE	$(Mn,Zn)Te2O5$	Tetragonal
ILESITE	$(Mn,Zn,Fe++)SO4.4H2O$	Monoclinic
REEDERITE-	$(Na,Al,Mn,Ca)15(Y,Ce,Nd,La)2(CO3)9(SO3F)(Cl,F)$	Hexagonal
MAXWELLITE	$(Na,Ca)(Fe+++,Al,Mg)(AsO4)(F,O)$	Monoclinic
ANDESINE	$(Na,Ca)(Si,Al)4O8$	Triclinic
OLIGOCLASE	$(Na,Ca)(Si,Al)4O8$	Triclinic
MONTMORILLONITE	$(Na,Ca)0,3(Al,Mg)2Si4O10(OH)2.nH2O$	Monoclinic
BURANGAITE	$(Na,Ca)2(Fe++,Mg)2Al10(PO4)8(OH,O)12.4H2O$	Monoclinic
LAVENITE	$(Na,Ca)2(Mn,Fe++)(Zr,Ti)Si2O7(O,OH,F)$	Monoclinic
SEIDOZERITE	$(Na,Ca)2(Zr,Ti,Mn)2Si2O7(O,F)2$	Monoclinic
HEULANDITE	$(Na,Ca)2-3Al3(Al,Si)2Si13O36.12H2O$	Monoclinic
LEUCOPHANITE	$(Na,Ca)2BeSi2(O,OH,F)7$	Triclinic ps Orthorhombic
FERROHAGENDORFITE	$(Na,Ca)2Fe++(Fe++,Fe+++)2(PO4)3$	Monoclinic
PYROCHLORE	$(Na,Ca)2Nb2O6(OH,F)$	Isometric
MICROLITE	$(Na,Ca)2Ta2O6(O,OH,F)$	Isometric
RINKITE (RINKOLITE)	$(Na,Ca)3(Ca,Ce)4Ti4(Si2O7)2(O,F)4$	Monoclinic
JANHAUGITE	$(Na,Ca)3(Mn++,Fe++)3(Ti++++,Zr,Nb)2(Si2O7)2O2(OH,F)2$	Monoclinic
BURBANKITE	$(Na,Ca)3(Sr,Ba,Ce)3(CO3)5$	Hexagonal
OKANOGANITE-	$(Na,Ca)3(Y,Ce)12Si6B2O27F14$	Trigonal
SODIUMBETPAKDALITE	$(Na,Ca)3Fe+++2(As2O4)(MoO4)6.15H2O$	Monoclinic
PETERSENITE-(Ce)	$(Na,Ca)4(Ce,La,Nd,Sr,Pr,Sm,Ba)2(CO3)5$	Monoclinic
HAUYNE	$(Na,Ca)4-8Al6Si6(O,S)24(SO4,Cl)1-2$	Isometric
FRANZINITE	$(Na,Ca)7(Si,Al)12O24(SO4,CO3,OH,Cl)3.H2O$	Hexagonal
LAZURITE	$(Na,Ca)7-8(Al,Si)12(O,S)24[(SO4),Cl2,(OH)2]$	
RECTORITE (MICA-SMECTITE)	$(Na,Ca)Al4(Si,Al)8O20(OH)4.2H2O$	Monoclinic
JERVISITE	$(Na,Ca,Fe++)(Sc,Mg,Fe++)Si2O6$	Monoclinic
SODIUMKOMAROVITE	$(Na,Ca,H)4Nb4Si4O12O8(OH,F)4.2H2O$	Orthorhombic
NENADKEVICHITE	$(Na,Ca,K)(Nb,Ti)Si2O6(O,OH).2H2O$	Orthorhombic
PANETHITE	$(Na,Ca,K)2(Mg,Fe++,Mn)2(PO4)2$	Monoclinic
VISHNEVITE	$(Na,Ca,K)6(Si,Al)12O24[(SO4),(CO3),Cl2]2-4.nH2O$	Hexagonal
MICROSOMMITE	$(Na,Ca,K)7-8(Si,Al)12O24(Cl,SO4,CO3)2-3$	Hexagonal
AFGHANITE	$(Na,Ca,K)8(Si,Al)12O24(SO4,Cl,CO3)3.H2O$	Hexagonal
TOUNKITE	$(Na,Ca,K)8Al6Si6O24(SO4)2Cl.H2O$	Hexagonal
SACROFANITE	$(Na,Ca,K)9(Si,Al)12O24[(OH)2,(SO4),(CO3),Cl2)]3.nH2O$	Hexagonal

Mineral Data

Mineral Name	Formula	Crystal System
MINAMIITE	(Na,Ca,K)Al3(SO4)2(OH)6	Trigonal
HERSCHELITE	(Na,Ca,K)AlSi2O6.3H2O	Trigonal
FERROWYLLIEITE	(Na,Ca,Mn)(Fe++,Mn)(Fe++,Fe+++,Mg)Al(PO4)3	Monoclinic
ROSEMARYITE	(Na,Ca,Mn++)(Mn++,Fe++)(Fe+++,Fe++,Mg)Al(PO4)3	Monoclinic
WYLLIEITE	(Na,Ca,Mn++)(Mn++,Fe++)(Fe++,Fe+++,Mg)Al(PO4)3	Monoclinic
CLARKEITE	(Na,Ca,Pb)2U2(O,OH)7	Orthorhombic
BEIDELLITE	(Na,Ca0,5)0,3Al2(Si,Al)4O10(OH)2.nH2O	Monoclinic
ILMAJOKITE	(Na,Ce,Ba)2TiSi3O5(OH)10.nH2O	Monoclinic
NATROBISTANTITE	(Na,Cs)Bi(Ta,Nb,Sb)4O12	Isometric
BRAMMALLITE	(Na,H3O)(Al,Mg,Fe)2(Si,Al)4O10[(OH)2,H2O]	Monoclinic
FLUORRICHTERITE	(Na,K)(Ca,Na)2(Mg,Fe)5Si8O22(F,OH,O)2	Monoclinic
POVONDRAITE	(Na,K)(Fe+++,Fe++)3(Fe,Mg,Al)6(BO3)3Si6O18(OH)4	Trigonal
WONESITE	(Na,K)(Mg,Fe,Al)6(Si,Al)8O20(OH,F)4	Monoclinic
MANJIROITE	(Na,K)(Mn++++,Mn++)8O16.nH2O	Tetragonal
DELHAYELITE	(Na,K)10Ca5Al6Si32O80(Cl2,F2,SO4)3.18H2O	Orthorhombic
TIETTAITE	(Na,K)17Fe+++TiSi16O29(OH)30.2H2O	Orthorhombic
BARYTOLAMPROPHYLLITE	(Na,K)2(Ba,Ca,Sr)2(Ti,Fe)3(SiO4)4(O,OH)2	Monoclinic
ROEDDERITE	(Na,K)2(Mg,Fe++)5Si12O30	Hexagonal
ZEMKORITE	(Na,K)2Ca(CO3)2	Hexagonal
LEMOYNITE	(Na,K)2CaZr2Si10O26.5-6H2O	Monoclinic
SODIUMPHARMACOSIDERITE	(Na,K)2Fe+++4(AsO4)3(OH)5.7H2O	Tetragonal ps Cubic
FERRIERITE	(Na,K)2Mg(Si,Al)18O36(OH).9H2O	Orthorhombic
ZIMBABWEITE	(Na,K)2PbAs+++4(Ta,Nb,Ti)4O18	Orthorhombic
DELINDEITE	(Na,K)3(Ba,Ca)4(Ti,Fe,Al)6Si8O26(OH)4	Monoclinic
NAFERTISITE	(Na,K)3(Fe++,Fe+++,Mg)9-10Ti2(Si,Fe+++,Al)12O34(O,OH)9	Monoclinic
YAGIITE	(Na,K)3Mg4(Al,Mg)6(Si,Al)24O60	Hexagonal
SPODIOPHYLLITE	(Na,K)4(Mg,Fe++)3(Fe+++,Al)2(Si8O24)	Monoclinic
REYERITE	(Na,K)4Ca14Si22Al2O58(OH)8.6H2O	Trigonal
MAGNESIUMASTROPHYLLITE	(Na,K)4Mg2(Fe++,Fe+++,Mn)5Ti2Si8O24(O,OH,F)7	Monoclinic
SAZYKINAITE-	(Na,K)5Y(Zr,Ti)Si6O18.6H2O	Trigonal
SHAFRANOVSKITE	(Na,K)6(Mn++,Fe++)3Si9O24.6H2O	Trigonal
MONTEREGIANITE-	(Na,K)6(Y,Ca)2Si16O38.10H2O	Monoclinic
QUADRIDAVYNE	(Na,K)6Ca2Al6Si6O24Cl4	Hexagonal
CANASITE	(Na,K)6Ca5Si12O30(OH,F)4	Monoclinic
BYSTRITE	(Na,K)7Ca(Si6Al6)O24S1,5.H2O	Trigonal
ANORTHOCLASE	(Na,K)AlSi3O8	Triclinic
NEPHELINE	(Na,K)AlSiO4	Hexagonal
MILLISITE	(Na,K)CaAl6(PO4)4(OH)9.3H2O	Tetragonal
TUHUALITE	(Na,K)Fe++Fe+++Si6O15	Orthorhombic
UNGARETTITE	(Na,K)Na2(Mn++,Mg)2Mn+++Si8O22O2	Monoclinic
FLUORFERROLEAKEITE	(Na,K)Na2Li(Fe++,Mn++,Mg)2Fe2+++Si8O22(O	Monoclinic

Mineral Name	Formula	Crystal System
(FLUOR-FERRO-LEAKEITE)	H,F)2	
THORNASITE	(Na,K)ThSi11(O,F,OH)25.8H2O	Trigonal
BANNERMANITE	(Na,K)xV++++xV+++++6-xO15	Monoclinic
EGGLETONITE	(Na,K,Ca)2(Mn,Fe)8(Si,Al)12O29(OH)7.11H2O	Monoclinic
CLINOPTILOLITE	(Na,K,Ca)2-3Al3(Al,Si)2Si13O36.12H2O	Monoclinic
BARRERITE	(Na,K,Ca)2Al2Si7O18.7H2O	Orthorhombic
GIUSEPPETTITE	(Na,K,Ca)7-8(Si,Al)12O24(SO4,Cl)1-2	Hexagonal
RANKAMAITE	(Na,K,Pb,Li)3(Ta,Nb,Al)11(O,OH)30	Orthorhombic
NATROMONTEBRASITE	(Na,Li)Al(PO4)(OH,F)	Triclinic
NATRONAMBULITE	(Na,Li)Mn4Si5O14(OH)	Triclinic
MURATAITE-	(Na,Y)4(Zn,Fe++)3(Ti,Nb)6O18(F,OH)4	Isometric
SODIUMURANOSPINITE	(Na2,Ca)(UO2)2(AsO4)2.5H2O	Tetragonal
GMELINITE	(Na2,Ca)Al2Si4O12.6H2O	Hexagonal
FAUJASITE	(Na2,Ca)Al2Si4O12.8H2O	Isometric
SODIUMDACHIARDITE	(Na2,Ca,K2)4-5Al8Si40O96.26H2O	Monoclinic
GREGORYITE	(Na2,K2,Ca)CO3	Hexagonal
KHANNESHITE	(NaCa)3(Ba,Sr,Ce,Ca)3(CO3)5	Hexagonal
ASHANITE	(Nb,Ta,U,Fe,Mn)4O8	Orthorhombic
NIOBOAESCHYNITE-(Nd)	(Nd,Ce)(Nb,Ti)2(O,OH)6	Orthorhombic
FERGUSONITE-(Nd)	(Nd,Ce)(Nb,Ti)O4	Monoclinic
FLORENCITE-(Nd)	(Nd,Ce)Al3(PO4)2(OH)6	Triclinic
FERGUSONITE-BETA-(Nd)	(Nd,Ce)NbO4	Monoclinic
AESCHYNITE-(Nd)	(Nd,Ce,Ca)(Ti,Nb)2(O,OH)6	Orthorhombic
RHABDOPHANE-(Nd)	(Nd,Ce,La)PO4.H2O	Hexagonal
MONAZITE-(Nd)	(Nd,Ce,La,Pr,Sm,Gd)(P,Si)O4	Monoclinic
HYDROXYLBASTNAESITE-(Nd)	(Nd,La)(CO3)(OH,F)	Hexagonal
LANTHANITE-(Nd)	(Nd,La)2(CO3)3.8H2O	Orthorhombic
ARSENOFLORENCITE-(Nd)	(Nd,La,Ce,Ba)(Al,Fe+++)3(AsO4,PO4)2(OH)6	Trigonal
AGARDITE-(Nd)	(Nd,La,Ce,Ca)Cu6(AsO4)3(OH)6.3H2O	Hexagonal
FRANCOISITE-(Nd)	(Nd,Y,Sm,Ce)(UO2)3(PO4)2O(OH).6H2O	Monoclinic
GODOVIKOVITE	(NH4)(Al,Fe+++)(SO4)2	Hexagonal
LONECREEKITE	(NH4)(Fe+++,Al)(SO4)2.12H2O	Isometric
NIAHITE	(NH4)(Mn++,Mg,Ca)PO4.H2O	Orthorhombic
URAMPHITE	(NH4)(UO2)(PO4).3H2O	Orthorhombic
OXAMMITE	(NH4)2(C2O4).H2O	Orthorhombic
CLAIRITE	(NH4)2(Fe+++,Mn+++)3(SO4)4(OH)3.3H2O	Triclinic
NICKELBOUSSINGAULTITE	(NH4)2(Ni,Mg)(SO4)2.6H2O	Monoclinic
AMMONIOBORITE	(NH4)2B10O16.5H2O	Monoclinic
MUNDRABILLAITE	(NH4)2Ca(HPO4)2.H2O	Monoclinic
SWAKNOITE	(NH4)2Ca(HPO4)2.H2O	Orthorhombic
KOKTAITE	(NH4)2Ca(SO4)2.H2O	Monoclinic
MOHRITE	(NH4)2Fe++(SO4)2.6H2O	Monoclinic
PHOSPHAMMITE	(NH4)2HPO4	Monoclinic
BOUSSINGAULTITE	(NH4)2Mg(SO4)2.6H2O	Monoclinic
EFREMOVITE	(NH4)2Mg2(SO4)3	Isometric

Mineral Data

Mineral Name	Formula	Crystal System
HANNAYITE	(NH4)2Mg3H4(PO4)4.8H2O	Triclinic
SCHERTELITE	(NH4)2MgH2(PO4)2.4H2O	Orthorhombic
BARARITE	(NH4)2SiF6	Hexagonal
CRYPTOHALITE	(NH4)2SiF6	Isometric
MASCAGNITE	(NH4)2SO4	Orthorhombic
LETOVICITE	(NH4)3H(SO4)2	Triclinic
TSCHERMIGITE	(NH4)Al(SO4)2.12H2O	Isometric
AMMONIOALUNITE	(NH4)Al3(SO4)2(OH)6	Trigonal
BUDDINGTONITE	(NH4)AlSi3O8.0,5H2O	Monoclinic
LARDERELLITE	(NH4)B5O6(OH)4	Monoclinic
BARBERIITE	(NH4)BF4	Orthorhombic
SABIEITE	(NH4)Fe+++(SO4)2	Trigonal
AMMONIOJAROSITE	(NH4)Fe+++3(SO4)2(OH)6	Trigonal
TESCHEMACHERITE	(NH4)HCO3	Orthorhombic
DITTMARITE	(NH4)Mg(PO4).H2O	Orthorhombic
STRUVITE	(NH4)MgPO4.6H2O	Orthorhombic
SPHENISCIDITE	(NH4,K)(Fe+++,Al)2(PO4)2(OH).2H2O	Monoclinic
KREMERSITE	(NH4,K)2Fe+++Cl5.H2O	Orthorhombic
TOBELITE	(NH4,K)Al2(Si3Al)O10(OH)2	Monoclinic
AMMONIOLEUCITE	(NH4,K)AlSi2O6	Tetragonal
BIPHOSPHAMMITE	(NH4,K)H2PO4	Tetragonal
LECONTITE	(NH4,K)Na(SO4).2H2O	Orthorhombic
JAMBORITE	(Ni++,Ni+++,Fe)(OH)2(OH,S,H2O)	Hexagonal
COMBLAINITE	(Ni++x,Co+++1-x)(OH)2(CO3)(1-x)/2.yH2O	Trigonal
SIEGENITE	(Ni,Co)3S4	Isometric
VOZHMINITE	(Ni,Co)4(As,Sb)S2	Hexagonal
NICKELSKUTTERUDITE	(Ni,Co)As3-x	Isometric
JOLLIFFEITE	(Ni,Co)AsSe	Isometric
AHLFELDITE	(Ni,Co)SeO3.2H2O	Monoclinic
PENROSEITE	(Ni,Co,Cu)Se2	Isometric
NICHROMITE	(Ni,Co,Fe++)(Cr,Fe+++,Al)2O4	Isometric
CARRBOYDITE	(Ni,Cu)14Al9(SO4,CO3)6(OH)43.7H2O	Hexagonal
HYDROMBOBOMKULITE	(Ni,Cu)Al4[(NO3)2,(SO4)](OH)12.13-14H2O	Monoclinic
MBOBOMKULITE	(Ni,Cu)Al4[(NO3)2,(SO4)](OH)12.3H2O	Monoclinic
NICKELALUMITE	(Ni,Cu)Al4[(SO4),(NO3)2](OH)12.3H2O	Monoclinic
TAENITE	(Ni,Fe)	Isometric
GODLEVSKITE	(Ni,Fe)7S6	Orthorhombic
PERRYITE	(Ni,Fe)8(Si,P)3	Hexagonal
ARUPITE	(Ni,Fe++)3(PO4)2.8H2O	Monoclinic
DWORNIKITE	(Ni,Fe++)SO4.H2O	Monoclinic
LIEBENBERGITE	(Ni,Mg)2SiO4	Orthorhombic
WILLEMSEITE	(Ni,Mg)3Si4O10(OH)2	Monoclinic
PIMELITE	(Ni,Mg)3Si4O10(OH)2.H2O	Monoclinic
FALCONDOITE	(Ni,Mg)4Si6O15(OH)2.6H2O	Orthorhombic
WIDGIEMOOLTHALITE	(Ni,Mg)5(CO3)4(OH)2.4-5H2O	Monoclinic
NICKELHEXAHYDRITE	(Ni,Mg,Fe++)(SO4).6H2O	Monoclinic

Mineral Data

Mineral Name	Formula	Crystal System
BRINDLEYITE (NIMESITE)	(Ni,Mg,Fe++)2Al(SiAl)O5(OH)4	Monoclinic
NIMITE	(Ni,Mg,Fe++)5Al(Si3Al)O10(OH)8	Monoclinic
GASPEITE	(Ni,Mg,Fe++)CO3	Trigonal
KEYSTONEITE	(Ni,Mg,Fe,Mn)3Te++++3O9.5H2O	Hexagonal
HEXATESTIBIOPANICKELITE	(Ni,Pd)(Te,Sb)	Hexagonal
OSMIUM	(Os,Ir)	Hexagonal
OMEIITE	(Os,Ru)As2	Orthorhombic
OSARSITE	(Os,Ru)AsS	Monoclinic
INCAITE	(Pb,Ag)4Sn4FeSb2S15	Monoclinic
WEIBULLITE	(Pb,Ag)6Bi8(S,Se)18	Orthorhombic
ASSELBORNITE	(Pb,Ba)(UO2)6(BiO)4(AsO4)2(OH)12.3H2O	Isometric
FERRAZITE	(Pb,Ba)3(PO4)2.8H2O	Triclinic
MONIMOLITE	(Pb,Ca)3Sb2O8	Isometric
KUKSITE	(Pb,Ca)3Zn3Te++++++O6(PO4,VO4)2	Orthorhombic
WOLSENDORFITE (WOELSENDORFITE)	(Pb,Ca)U2O7.2H2O	Orthorhombic
HANCOCKITE	(Pb,Ca,Sr)2(Al,Fe+++)3(SiO4)3(OH)	Monoclinic
PLUMBOMICROLITE	(Pb,Ca,U)2Ta2O6(OH)	Isometric
SHADLUNITE	(Pb,Cd)(Fe,Cu)8S8	Isometric
FERRISURITE	(Pb,Cu)2-3(CO3)1,5-2(OH,F)0,5-1[(Fe,Al)2Si4O10(OH)2].nH2O	Monoclinic
FORNACITE	(Pb,Cu)3[(Cr,As)O4]2(OH)	Monoclinic
PROUDITE	(Pb,Cu)8Bi9-10(S,Se)22	Monoclinic
XINGZHONGITE	(Pb,Cu,Fe)(Ir.Pt,Rh)2S4	Isometric
SAKHAROVAITE	(Pb,Fe)(Bi,Sb)2S4	Orthorhombic
MOROZEVICZITE	(Pb,Fe)3Ge1-xS4	Isometric
PARKINSONITE	(Pb,Mo)O8Cl2	Tetragonal
HINSDALITE	(Pb,Sr)Al3(PO4)(SO4)(OH)6	Trigonal
HATCHITE	(Pb,Tl)2AgAs2S5	Triclinic
HUTCHINSONITE	(Pb,Tl)2As5S9	Orthorhombic
RATHITE	(Pb,Tl)3As5S10	Monoclinic
PLUMBOBETAFITE	(Pb,U,Ca)(Ti,Nb)2O6(OH,F)	Isometric
PLUMBOPYROCHLORE	(Pb,Y,U,Ca)2-xNb2O6(OH)	Isometric
TELARGPALITE	(Pd,Ag)3Te	Isometric
VASILITE	(Pd,Cu)16(S,Te)7	Isometric
STANNOPALLADINITE	(Pd,Cu)3Sn2	Hexagonal
OOSTERBOSCHITE	(Pd,Cu)7Se3	Orthorhombic
TAIMYRITE	(Pd,Cu,Pt)3Sn	Orthorhombic
ATHENEITE	(Pd,Hg)3As	Hexagonal
VYSOTSKITE	(Pd,Ni)S	Tetragonal
SUDBURYITE	(Pd,Ni)Sb	Hexagonal
MERENSKYITE	(Pd,Pt)(Te,Bi)2	Trigonal
VINCENTITE	(Pd,Pt)3(As,Sb,Te)	
ATOKITE	(Pd,Pt)3Sn	Isometric
OULANKAITE	(Pd,Pt)5(Cu,Fe)4SnTe2S2	Tetragonal
MICHENERITE	(Pd,Pt)BiTe	Isometric

Mineral Data

Mineral Name	Formula	Crystal System
ZVYAGINTSEVITE	(Pd,Pt,Au)3(Pb,Sn)	Isometric
KHARAELAKHITE	(Pt,Cu,Pb,Fe,Ni)9S8	Orthorhombic
CRERARITE	(Pt,Pb)Bi3(S,Se)4-x(x=0,7)	Isometric
MONCHEITE	(Pt,Pd)(Te,Bi)2	Trigonal
ISOFERROPLATINUM	(Pt,Pd)3(Fe,Cu)	Isometric
RUSTENBURGITE	(Pt,Pd)3Sn	Isometric
GENKINITE	(Pt,Pd)4Sb3	Tetragonal
BRAGGITE	(Pt,Pd,Ni)S	Tetragonal
COOPERITE	(Pt,Pd,Ni)S	Tetragonal
PLATARSITE	(Pt,Rh,Ru)AsS	Isometric
BOWIEITE	(Rh,Ir,Pt)1,77S3	Orthorhombic
RHODIUM	(Rh,Pt)	Isometric
HOLLINGWORTHITE	(Rh,Pt,Pd)AsS	Isometric
RUTHENIUM	(Ru,Ir,Os)	Hexagonal
RUTHENARSENITE	(Ru,Ni)As	Orthorhombic
ANDUOITE	(Ru,Os)As2	Orthorhombic
STIBIOBETAFITE	(Sb+++,Ca)2(Ti,Nb,Ta)2(O,OH)7	Isometric
STIBIOMICROLITE	(Sb,Ca,Na)2(Ta,Nb)2(O,OH)7	
THORTVEITITE	(Sc,Y)2Si2O7	Monoclinic
SELENTELLURIUM	(Se,Te)	Trigonal
STANNOMICROLITE (SUKULAITE)	(Sn++,Fe++)(Ta,Nb,Sn++++)2(O,OH)7	Isometric
VARLAMOFFITE	(Sn,Fe)(O,OH)2	Isometric
BREWSTERITE	(Sr,Ba,Ca)Al2Si6O16.5H2O	Monoclinic
STENONITE	(Sr,Ba,Na)2Al(CO3)F5	Monoclinic
PALERMOITE	(Sr,Ca)(Li,Na)2Al4(PO4)4(OH)4	Orthorhombic
GOEDKENITE	(Sr,Ca)2Al(PO4)2(OH)	Monoclinic
STRONTIOGINORITE	(Sr,Ca)2B14O23.8H2O	Monoclinic
STRONTIUMAPATITE	(Sr,Ca)5(PO4)3(OH,F)	Hexagonal
STRONTIODRESSERITE	(Sr,Ca)Al2(CO3)2(OH)4.H2O	Orthorhombic
SLAWSONITE	(Sr,Ca)Al2Si2O8	Monoclinic
DAQINGSHANITE-(Ce)	(Sr,Ca,Ba)3(Ce,La)(PO4)(CO3)3-x(OH,F)x	Hexagonal
ARSENOGOYAZITE (WEILERITE)	(Sr,Ca,Ba)Al3(AsO4,PO4)(OH)5F.H2O	Trigonal
KEMMLITZITE	(Sr,Ce)Al3(AsO4)(SO4)(OH)6	Trigonal
BELOVITE	(Sr,Ce,Na,Ca)5(PO4)3(OH)	Hexagonal
STRONTIOCHEVKINITE	(Sr,La,Ce,Ca)4(Fe++,Fe+++)(Ti+++,Zr)3Si4O12O10	Monoclinic
CRICHTONITE	(Sr,La,Ce,Y)(Ti,Fe+++,Mn)21O38	Trigonal
KIMROBINSONITE	(Ta,Nb)(OH)5-2x(O,CO3)x	Isometric
BEHIERITE	(Ta,Nb)BO4	Tetragonal
IXIOLITE	(Ta,Nb,Sn,Fe,Mn)4O8	Monoclinic
WODGINITE	(Ta,Nb,Sn,Mn,Fe)16O32	Monoclinic
KIVUITE	(Th,Ca,Pb)H2(UO2)4(PO4)2(OH)8.7H2O	Orthorhombic
EYLETTERSITE	(Th,Pb)1-xAl3(PO4,SiO4)2(OH)6	Trigonal
GRAYITE	(Th,Pb,Ca)PO4.H2O	ps Hexagonal
THORITE	(Th,U)SiO4	Tetragonal

Mineral Data

Mineral Name	Formula	Crystal System
THORUTITE	(Th,U,Ca)Ti2(O,OH)6	Monoclinic
ILMENORUTILE	(Ti,Nb,Fe+++)3O6	Tetragonal
STRUVERITE (STRUEVERITE)	(Ti,Ta,Fe+++)3O6	Tetragonal
KHAMRABAEVITE	(Ti,V,Fe)C	Isometric
SRILANKITE	(Ti,Zr)O2	Orthorhombic
STALDERITE	(Tl,Cu)(Zn,Fe,Hg)AsS3	Tetragonal
CHALCOTHALLITE	(Tl,K)2Cu5,5(Fe,Ag)SbS4	Orthorhombic ps Tetragonal
THALFENISITE	(Tl,K)6Na(Fe,Ni,Cu)25S26Cl	Isometric
DORALLCHARITE	(Tl,K)Fe+++3(SO4)2(OH)6	Orthorhombic
CHABOURNEITE	(Tl,Pb)21(Sb,As)91S147	Triclinic
URANPYROCHLORE	(U,Ca,Ca)2(Nb,Ta)2O6(OH,F)	Isometric
NINGYOITE	(U,Ca,Ce)2(PO4)2.1-2H2O	Orthorhombic ps Hexagonal
URANMICROLITE (DJALMAITE)	(U,Ca,Ce)2(Ta,Nb)2O6(OH,F)	Isometric
BRANNERITE	(U,Ca,Y,Ce)(Ti,Fe)2O6	Monoclinic
ISHIKAWAITE	(U,Fe,Y,Ca)(Nb,Ta)O4	Orthorhombic
URANOPOLYCRASE	(U,Y)(Ti,Nb,Ta)2O6	Orthorhombic
JOLIOTITE	(UO2)(CO3).nH2O,(n=2?)	Orthorhombic
IANTHINITE	(UO2).5(UO3).10H2O	Orthorhombic
SODDYITE	(UO2)2SiO4.2H2O	Orthorhombic
TROGERITE (TROEGERITE)	(UO2)3(AsO4)2.12H2O	Tetragonal
HAYNESITE	(UO2)3(SeO3)2(OH)2.5H2O	Orthorhombic
URANOPILITE	(UO2)6(SO4)(OH)10.12H2O	Monoclinic
METAURANOPILITE	(UO2)6(SO4)(OH)10.5H2O	Monoclinic
UMOHOITE	(UO2)MoO4.4H2O	Monoclinic/Orthorhombic
SCHMITTERITE	(UO2)TeO3	Orthorhombic
MONTROSEITE	(V+++,Fe+++)O(OH)	Orthorhombic
NOLANITE	(V+++,Fe++,Fe+++,Ti)10O14(OH)2	Hexagonal
ALUMOTUNGSTITE	(W,Al)(O,OH)3	Cubic ps Cubic
CALCYBEBOROSILITE-	(Y,Ca)2(B,Be)2Si2O8(OH)2	Monoclinic
BRITHOLITE- (ABUKUMALITE)	(Y,Ca)5(SiO4,PO4)3(OH,F)	Hexagonal
AGARDITE-	(Y,Ca)Cu6(AsO4)3(OH)6.3H2O	Hexagonal
EUXENITE-	(Y,Ca,Ce,U,Th)(Nb,Ta,Ti)2O6	Orthorhombic
POLYCRASE-	(Y,Ca,Ce,U,Th)(Ti,Nb,Ta)2O6	Orthorhombic
AESCHYNITE-	(Y,Ca,Fe,Th)(Ti,Nb)2(O,OH)6	Orthorhombic
TRITOMITE- (SPENCITE)	(Y,Ca,La,Fe++)5(Si,B,Al)3(O,OH,F)13	Trigonal
BASTNASITE- (BASTNAESITE-)	(Y,Ce)(CO3)F	Hexagonal
SYNCHYSITE- (DOVERITE)	(Y,Ce)Ca(CO3)2F	Monoclinic ps Hexagonal
TANTEUXENITE-	(Y,Ce,Ca)(Ta,Nb,Ti)2(O,OH)6	Orthorhombic
TANTALAESCHYNITE-	(Y,Ce,Ca)(Ta,Ti,Nb)2O6	Orthorhombic
ALLANITE- (YTTROORTHITE)	(Y,Ce,Ca)2(Al,Fe+++)3(SiO4)3(OH)	Monoclinic
LORANSKITE-	(Y,Ce,Ca)ZrTaO6	Orthorhombic
PETERSITE-	(Y,Ce,Nd,Ca)Cu6(PO4)3(OH)6.3H2O	Hexagonal
SAMARSKITE-	(Y,Ce,U,Fe+++)3(Nb,Ta,Ti)5O16	Monoclinic

Mineral Name	Formula	Crystal System
BIJVOETITE-	(Y,Dy)2(UO2)4(CO3)4(OH)6.11H2O	Orthorhombic
YFTISITE-	(Y,Dy,Er)4(Ti,Sn)O(SiO4)2(F,OH)6	Orthorhombic
TRIMOUNSITE-	(Y,Dy,Er,Yb,Gd,Ho,Tb,Sm)2Ti2SiO9	Monoclinic
YTTROPYROCHLORE-	(Y,Na,Ca,U)1-2(Nb,Ta,Ti)2(O,OH)7	Isometric
KAMOTOITE-	(Y,Nd,Gd)2U++++++4(CO3)3.14,5H2O	Monoclinic
YTTRIALITE-	(Y,Th)2Si2O7	Hexagonal
YTTROCRASITE-	(Y,Th,Ca,U)(Ti,Fe+++)2(O,OH)6	Orthorhombic
KOBEITE-	(Y,U)(Ti,Nb)2(O,OH)6	Amorphous
YTTROBETAFITE-	(Y,U,Ce)2(Ti,Nb,Ta)2O6(OH)	Isometric
YTTROCOLUMBITE-	(Y,U,Fe++)(Nb,Ta)O4	Orthorhombic
YTTROTANTALITE-	(Y,U,Fe++)(Ta,Nb)O4	Orthorhombic
KEIVIITE-	(Y,Yb)2Si2O7	Monoclinic
HINGGANITE-	(Y,Yb,Er)BeSiO4(OH)	Monoclinic
KULIOKITE-	(Y,Yb,Er,Dy,Lu,Gd,Tm,Ho)4Al(SiO4)2(OH)2F5	Triclinic
KEIVIITE-(Yb)	(Yb,Y)2Si2O7	Monoclinic
HINGGANITE-(Yb)	(Yb,Y)BeSiO4(OH)	Monoclinic
FRAIPONTITE	(Zn,Al)3(Si,Al)2O5(OH)4	Monoclinic
GUARINOITE	(Zn,Co,Ni)6(SO4)(OH,Cl)10.5H2O	Hexagonal
GLAUCOCERINITE	(Zn,Cu)10Al6(SO4)3(OH)32.18H2O	Hexagonal
ZINCROSASITE	(Zn,Cu)2(CO3)(OH)2	Monoclinic
BECHERERITE	(Zn,Cu)3Zn(S,Si)(O,OH)4(OH)13	Trigonal
NAMUWITE	(Zn,Cu)4(SO4)(OH)6.4H2O	Hexagonal
AURICHALCITE	(Zn,Cu)5(CO3)2(OH)6	Monoclinic
FAUSTITE	(Zn,Cu)Al6(PO4)4(OH)8.4H2O	Triclinic
ZINCMELANTERITE	(Zn,Cu,Fe++)SO4.7H2O	Monoclinic
SPHALERITE (BLENDE)	(Zn,Fe)S	Isometric
WURTZITE	(Zn,Fe)S	Hexagonal
GERDTREMMELITE	(Zn,Fe++)(Al,Fe+++)2(AsO4)(OH)5	Triclinic
BIANCHITE	(Zn,Fe++)(SO4).6H2O	Monoclinic
ZEMANNITE	(Zn,Fe++)2(Te++++O3)3NaxH2-x.nH2O	Hexagonal
SANMARTINITE	(Zn,Fe++)WO4	Monoclinic
METAKOTTIGITE (METAKOETTIGITE)	(Zn,Fe+++,Fe++)3(AsO4)2.8(H2O,OH)	Triclinic
BAILEYCHLORE	(Zn,Fe++,Al,Mg)6(Si,Al)4O10(OH)8	Triclinic
DIETRICHITE	(Zn,Fe++,Mn)Al2(SO4)4.22H2O	Monoclinic
CHALCOPHANITE	(Zn,Fe++,Mn++)Mn++++3O7.3H2O	Trigonal
ECANDREWSITE	(Zn,Fe++,Mn++)TiO3	Trigonal
POLHEMUSITE	(Zn,Hg)S	Tetragonal ps Cubic
BOYLEITE	(Zn,Mg)SO4.4H2O	Monoclinic
NIGERITE	(Zn,Mg,Fe++)(Sn,Zn)2(Al,Fe+++)12O22(OH)2	Trigonal
ZINCOBOTRYOGEN	(Zn,Mg,Mn)Fe+++(SO4)2(OH).7H2O	Monoclinic
SCLARITE	(Zn,Mg,Mn++)4Zn3(CO3)2(OH)10	Monoclinic
ZINCITE	(Zn,Mn)O	Hexagonal
GUNNINGITE	(Zn,Mn)SO4.H2O	Monoclinic
WOODRUFFITE	(Zn,Mn++)Mn++++3O7.1-2H2O	Monoclinic
FRANKLINITE	(Zn,Mn++,Fe++)(Fe+++,Mn+++)2O4	Isometric

Mineral Data

Mineral Name	Formula	Crystal System
ALVANITE	(Zn,Ni)Al4(VO3)2(OH)12.2H2O	Monoclinic
KURUMSAKITE	(Zn,Ni,Cu)8Al8V2Si5O35.27H2O	Orthorhombic
TAZHERANITE	(Zr,Ti,Ca)O2	Isometric
AMBER (SUCCINITE,BERNSTEIN)	[C,H,O]	Amorphous
VIAENEITE	11FeS2.(Fe,Pb)(S2O3)	Monoclinic
KITAIBELITE	15Bi2S3.5Ag2S.PbS	
GIESSENITE	2(Cu2Pb26(Bi,Sb)20S57)	Monoclinic
KILLALAITE	2Ca3Si2O7.H2O	Monoclinic
BASSANITE	2CaSO4.H2O	Monoclinic ps Hexagonal
ORICKITE	2CuFeS2.H2O	Hexagonal
BARSTOWITE	3PbCl2.PbCO3.H2OorPb4Cl6CO3.H2O	Monoclinic
SILHYDRITE	3SiO2.H2O	Orthorhombic
VALLERIITE	4(Fe,Cu)S.3(Mg,Al)(OH)2	Hexagonal
HAAPALAITE	4(Fe,Ni)S.3(Mg,Fe++)(OH)2	Hexagonal
AKDALAITE	4Al2O3.H2O	Hexagonal
FERRIHYDRITE	5Fe+++2O3.9H2O	Hexagonal
TOCHILINITE	6Fe0,9S.5(Mg,Fe++)(OH)2	Triclinic
SANTANAITE	9PbO.2PbO2.CrO3	Hexagonal
BLAKEITE	Aferrictellurite	
SILVER-2H	Ag	Hexagonal
SILVER-3C (ARGENT)	Ag	Isometric
SILVER-4H	Ag	Hexagonal
ARGENTOPENTLANDITE	Ag(Fe,Ni)8S8	Isometric
ARAMAYOITE	Ag(Sb,Bi)S2	Triclinic
SCHACHNERITE	Ag1,1Hg0,9	Hexagonal
ALLARGENTUM	Ag1-xSbx	Hexagonal
PEARCEITE	Ag16As2S11	Monoclinic
ROSHCHINITE	Ag19Pb10Sb51S96orPb(Ag,Cu)2(Sb,As)5S10	Orthorhombic
OURAYITE	Ag25Pb30Bi41S104	Orthorhombic
DERVILLITE	Ag2AsS2	Monoclinic
TOYOHAITE	Ag2FeSn3S8	Tetragonal
HOCARTITE	Ag2FeSnS4	Tetragonal
MOSCHELLANDSBERGITE	Ag2Hg3	Isometric
IMITERITE	Ag2HgS2	Monoclinic
STERRYITE	Ag2Pb10(Sb,As)12S29	Orthorhombic
ACANTHITE	Ag2S	Monoclinic
STETEFELDTITE	Ag2Sb2(O,OH)7	Isometric
NAUMANNITE	Ag2Se	Orthorhombic ps Cubic
HESSITE	Ag2Te	Monoclinic
PIRQUITASITE	Ag2ZnSnS4	Tetragonal
PROUSTITE	Ag3AsS3	Trigonal
XANTHOCONITE	Ag3AsS3	Monoclinic
UYTENBOGAARDTITE	Ag3AuS2	Tetragonal
FISCHESSERITE	Ag3AuSe2	Isometric
PETZITE	Ag3AuTe2	Isometric

Mineral Data

Mineral Name	Formula	Crystal System
MUMMEITE	Ag3CuPbBi6S13orAg2Cu2Pb2Bi6S13	Monoclinic
JALPAITE	Ag3CuS2	Tetragonal
LUANHEITE	Ag3Hg	Hexagonal
PARASCHACHNERITE	Ag3Hg2	Orthorhombic
SCHIRMERITE	Ag3Pb3Bi9S18toAg3Pb6Bi7S18	Orthorhombic
RAMDOHRITE	Ag3Pb6Sb11S24	Monoclinic
DYSCRASITE	Ag3Sb	Orthorhombic
PYRARGYRITE	Ag3SbS3	Trigonal
PYROSTILPNITE	Ag3SbS3	Monoclinic
SAMSONITE	Ag4MnSb2S6	Monoclinic
SOPCHEITE	Ag4Pd3Te4	Orthorhombic
AGUILARITE	Ag4SeS	Orthorhombic
CERVELLEITE	Ag4TeS	Isometric
BORODAEVITE	Ag5(Bi,Sb)9S16	Monoclinic
VIKINGITE	Ag5Pb8Bi13S30	Monoclinic
SELENOSTEPHANITE	Ag5Sb(Se,S)4	Orthorhombic
STEPHANITE	Ag5SbS4	Orthorhombic
ARCUBISITE	Ag6CuBiS4	
BILLINGSLEYITE	Ag7AsS6	Isometric
ESKIMOITE	Ag7Pb10Bi15S36	Monoclinic
TREASURITE	Ag7Pb6Bi15S32	Monoclinic
STUTZITE (STUETZITE)	Ag7Te4	Hexagonal
BENLEONARDITE	Ag8(Sb,As)Te2S3	Tetragonal
ARGYRODITE	Ag8GeS6	Orthorhombic ps Cubic
CANFIELDITE	Ag8SnS6	Orthorhombic ps Cubic
EUGENITE	Ag9Hg2	Isometric
TSNIGRIITE	Ag9SbTe3(S,Se)3	Monoclinic
SMITHITE	AgAsS2	Monoclinic
TRECHMANNITE	AgAsS2	Trigonal
MATILDITE	AgBiS2	Hexagonal
SCHAPBACHITE	AgBiS2	Isometric
BOHDANOWICZITE	AgBiSe2	Hexagonal
VOLYNSKITE	AgBiTe2	Orthorhombic
BROMARGYRITE (BROMYRITE)	AgBr	Isometric
CHLORARGYRITE (CERARGYRITE)	AgCl	Isometric
CAMERONITE	AgCu7Te10	Tetragonal
STROMEYERITE	AgCuS	Orthorhombic
ARGENTOJAROSITE	AgFe+++3(SO4)2(OH)6	Trigonal
ARGENTOPYRITE	AgFe2S3	Orthorhombic
STERNBERGITE	AgFe2S3	Orthorhombic
LAFFITTITE	AgHgAsS3	Monoclinic
IODARGYRITE (IODYRITE)	AgI	Hexagonal
PADERAITE	AgPb2Cu6Bi11S22	Monoclinic
UCHUCCHACUAITE	AgPb3MnSb5S12	Monoclinic ps Orthorhombic

Mineral Name	Formula	Crystal System
ZOUBEKITE	AgPb4Sb4S10	Orthorhombic
CUPROPAVONITE	AgPbCu2Bi5S10	Monoclinic
FREIESLEBENITE	AgPbSbS3	Monoclinic
MIARGYRITE	AgSbS2	Monoclinic
EMPRESSITE	AgTe	Orthorhombic
ALUMINIUM (ALUMINUM)	Al	Isometric
CADWALADERITE	Al(OH)2Cl.4H2O	Amorphous
BAYERITE	Al(OH)3	Monoclinic
DOYLEITE	Al(OH)3	Triclinic
GIBBSITE (HYDRARGILLITE)	Al(OH)3	Monoclinic
NORDSTRANDITE	Al(OH)3	Triclinic
JURBANITE	Al(SO4)(OH).5H2O	Monoclinic
KHADEMITE (ROSTITE)	Al(SO4)F.H2O	Orthorhombic
THREADGOLDITE	Al(UO2)2(PO4)2(OH).8H2O	Monoclinic
METAVANURALITE	Al(UO2)2V2O8(OH).8H2O	Triclinic
MUNDITE	Al(UO2)3(PO4)2(OH)3,5.5H2O	Orthorhombic
UPALITE	Al(UO2)3(PO4)2O(OH).7H2O	Monoclinic
VASHEGYITE	Al11(PO4)9(OH)6.38H2OorAl6(PO4)5(OH)3.23H2O	Orthorhombic
ZAHERITE	Al12(SO4)5(OH)26.20H20	Triclinic
SATPAEVITE	Al12V++++2V+++++6O37.30H2O	Orthorhombic
ZUNYITE	Al13Si5O20(OH,F)18Cl	Isometric
HYDROSCARBROITE	Al14(CO3)3(OH)36.nH2O	Triclinic
BULACHITE	Al2(AsO4)(OH)3.3H2O	Orthorhombic
AUGELITE	Al2(PO4)(OH)3	Monoclinic
BOLIVARITE	Al2(PO4)(OH)3.4-5H2O	Amorphous
SENEGALITE	Al2(PO4)(OH)3.H2O	Orthorhombic
SANJUANITE	Al2(PO4)(SO4)(OH).9H2O	Monoclinic
METASCHODERITE	Al2(PO4)(VO4).6H2O	Monoclinic
SCHODERITE	Al2(PO4)(VO4).8H2O	Monoclinic
FLUELLITE	Al2(PO4)F2(OH).7H2O	Orthorhombic
METAALUMINITE	Al2(SO4)(OH)4.5H2O	Monoclinic
ALUMINITE	Al2(SO4)(OH)4.7H2O	Monoclinic ps Orthorhombic
ALUNOGEN	Al2(SO4)3.17H2O	Triclinic
FURONGITE	Al2(UO2)(PO4)3(OH)2.8H2O	Triclinic
PHURALUMITE	Al2(UO2)3(PO4)2(OH)6.10H2O	Monoclinic
ALUMINOCOPIAPITE	Al2/3Fe+++4(SO4)6O(OH)2.20H2O	Triclinic
MELLITE	Al2[C6(COO)6].16H2O	Tetragonal
CORUNDUM	Al2O3	Trigonal
ALLOPHANE	Al2O3.(SiO2)1.3-2.(H2O)2.5-3	Amorphous
DICKITE	Al2Si2O5(OH)4	Monoclinic
HALLOYSITE	Al2Si2O5(OH)4	Monoclinic
KAOLINITE	Al2Si2O5(OH)4	Triclinic
NACRITE	Al2Si2O5(OH)4	Monoclinic
PIANLINITE	Al2Si2O6(OH)2	Amorphous
PYROPHYLLITE	Al2Si4O10(OH)2	Monoclinic

Mineral Data

Mineral Name	Formula	Crystal System
IMOGOLITE	Al2SiO3(OH)4	
TOPAZ	Al2SiO4(F,OH)2	Orthorhombic
SILLIMANITE	Al2SiO5=Al[6]Al[4]OSiO4	Orthorhombic
ANDALUSITE	Al2SiO5=Al[6]Al[5]OSiO4	Orthorhombic
KYANITE (DISTHENE)	Al2SiO5=Al[6]Al[6]OSiO4	Triclinic
EVANSITE	Al3(PO4)(OH)6.6H2O	Amorphous
WAVELLITE	Al3(PO4)2(OH,F)3.5H2O	Orthorhombic
KINGITE	Al3(PO4)2(OH,F)3.9H2O	Triclinic
MOREAUITE	Al3(UO2)(PO4)3(OH)2.13H2O	Monoclinic
TRIANGULITE	Al3(UO2)4(PO4)4(OH)5.5H2O	Triclinic
TROLLEITE	Al4(PO4)3(OH)3	Monoclinic
VANTASSELITE	Al4(PO4)3(OH)3.9H2O	Orthorhombic
HYDROBASALUMINITE	Al4(SO4)(OH)10.12-36H2O	Monoclinic
BASALUMINITE	Al4(SO4)(OH)10.5H2O	Hexagonal
FELSOBANYAITE (FELSOBANYITE)	Al4(SO4)(OH)10.5H2O	Orthorhombic
METAALUNOGEN	Al4(SO4)6.27H2O	Monoclinic
SIMPSONITE	Al4(Ta,Nb)3(O,OH,F)14	Trigonal
SCARBROITE	Al5(OH)13(CO3).5H2O	Hexagonal
HOTSONITE	Al5(PO4)(SO4)(OH)10.8H2O	Triclinic
KRIBERGITE	Al5(PO4)3(SO4)(OH)4.2H2O	Triclinic
BAHIANITE	Al5Sb+++++3O14(OH)2	Monoclinic
HOLTITE	Al6(Al,Ta)(Si,Sb)3BO15(O,OH)2	Orthorhombic
PLANERITE	Al6(PO4)2(PO3OH)2(OH)8.4H2O	Triclinic
DUMORTIERITE	Al6,5-7(BO3)(SiO4)3(O,OH)3	Orthorhombic
JEREMEJEVITE	Al6B5O15(F,OH)3	Hexagonal
MULLITE	Al6Si2O13	Orthorhombic
ALARSITE	AlAsO4	Trigonal
MANSFIELDITE	AlAsO4.2H2O	Orthorhombic
CHLORALUMINITE	AlCl3.6H2O	Trigonal
ZHARCHIKHITE	AlF(OH)2	Monoclinic
ROSENBEGITE	AlF3.3H2O	Tetragonal
BOHMITE (BOEHMITE)	AlO(OH)	Orthorhombic
DIASPORE	AlO(OH)	Orthorhombic
BERLINITE	AlPO4	Trigonal
METAVARISCITE	AlPO4.2H2O	Monoclinic
VARISCITE	AlPO4.2H2O	Orthorhombic
ALUMOTANTITE	AlTaO4	Orthorhombic
STEIGERITE	AlVO4.3H2O	Monoclinic
SCRUTINYITE	àPbO2	Orthorhombic
ARSENIC	As	Trigonal
ARSENOLAMPRITE	As	Orthorhombic
JEROMITE	As(S,Se)2	Amorphous
LAPHAMITE	As2(Se,S)3	Monoclinic
ARSENOLITE	As2O3	Isometric
CLAUDETITE	As2O3	Monoclinic

Mineral Data

Mineral Name	Formula	Crystal System
ORPIMENT	As2S3	Monoclinic
DURANUSITE	As4S	Orthorhombic
DIMORPHITE	As4S3	Orthorhombic
UZONITE (USONITE)	As4S5	Monoclinic
ALACRANITE	As8S9	Monoclinic
PARAREALGAR	AsS	Monoclinic
REALGAR	AsS	Monoclinic
GETCHELLITE	AsSbS3	Monoclinic
TOMICHITE	AsTi3(V,Fe)4O13(OH)	Monoclinic
GOLD (OR)	Au	Isometric
ANYUIITE	Au(Pb,Sb)2	Tetragonal
MALDONITE	Au2Bi	Isometric
BILIBINSKITE	Au3Cu2PbTe2	ps Cubic
BEZSMERTNOVITE	Au4Cu(Te,Pb)	Orthorhombic
PETROVSKAITE	AuAg(S,Se)	Monoclinic
TETRAAURICUPRIDE	AuCu	Tetragonal
NAGYAGITE	AuPb(Sb,Bi)Te2-3S6	Orthorhombic
BUCKHORNITE	AuPb2BiTe2S3	Orthorhombic
AUROSTIBITE	AuSb2	Isometric
AUROANTIMONATE	AuSbO3	
CALAVERITE	AuTe2	Monoclinic
KRENNERITE	AuTe2	Orthorhombic
BARIUMALUMOPHARMACOS IDERITE	Ba(Al,Fe+++)4(AsO4)3(OH)5.5H2O	Tetragonal ps Cubic
KRASNOVITE	Ba(Al,Mg)[(OH)2/(PO4,CO3)].H2O	Orthorhombic
CURETONITE	Ba(Al,Ti++++)(PO4)4(O,OH)F	Monoclinic
CORDYLITE-(Ce)	Ba(Ce,La)2(CO3)3F2	Hexagonal
HASHEMITE	Ba(Cr,S)O4	Orthorhombic
BAFERTISITE	Ba(Fe++,Mn)2TiSi2O7(O,OH)2	Monoclinic
KULANITE	Ba(Fe++,Mn,Mg)2Al2(PO4)3(OH)3	Triclinic
PENIKISITE	Ba(Mg,Fe++)2Al2(PO4)3(OH)3	Triclinic
HOLLANDITE	Ba(Mn++++,Mn++)8O16	Monoclinic ps Tetragonal
PERLOFFITE	Ba(Mn,Fe++)2Fe+++2(PO4)3(OH)3	Monoclinic
HEJTMANITE	Ba(Mn,Fe++)2TiO(Si2O7)(OH,F)2	Monoclinic
NITROBARITE	Ba(NO3)2	Isometric
PABSTITE	Ba(Sn,Ti)Si3O9	Hexagonal
BARIOMICROLITE (RIJKEBOERITE)	Ba(Ta,Nb)2(O,OH)7	Isometric
ANKANGITE	Ba(Ti,V+++,Cr+++)8O16	Tetragonal
HEINRICHITE	Ba(UO2)2(AsO4)2.10-12H2O	Tetragonal
METAHEINRICHITE	Ba(UO2)2(AsO4)2.8H2O	Tetragonal
URANOCIRCITE	Ba(UO2)2(PO4)2.12H2O	Tetragonal
METAURANOCIRCITE	Ba(UO2)2(PO4)2.6-8H2O	Monoclinic
GUILLEMINITE	Ba(UO2)3(SeO3)2O2.3H2O	Orthorhombic
PROTASITE	Ba(UO2)3O3(OH)2.3H2O	Monoclinic ps Hexagonal
BILLIETITE	Ba(UO2)6O4(OH)6.8H2O	Orthorhombic

Mineral Name	Formula	Crystal System
CAPPELENITE-	Ba(Y,Ce)6Si3B6O24F2	Trigonal
HAWTHORNEITE	Ba[Ti3Cr4Fe4Mg]O19	Hexagonal
WALTHIERITE	Ba0,5Al3(SO4)2(OH)6	Trigonal
MUIRITE	Ba10Ca2MnTiSi10O30(OH,Cl,F)10	Tetragonal
GAMAGARITE	Ba2(Fe+++,Mn+++)(VO4)2(OH)	Monoclinic
VERPLANCKITE	Ba2(Mn,Fe++,Ti)Si2O6(O,OH,Cl,F)2.3H2O	Hexagonal
SUZUKIITE	Ba2(VO2)Si4O12	Orthorhombic
PELLYITE	Ba2Ca(Fe++,Mg)2Si6O17	Orthorhombic
USOVITE	Ba2CaMgAl2F14	Monoclinic
ZHONGHUACERITE-(Ce)	Ba2Ce(CO3)3F	Trigonal
ILIMAUSSITE-(Ce)	Ba2Na4CeFe+++Nb2Si8O28.5H2O	Hexagonal
JOAQUINITE-(Ce)	Ba2NaCe2Fe++(Ti,Nb)2Si8O26(OH,F).H2O	Monoclinic
ORTHOJOAQUINITE-(Ce)	Ba2NaCe2Fe++Ti2Si8O26(O,OH).H2O	Orthorhombic
FRESNOITE	Ba2TiSi2O8	Tetragonal
BELKOVITE	Ba3(Nb,Ti)6(Si2O7)2O12	Hexagonal
CEBAITE-(Nd)	Ba3(Nd,Ce)2(CO3)5F2	Monoclinic
CEBAITE-(Ce)	Ba3Ce2(CO3)5F2	Monoclinic
MCKELVEYITE-	Ba3Na(Ca,U)Y(CO3)6.3H2O	Triclinic psTrigonal
GARRELSITE	Ba3NaSi2B7O16(OH)4	Monoclinic
TARAMELLITE	Ba4(Fe+++,Ti,Fe++,Mg)4(B2Si2O27)O2Clx	Orthorhombic
JONESITE	Ba4(K,Na)2Ti4Al2Si10O36.6H2O	Orthorhombic
TITANTARAMELLITE	Ba4(Ti,Fe+++,Fe++,Mg)4(B2Si8O27)O2Clx	Orthorhombic
BAOTITE	Ba4(Ti,Nb)8Si4O28Cl	Tetragonal
NAGASHIMALITE	Ba4(V+++,Ti)4Si8B2O27Cl(O,OH)2	Orthorhombic
WENKITE	Ba4Ca6(Si,Al)20O39(OH)2(SO4)3.nH2O	Hexagonal
ALFORSITE	Ba5(PO4)3Cl	Hexagonal
TRASKITE	Ba9Fe++2Ti2(SiO3)12(OH,Cl,F)6.6H2O	Hexagonal
HYDRODRESSERITE	BaAl2(CO3)2(OH)4.3H2O	Triclinic
DRESSERITE	BaAl2(CO3)2(OH)4.H2O	Orthorhombic
JAGOWERITE	BaAl2(PO4)2(OH)2	Triclinic
CYMRITE	BaAl2Si2(O,OH)8.H2O	Monoclinic
CELSIAN	BaAl2Si2O8	Monoclinic
PARACELSIAN	BaAl2Si2O8	Monoclinic
EDINGTONITE	BaAl2Si3O10.4H2O	Orthorhombic/Tetragonal
GORCEIXITE	BaAl3(PO4)(PO3OH)(OH)6	Monoclinic psTrigonal
ARSENOGORCEIXITE	BaAl3AsO3(OH)(AsO4,PO4)(OH,F)6	Trigonal
BABEFPHITE	BaBe(PO4)(F,O)	Tetragonal
BARYLITE	BaBe2Si2O7	Orthorhombic ps Hexagonal
ALSTONITE	BaCa(CO3)2	Triclinic ps Orthorhombic
BARYTOCALCITE	BaCa(CO3)2	Monoclinic
PARALSTONITE	BaCa(CO3)2	Trigonal
ARMENITE	BaCa2Al6Si9O30.2H2O	Orthorhombic
WALSTROMITE	BaCa2Si3O9	Triclinic
MACDONALDITE	BaCa4Si16O36(OH)2.10H2O	Orthorhombic
HUANGHOITE-(Ce) (HUANGHEITE)	BaCe(CO3)2F	Trigonal

Mineral Name	Formula	Crystal System
WITHERITE	BaCO3	Orthorhombic
EFFENBERGERITE	BaCuSi4O10	Tetragonal
FRANKDICKSONITE	BaF2	Isometric
ANDREMEYERITE	BaFe(Fe++,Mn,Mg)Si2O7	Monoclinic
DUSSERTITE	BaFe+++3(AsO4)2(OH)5	Trigonal
BARIUMPHARMACOSIDERITE	BaFe+++4(AsO4)3(OH)5.5H2O	Tetragonal ps Cubic
GILLESPITE	BaFe++Si4O10	Tetragonal
NORSETHITE	BaMg(CO3)2	Trigonal
BALIPHOLITE	BaMg2LiAl3Si4O12(OH,F)8	Orthorhombic
NOELBENSONITE	BaMn+++2Si2O7(OH)2.H2O	Orthorhombic
ORTHOERICSSONITE	BaMn2(Fe+++O)Si2O7(OH)	Orthorhombic
ERICSSONITE	BaMn2Fe+++OSi2O7(OH)	Monoclinic
BANALSITE	BaNa2Al4Si4O16	Orthorhombic
TIENSHANITE	BaNa2MnTiB2Si6O20	Hexagonal
LEUCOSPHENITE	BaNa4Ti2B2Si10O30	Monoclinic
BORNEMANITE	BaNa4Ti2NbSi4O17(F,OH).Na3PO4	Orthorhombic
BAIYUNEBOITE-(Ce)	BaNaCe2(CO3)4F	Hexagonal
KRAUSKOPFITE	BaSi2O4(OH)2.2H2O	Monoclinic
SANBORNITE	BaSi2O5	Orthorhombic
BARITE (BARYTE,BARYTINE)	BaSO4	Orthorhombic
PARABARIOMICROLITE	BaTa4O10(OH)2.2H20	Trigonal
REDLEDGEITE	BaTi6Cr+++2O16.H2O	Tetragonal
MANNARDITE	BaTi6V+++2O16.H2O	Tetragonal
BENITOITE	BaTiSi3O9	Hexagonal
BAURANOITE	BaU2O7.4-5H2O	
BAZIRITE	BaZrSi3O9	Hexagonal
KOMKOVITE	BaZrSi3O9.3H2O	Trigonal
BEHOITE	Be(OH)2	Orthorhombic
CLINOBEHOITE	Be(OH)2	Monoclinic
BEARSITE	Be2(AsO4)(OH).4H2O	Monoclinic
BERBORITE-1T	Be2(BO3)(OH,F).H2O	Trigonal
BERBORITE-2T	Be2(BO3)(OH,F).H2O	Trigonal
BERBORITE-3H	Be2(BO3)(OH,F).H2O	Hexagonal
MORAESITE	Be2(PO4)(OH).4H2O	Monoclinic
HAMBERGITE	Be2BO3(OH)	Orthorhombic
PHENAKITE	Be2SiO4	Trigonal
BAZZITE	Be3(Sc,Al)2Si6O18	Hexagonal
BERYL	Be3Al2Si6O18	Hexagonal
BERYLLITE	Be3SiO4(OH)2.H2O	Orthorhombic
BERTRANDITE	Be4Si2O7(OH)2	Orthorhombic
CHRYSOBERYL	BeAl2O4	Orthorhombic
EUCLASE	BeAlSiO4(OH)	Monoclinic
BROMELLITE	BeO	Hexagonal
BISMUTH	Bi	Trigonal
ZAIRITE	Bi(Fe+++,Al)3(PO4)2(OH)6	Trigonal

Mineral Data

Mineral Name	Formula	Crystal System
BISMUTOCOLUMBITE	Bi(Nb,Ta)O4	Tetragonal
INGODITE	Bi(S,Te)	Hexagonal
BISMUTOSTIBICONITE	Bi(Sb+++++,Fe+++)2O7	Isometric
NEVSKITE	Bi(Se,S)	Trigonal
BISMUTOTANTALITE	Bi(Ta,Nb)O4	Orthorhombic
SULPHOTSUMOITE	Bi(Te,S)	Trigonal
ORTHOWALPURGITE	Bi(UO2)(AsO4)2O4	Orthorhombic
MRAZEKITE	Bi+++2Cu++3(PO4)2O2(OH)2.2H2O	Monoclinic
PINGGUITE	Bi+++6Te++++2O13	Orthorhombic
SILLENITE	Bi12SiO20	Isometric
ARSENOBISMITE	Bi2(AsO4)(OH)3	
ATELESTITE	Bi2(AsO4)O(OH)	Monoclinic
BISMUTITE	Bi2(CO3)O2	Tetragonal
SMRKOVECITE	Bi2(PO4)O(OH)	Monoclinic
PARAGUANAJUATITE	Bi2(Se,S)3	Trigonal
PAULKELLERITE	Bi2Fe+++(PO4)O2(OH)2	Monoclinic
KOECHLINITE	Bi2MoO6	Orthorhombic
CANNONITE	Bi2O(OH)2SO4	Monoclinic
BISMITE	Bi2O3	Monoclinic
SPHAEROBISMOITE	Bi2O3	Tetragonal
BISMUTHINITE	Bi2S3	Orthorhombic
SKIPPENITE	Bi2Se2(Te,S)	Trigonal
GUANAJUATITE	Bi2Se3	Orthorhombic
MONTANITE	Bi2Te++++++O6.2H2O	Monoclinic
SMIRNITE	Bi2Te++++O5	Orthorhombic
TETRADYMITE	Bi2Te2S	Trigonal
KAWAZULITE	Bi2Te2Se	Trigonal
TELLUROBISMUTHITE	Bi2Te3	Trigonal
URANOSPHAERITE	Bi2U2O9.3H2O	Monoclinic
RUSSELLITE	Bi2WO6	Tetragonal
PREISINGERITE	Bi3(AsO4)2O(OH)	Triclinic
SCHUMACHERITE	Bi3[(V,As,P)O4]2O(OH)	Triclinic
SZTROKAYITE	Bi3TeS2	
IKUNOLITE	Bi4(S,Se)3	Trigonal
JOSEITE-A	Bi4(S,Te)3	Trigonal
LAITAKARITE	Bi4(Se,S)3	Trigonal
EULYTITE	Bi4(SiO4)3	Isometric
JOSEITE-B	Bi4(Te,S)3	Trigonal
WALPURGITE	Bi4(UO2)(AsO4)2O4.2H2O	Triclinic
PILSENITE	Bi4Te3	Trigonal
PROTOJOSEITE	Bi4TeS2	Hexagonal
YECORAITE	Bi5Fe+++3(Te++++O3)(Te++++++O4)2O9.9H2O	
CHILUITE (CHILUNITE)	Bi6Te++++++2Mo++++++2O21	Hexagonal
HEDLEYITE	Bi7Te3	Trigonal
VIHORLATITE	Bi8+x(Se,Te,S)11-x	Trigonal
WAYLANDITE	BiAl3(PO4)2(OH)6	Trigonal

Mineral Name	Formula	Crystal System
ROOSEVELTITE	BiAsO4	Monoclinic
TETRAROOSEVELTITE	BiAsO4	Tetragonal
MIXITE	BiCu6(AsO4)3(OH)6.3H2O	Hexagonal
GANANITE	BiF3	Isometric
BISMUTOFERRITE	BiFe+++2(SiO4)2(OH)	Monoclinic
DAUBREEITE	BiO(OH,Cl)	Tetragonal
BISMOCLITE	BiOCl	Tetragonal
ZAVARITSKITE	BiOF	Tetragonal
XIMENGITE	BiPO4	Hexagonal
TSUMOITE	BiTe	Trigonal
CLINOBISVANITE	BiVO4	Monoclinic
DREYERITE	BiVO4	Tetragonal
PUCHERITE	BiVO4	Orthorhombic
CHAOITE	C	Hexagonal
DIAMOND	C	Isometric
GRAPHITE	C	Hexagonal/Trigonal
LONSDALEITE	C	Hexagonal
TINNUNKULITE	C10H12N6O8	
FLAGSTAFFITE	C10H22O3	Orthorhombic
RAVATITE	C14H10	Orthorhombic
SIMONELLITE	C19H24	Orthorhombic
REFIKITE	C19H31COOH	Orthorhombic
FICHTELITE	C19H34	Monoclinic
HARTITE	C20H34	Triclinic
DINITE	C20H36	Orthorhombic
IDRIALITE	C22H14	Orthorhombic
KARPATITE (CORONENE)	C24H12	Monoclinic
GUANINE	C5H3(NH2)N4O	Monoclinic
URICITE	C5H4N4O3	Monoclinic
KLADNOITE	C6H4(CO)2NH	Monoclinic
TSCHERNICHITE	Ca(Al2Si6O16).8H2O	Tetragonal/Monoclinic
WEDDELLITE (CALCIUMOXALATE)	Ca(C2O4).2H2O	Tetragonal
WHEWELLITE	Ca(C2O4).H2O	Monoclinic
SYNCHYSITE-(Ce)	Ca(Ce,La)(CO3)2F	Monoclinic ps Hexagonal
DISSAKISITE-(Ce)	Ca(Ce,La)(Mg,Fe++)(Al,Fe+++)2Si3O12(OH)	Monoclinic
PARISITE-(Ce)	Ca(Ce,La)2(CO3)3F2	Trigonal
ORTHOSERPIERITE	Ca(Cu,Zn)4(SO4)2(OH)6.3H2O	Orthorhombic
SERPIERITE	Ca(Cu,Zn)4(SO4)2(OH)6.3H2O	Monoclinic
CAFETITE	Ca(Fe+++,Al)2Ti4O12.4H2O	Orthorhombic
FERROBUSTAMITE	Ca(Fe++,Ca,Mn)Si2O6	Triclinic
HARRISONITE	Ca(Fe++,Mg)6(PO4)2(SiO4)2	Trigonal
ANKERITE	Ca(Fe++,Mg,Mn)(CO3)2	Trigonal
WHITEITE-(CaFeMg)	Ca(Fe++,Mn++)Mg2Al2(PO4)4(OH)2.8H2O	Monoclinic
LAUTARITE	Ca(IO3)2	Monoclinic
BRUGGENITE	Ca(IO3)2.H2O	Monoclinic

Mineral Name	Formula	Crystal System
(BRUEGGENITE)		
ALLANITE-(La)	Ca(La,Ce)(Fe++,Mn++)(Al,Fe+++)2(SiO4)(Si2O7)O(OH)	Monoclinic
LIDDICOATITE	Ca(Li,Al)3Al6(BO3)3Si6O18(O,OH,F)4	Trigonal
CLINTONITE	Ca(Mg,Al)3(Al3Si)O10(OH)2	Monoclinic
CLINOKURCHATOVITE	Ca(Mg,Fe++,Mn)B2O5	Monoclinic
KURCHATOVITE	Ca(Mg,Mn,Fe++)B2O5	Orthorhombic
BRAUNITEII	Ca(Mn+++,Fe+++)14SiO24	Tetragonal
ROSCHERITE	Ca(Mn++,Fe++)5Be4(PO4)6(OH)4.6H2O	Monoclinic/Triclinic
KUTNOHORITE (KUTNAHORITE)	Ca(Mn,Mg,Fe++)(CO3)2	Trigonal
KECKITE	Ca(Mn,Zn)2Fe+++3(PO4)4(OH)3.2H2O	Monoclinic
CALCIOANCYLITE-(Nd)	Ca(Nd,Ce,Gd,Y)3(CO3)4(OH)3.H2O	Orthorhombic
PARISITE-(Nd)	Ca(Nd,Ce,La)2(CO3)3F2	Trigonal
SYNCHYSITE-(Nd)	Ca(Nd,La)(CO3)2F	Monoclinic ps Hexagonal
SHABAITE-(Nd)	Ca(Nd,Y)2(UO2)(CO3)4(OH)2.6H2O	Monoclinic
NICKELAUSTINITE	Ca(Ni,Zn)(AsO4)(OH)	Orthorhombic
NITROCALCITE	Ca(NO3)2.4H2O	Monoclinic
PORTLANDITE	Ca(OH)2	Hexagonal
CASCANDITE	Ca(Sc,Fe++)Si3O8(OH)	Triclinic
RYNERSONITE	Ca(Ta,Nb)2O6	Orthorhombic
METAZELLERITE	Ca(UO2)(CO3)2.3H2O	Orthorhombic
ZELLERITE	Ca(UO2)(CO3)2.5H2O	Orthorhombic
URANOSPINITE	Ca(UO2)2(AsO4)2.10H2O	Tetragonal
METAURANOSPINITE	Ca(UO2)2(AsO4)2.8H2O	Tetragonal
AUTUNITE	Ca(UO2)2(PO4)2.10-12H2O	Tetragonal
METAAUTUNITE (METAAUTUNITE)	Ca(UO2)2(PO4)2.2-6H2O	Tetragonal
HAIWEEITE	Ca(UO2)2Si6O15.5H2O	Monoclinic
METAHAIWEEITE	Ca(UO2)2Si6O15.nH2O	Monoclinic
URANOPHANE (URANOTILE)	Ca(UO2)2SiO3(OH)2.5H2O	Monoclinic
URANOPHANEBETA	Ca(UO2)2SiO3(OH)2.5H2O	Monoclinic
RAUVITE	Ca(UO2)2V+++++10O28.16H2O	
METATYUYAMUNITE	Ca(UO2)2V2O8.3H2O	Orthorhombic
TYUYAMUNITE	Ca(UO2)2V2O8.5-8H2O	Orthorhombic
URANCALCARITE	Ca(UO2)3(CO3)(OH)6.3H2O	Orthorhombic
CALCURMOLITE	Ca(UO2)3(MoO4)3(OH)2.11H2O	Monoclinic
ARSENURANYLITE	Ca(UO2)4(AsO4)2(OH)4.6H2O	Orthorhombic
RABEJACITE	Ca(UO2)4(SO4)2(OH)6.6H2O	Orthorhombic
SHARPITE	Ca(UO2)6(CO3)5(OH)4.6H2O	Orthorhombic
BECQUERELITE	Ca(UO2)6O4(OH)6.8H2O	Orthorhombic
VANADOMALAYAITE	Ca(V++++,Ti++++)OSiO4	Monoclinic
CAVANSITE	Ca(VO)Si4O10.4H2O	Orthorhombic
PENTAGONITE	Ca(VO)Si4O10.4H2O	Orthorhombic
SARYARKITE-	Ca(Y,Th)Al5(SiO4)2(PO4,SO4)2(OH)7.6H2O	Hexagonal
FERRILOTHARMEYERITE	Ca(Zn,Cu++)(Fe+++,Zn)[(As+++++O3(OH)2](OH)3	Monoclinic

Mineral Name	Formula	Crystal System
PETEDUNNITE	Ca(Zn,Mn++,Fe++,Mg)Si2O6	Monoclinic
TRIMERITE	Ca,Mn2Be3(SiO4)3	Monoclinic
HEXAHYDROBORITE	Ca[B(OH)4]2.2H2O	Monoclinic
CALCLACITE	Ca[Cl2/CH3COO].10H2O	Monoclinic/Triclinic
ALIETTITE	Ca0,2Mg6(Si,Al)8O20(OH)4.4H2O	
VOLKONSKOITE (VOLCHONSKOITE)	Ca0,3(Cr+++,Mg,Fe+++)2(Si,Al)4O10(OH)2.4H2O	Monoclinic
HUANGITE	Ca0,5Al3(SO4)2(OH)6	Trigonal
BELYANKINITE	Ca1-2(Ti,Zr,Nb)5O12.9H2O	Amorphous
TVEITITE-	Ca1-xYxF2+x	Monoclinic ps Cubic
RUSTUMITE	Ca10(Si2O7)2(SiO4)Cl2(OH)2	Monoclinic
HYDROXYLELLESTADITE	Ca10(SiO4)3(SO4)3(OH,Cl,F)2	Monoclinic ps Hexagonal
VISEITE	Ca10Al24(SiO4)6(PO4)7O22F3.72H2O	Isometric
OYELITE	Ca10B2Si8O29.12H2O	Orthorhombic
VESUVIANITE (IDOCRASE)	Ca10Mg2Al4(SiO4)5(Si2O7)2(OH)4	Tetragonal
JUANITE	Ca10Mg4Al2Si11O39.4H2O	Orthorhombic
JASMUNDITE	Ca11(SiO4)4O2S	Tetragonal
MAYENITE	Ca12Al14O33	Isometric
TACHARANITE	Ca12Al2Si18O51.18H2O	Monoclinic
WAWAYANDAITE	Ca12Mn4B2Be18Si12O46(OH,Cl)30[Ortho.]	Monoclinic
TUNGUSITE	Ca14Fe++9(Si4O10)6(OH)22	Triclinic
SAMFOWLERITE	Ca14Mn++3Zn2(Be,Zn)2Be6(SiO4)6(Si2O7)4(OH,F)6	Monoclinic
NIOCALITE	Ca14Nb2(Si2O7)4O6F2	Monoclinic
BROWNMILLERITE	Ca2(Al,Fe+++)2O5	Orthorhombic
PIEMONTITE	Ca2(Al,Mn,Fe)3(SiO4)3(OH)=Ca2(Mn,Fe)Al2(SiO4)(Si2O7)O(OH)	Monoclinic
AMINOFFITE	Ca2(Be,Al)Si2O7(OH).H2O	Tetragonal
RONTGENITE-(Ce) (ROENTGENITE-(Ce))	Ca2(Ce,La)3(CO3)5F3	Trigonal
ROSELITE	Ca2(Co,Mg)(AsO4)2.2H2O	Monoclinic
BRENKITE	Ca2(CO3)F2	Orthorhombic
EPIDOTE	Ca2(Fe+++,Al)3(SiO4)3(OH)=Ca2(Fe,Al)Al2(SiO4)(Si2O7)O(OH)	Monoclinic
FRANKLINFURNACEITE	Ca2(Fe+++Al)Mn+++Mn++3Zn2Si2O10(OH)8	Monoclinic
FERROTSCHERMAKITE	Ca2(Fe++,Mg)3Al2(Si7Al)O22(OH)2	Monoclinic
FERROFERRITSCHERMAKITE	Ca2(Fe++,Mg)3Fe+++2(Si7Al)O22(OH)2	Monoclinic
FERROHORNBLENDE	Ca2(Fe++,Mg)4Al(Si7Al)O22(OH,F)2	Monoclinic
FERROACTINOLITE	Ca2(Fe++,Mg)5Si8O22(OH)2	Monoclinic
HOMILITE	Ca2(Fe++,Mg)B2Si2O10	Monoclinic
PENOBSQUISITE	Ca2(Fe++,Mg)B9O13Cl(OH)6.4H2O	Monoclinic
MESSELITE	Ca2(Fe++,Mn)(PO4)2.2H2O	Triclinic
BABINGTONITE	Ca2(Fe++,Mn)Fe+++Si5O14(OH)	Triclinic
DIETZEITE	Ca2(IO3)2(CrO4)	Monoclinic
SHUISKITE	Ca2(Mg,Al)(Cr,Al)2(SiO4)(Si2O7)(OH)2.H2O	Monoclinic
SERENDIBITE	Ca2(Mg,Al)6(Si,Al,B)6O20	Triclinic

Mineral Data

Mineral Name	Formula	Crystal System
WENDWILSONITE	Ca2(Mg,Co)(AsO4)2.2H2O	Monoclinic
COLLINSITE	Ca2(Mg,Fe++)(PO4)2.2H2O	Triclinic
TSCHERMAKITE	Ca2(Mg,Fe++)3Al2(Si7Al)O22(OH)2	Monoclinic
MAGNESIOHORNBLENDE	Ca2(Mg,Fe++)4Al(Si7Al)O22(OH,F)2	Monoclinic
ACTINOLITE	Ca2(Mg,Fe++)5Si8O22(OH)2	Monoclinic
TREMOLITE	Ca2(Mg,Fe++)5Si8O22(OH)2	Monoclinic
CEBOLLITE	Ca2(Mg,Fe++,Al)Si2(O,OH)7	Orthorhombic
RHONITE (RHOENITE)	Ca2(Mg,Fe++,Fe+++,Ti)6(Si,Al)6O20	Triclinic
FEDOROVSKITE	Ca2(Mg,Mn)2B4O7(OH)6	Orthorhombic
DORRITE	Ca2(Mg2,Fe+++4)Al4Si2O20	Triclinic
PUMPELLYITE-(Mn)	Ca2(Mn++,Mg)(Al,Mn+++,Fe)2(SiO4)(Si2O7(OH)2.H2O	Monoclinic
FAIRFIELDITE	Ca2(Mn,Fe++)(PO4)2.2H2O	Triclinic
MANGANBABINGTONITE	Ca2(Mn,Fe++)Fe+++Si5O14(OH)	Triclinic
BRANDTITE	Ca2(Mn,Mg)(AsO4)2.2H2O	Monoclinic
CALCIOBETAFITE	Ca2(Nb,Ti)2(O,OH)7	Isometric
CASSIDYITE	Ca2(Ni,Mg)(PO4)2.2H2O	Triclinic
ISOCLASITE	Ca2(PO4)(OH).2H2O	Monoclinic
SPODIOSITE	Ca2(PO4)F	Orthorhombic
HANNEBACHITE	Ca2(SO3)2.H2O	Orthorhombic
RAPIDCREEKITE	Ca2(SO4)(CO3).4H2O	Orthorhombic
ARDEALITE	Ca2(SO4)(HPO4).4H2O	Monoclinic
LIEBIGITE	Ca2(UO2)(CO3)3.11H2O	Orthorhombic
PHURCALITE	Ca2(UO2)3O2(PO4)2.7H2O	Orthorhombic
KAINOSITE-	Ca2(Y,Ce)2Si4O12(CO3).H2O	Orthorhombic
OGDENSBURGITE	Ca2(Zn,Mn)Fe+++4(AsO4)4(OH)6.6H2O	Orthorhombic ps Hexagonal
ROGGIANITE	Ca2[Be(OH)2Al2Si4O13].<2.5H2O	Tetragonal
HARKERITE	Ca24Mg8Al2(SiO4)8(BO3)6(CO3)10.2H2O	Isometric
GEHLENITE	Ca2Al(AlSi)O7	Tetragonal
HYDROCALUMITE	Ca2Al(OH)6[Cl1-x(OH)x].3H2O	Monoclinic
BEARTHITE	Ca2Al(PO4)2(OH)	Monoclinic
STRATLINGITE (STRAETLINGITE)	Ca2Al[(OH)6AlSiO2-3(OH)4-3].2,5H2O	Trigonal
VERTUMNITE	Ca2Al[(OH)6AlSiO2-3(OH)4-3].2,5H2O	Monoclinic ps Hexagonal
ANORTHITE	Ca2Al2Si2O8	Triclinic
PREHNITE	Ca2Al2Si3O10(OH)2	Orthorhombic
BICCHULITE	Ca2Al2SiO6(OH)2	Isometric
KAMAISHILITE	Ca2Al2SiO6(OH)2	Tetragonal
MUKHINITE	Ca2Al2V+++(SiO4)3(OH)	Monoclinic
CLINOZOISITE	Ca2Al3(SiO4)3(OH)=Ca2AlAl2(SiO4)(Si2O7)O(OH)	Monoclinic
ZOISITE	Ca2Al3(SiO4)3(OH)=Ca2AlAl2(SiO4)(Si2O7)O(OH)	Orthorhombic
PARTHEITE	Ca2Al4Si4O15(OH)2.4H2O	Monoclinic
GISMONDINE	Ca2Al4Si4O16.9H2O	Monoclinic
CARLHINTZEITE	Ca2AlF7.H2O	Triclinic ps Monoclinic
CAHNITE	Ca2B(AsO4)(OH)4	Tetragonal

Mineral Name	Formula	Crystal System
GINORITE	Ca2B14O23.8H2O	Monoclinic
SOLONGOITE	Ca2B3O4(OH)4Cl	Monoclinic
HYDROCHLORBORITE	Ca2B4O4(OH)7Cl.7H2O	Monoclinic
EKATERINITE	Ca2B4O7(Cl,OH)2.2H2O	Hexagonal
TYRETSKITE-1Tc	Ca2B5O9(OH).H2O	Triclinic
HILGARDITE-1Tc	Ca2B5O9Cl.H2O	Triclinic
HILGARDITE-3Tc	Ca2B5O9Cl.H2O	Triclinic
HILGARDITE-4M	Ca2B5O9Cl.H2O	Monoclinic
HOWLITE	Ca2B5SiO9(OH)5	Monoclinic
COLEMANITE	Ca2B6O11.5H2O	Monoclinic
MEYERHOFFERITE	Ca2B6O6(OH)10.2H2O	Triclinic
INYOITE	Ca2B6O6(OH)10.8H2O	Monoclinic
URALOLITE	Ca2Be4(PO4)3(OH)3.5H2O	Monoclinic
GUGIAITE	Ca2BeSi2O7	Tetragonal
ROSELITEBETA (ROSELITEBETA)	Ca2Co(AsO4)2.2H2O	Triclinic
VOGLITE	Ca2Cu(UO2)(CO3)4.6H2O	Monoclinic
HENMILITE	Ca2Cu[B(OH)4]2(OH)4	Triclinic
KINOITE	Ca2Cu2Si3O8(OH)4	Monoclinic
RICHELSDORFITE	Ca2Cu5Sb(AsO4)4Cl(OH)6.6H2O	Monoclinic
SHUBNIKOVITE	Ca2Cu8(AsO4)6Cl(OH).7H2O	Orthorhombic
CLINOTYROLITE	Ca2Cu9[(As,S)O4]4(O,OH)10.10H2O	Monoclinic
JULGOLDITE-(Fe)	Ca2Fe++(Fe+++,Al)2(SiO4)(Si2O7)(OH)2.H2O	Monoclinic
ANAPAITE	Ca2Fe++(PO4)2.4H2O	Triclinic
SREBRODOLSKITE	Ca2Fe+++2O5	Orthorhombic
ARSENIOSIDERITE	Ca2Fe+++3(AsO4)3O2.3H2O	Monoclinic
YUKONITE	Ca2Fe+++3(AsO4)4(OH).12H2O	Amorphous
MITRIDATITE	Ca2Fe+++3(PO4)3O2.3H2O	Monoclinic
KOLFANITE	Ca2Fe+++3O2(AsO4)3.2H2O	Monoclinic
FERROALUMINOTSCHERMA KITE	Ca2Fe++3Al2(Si7Al)O22(OH)2	Monoclinic
PUMPELLYITE-(Fe) (FERROPUMPELLYITE)	Ca2Fe++Al2(SiO4)(Si2O7)(OH)2.H2O	Monoclinic
FERROAXINITE	Ca2Fe++Al2BO3Si4O12(OH)	Triclinic
TALMESSITE	Ca2Mg(AsO4)2.2H2O	Triclinic
CARBOBORITE	Ca2Mg(CO3)2B2(OH)8.4H2O	Monoclinic
SERGEEVITE	Ca2Mg11(CO3)27(HCO3)12(OH)4.6H2O	Trigonal
PUMPELLYITE-(Mg)	Ca2MgAl2(SiO4)(Si2O7)(OH)2.H2O	Monoclinic
MAGNESIOAXINITE	Ca2MgAl2BO3Si4O12(OH)	Triclinic
AKERMANITE	Ca2MgSi2O7	Tetragonal
PARABRANDTITE	Ca2Mn++(AsO4).2H2O	Triclinic
PARAROBERTSITE	Ca2Mn+++3(PO4)3O2.3H2O	Monoclinic
MACFALLITE	Ca2Mn+++3(SiO4)(Si2O7)(OH)3	Monoclinic
ROWEITE	Ca2Mn++2B4O7(OH)6	Orthorhombic
MANGANAXINITE	Ca2Mn++Al2BO3Si4O12(OH)	Triclinic
ORIENTITE	Ca2Mn++Mn+++2Si3O10(OH)4	Orthorhombic
INESITE	Ca2Mn7Si10O28(OH)2.5H2O	Triclinic

Mineral Data

Mineral Name	Formula	Crystal System
WELSHITE	Ca2Sb+++++Mg4Fe+++Si4Be2O20	Triclinic
SUOLUNITE	Ca2Si2O5(OH)2.H2O	Orthorhombic
BULTFONTEINITE	Ca2SiO2(OH,F)4	Triclinic
LARNITE	Ca2SiO4	Monoclinic
EAKERITE	Ca2SnAl2Si6O18(OH)2.2H2O	Monoclinic
HENDERSONITE	Ca2V++++V+++++8O24.8H2O	Orthorhombic
PINTADOITE	Ca2V2O7.9H2O	
PARANIITE-	Ca2Y(AsO4)(WO4)2	Tetragonal
GAITITE	Ca2Zn(AsO4)2.2H2O	Triclinic
ZINCROSELITE	Ca2Zn(AsO4)2.2H2O	Monoclinic
JUNGITE	Ca2Zn4Fe+++8(PO4)9(OH)9.16H2O	Orthorhombic
EHRLEITE	Ca2ZnBe(PO4)2(PO3OH).4H2O	Triclinic
HARDYSTONITE	Ca2ZnSi2O7	Tetragonal
PHAUNOUXITE	Ca3(AsO4).11H2O	Triclinic
RAUENTHALITE	Ca3(AsO4)2.10H2O	Monoclinic/Triclinic
TADZHIKITE-(Ce)	Ca3(Ce,Y)2(Ti,Al,Fe+++)B4Si4O22	Monoclinic
CHUKHROVITE-(Ce)	Ca3(Ce,Y)Al2(SO4)F13.10H2O	Isometric
HENRITERMIERITE	Ca3(Mn,Al)2(SiO4)2(OH)4	Tetragonal
NAGELSCHMIDTITE	Ca3(PO4)2.2(a-Ca2SiO4)	Hexagonal
ORSCHALLITE	Ca3(SO3)2(SO4).12H2O	Trigonal
UHLIGITE	Ca3(Ti,Al,Zr)9O20	Isometric
ASBECASITE	Ca3(Ti,Fe,Sn)(Be,B,Al)2(As+++,Sb+++)3O6(SiO4)2	Trigonal
GOLDMANITE	Ca3(V,Al,Fe+++)2(SiO4)3	Isometric
CHUKHROVITE-	Ca3(Y,Ce)Al2(SO4)F13.10H2O	Isometric
KIMZEYITE	Ca3(Zr,Ti)2(Si,Al,Fe+++)3O12	Isometric
BAGHDADITE	Ca3(Zr,Ti)Si2O9	Monoclinic
EARLANDITE	Ca3[C(OH)(CH2)2(COO)3]2.4H2O	Monoclinic
GROSSULAR	Ca3Al2(SiO4)3	Isometric
HIBSCHITE	Ca3Al2(SiO4)3-x(OH)4x(x=0,2-1,5)	Isometric
KATOITE	Ca3Al2(SiO4)3-x(OH)4xx=1,5-3	Isometric
CREEDITE	Ca3Al2(SO4)(F,OH)10.2H2O	Monoclinic
YAROSLAVITE	Ca3Al2F10(OH)2.H2O	Orthorhombic
PERHAMITE	Ca3Al7(SiO4)3(PO4)(OH)3.16.5H2O	Hexagonal
TAKEDAITE	Ca3B2O6	Trigonal
OLSHANSKYITE	Ca3B4(OH)18	Monoclinic
NIFONTOVITE	Ca3B6O6(OH)12.2H2O	Monoclinic
PARAFRANSOLEITE	Ca3Be2(PO4)2(PO3OH)2.4H2O	Triclinic
UVAROVITE	Ca3Cr2(SiO4)3	Isometric
LIEBAUITE	Ca3Cu5Si9O26	Monoclinic
ANDRADITE	Ca3Fe+++2(SiO4)3	Isometric
SCHAURTEITE	Ca3Ge++++(SO4)2(OH)6.3H2O	Hexagonal
HSIANGHUALITE	Ca3Li2Be3(SiO4)3F2	Isometric
SAKHAITE	Ca3Mg(BO3)2(CO3).nH2O,(n<1)	Isometric
MERWINITE	Ca3Mg(SiO4)2	Monoclinic
RABBITTITE	Ca3Mg3(UO2)2(CO3)6(OH)4.18H2O	Monoclinic

Mineral Name	Formula	Crystal System
DESPUJOLSITE	Ca3Mn++++(SO4)2(OH)6.3H2O	Hexagonal
JOURAVSKITE	Ca3Mn++++(SO4,CO3)2(OH)6.12H2O	Hexagonal
ROBERTSITE	Ca3Mn+++4(PO4)4(OH)6.3H2O	Monoclinic
INGERSONITE	Ca3Mn++Sb+++++4O14	Hexagonal
THAUMASITE	Ca3Si(CO3)(SO4)(OH)6.12H2O	Hexagonal
AFWILLITE	Ca3Si2O4(OH)6	Monoclinic
KILCHOANITE	Ca3Si2O7	Orthorhombic
RANKINITE	Ca3Si2O7	Monoclinic
FOSHALLASITE	Ca3Si2O7.3H2O	Monoclinic
ROSENHAHNITE	Ca3Si3O8(OH)2	Triclinic
NEKOITE	Ca3Si6O15.7H2O	Triclinic
HATRURITE	Ca3SiO5	Trigonal
YAFSOANITE	Ca3Te2Zn3O12	Isometric
SCHORLOMITE	Ca3Ti++++2(Fe+++2Si)O12	Isometric
MORIMOTOITE	Ca3TiFe++(SiO4)3	Isometric
PASCOITE	Ca3V10O28.17H2O	Monoclinic
CAYSICHITE-	Ca3Y4GdSi8O20(CO3)6(OH).2H20	Orthorhombic
AERINITE	Ca4(Al,Fe+++,Mg,Fe++)10Si12O36(CO3).12H2O	Monoclinic
STANFIELDITE	Ca4(Mg,Fe++,Mn)5(PO4)6	Monoclinic
JAFFEITE	Ca4(Si3O7)(OH)6	Hexagonal
YEELIMITE	Ca4Al6O12(SO4)	Isometric
MEIONITE	Ca4Al6Si6O24CO3	Tetragonal
TERTSCHITE	Ca4B10O19.20H2O	Monoclinic
PRICEITE	Ca4B10O19.7H2O	Triclinic
BAKERITE	Ca4B4(BO4)(SiO4)3(OH)3.H2O	Monoclinic
BAVENITE	Ca4Be2Al2Si9O26(OH)2	Orthorhombic
CALCIOFERRITE	Ca4Fe++(Fe+++,Al)4(PO4)6(OH)4.12H2O	Monoclinic
XANTHOXENITE	Ca4Fe+++2(PO4)4(OH)2.3H2O	Triclinic
ALBRECHTSCHRAUFITE	Ca4Mg(UO2)2(CO3)6F2.17H2O	Triclinic
MONTGOMERYITE	Ca4MgAl4(PO4)6(OH)4.12H2O	Monoclinic
TERUGGITE	Ca4MgAs2B12O22(OH)12.12H2O	Monoclinic
BORCARITE	Ca4MgB4O6(OH)6(CO3)2	Triclinic
IRHTEMITE	Ca4MgH2(AsO4)4.4H2O	Monoclinic
GAUDEFROYITE	Ca4Mn+++3-x(BO3)3(CO3)(O,OH)3	Hexagonal
KITTATINNYITE	Ca4Mn++2Mn+++4Si4O16(OH)8.18H2O	Hexagonal
ORLYMANITE	Ca4Mn++3Si8O20(OH)6.2H2O	Trigonal
WALLKILLDELLITE	Ca4Mn++6As+++++4O16(OH)8.18H2O	Hexagonal
ZODACITE	Ca4Mn++Fe+++4(PO4)6(OH)4.12H2O	Monoclinic
MONGOLITE	Ca4Nb6Si5O24(OH)10.5-6H2O	Tetragonal
FUKALITE	Ca4Si2O6(CO3)(OH,F)2	Orthorhombic
CUSPIDINE	Ca4Si2O7(F,OH)2	Monoclinic
TRABZONITE	Ca4Si3O10.2H2O	Monoclinic
ZEOPHYLLITE	Ca4Si3O8(OH,F)4.2H2O	Triclinic ps Hexagonal
FOSHAGITE	Ca4Si3O9(OH)2	Triclinic
SAINFELDITE	Ca5(AsO4)[AsO3(OH)]2.4H2O	Monoclinic
JOHNBAUMITE	Ca5(AsO4)3(OH)	Hexagonal

Mineral Data

Mineral Name	Formula	Crystal System
SVABITE	Ca5(AsO4)3F	Hexagonal
HYDROXYLAPATITE (HYDROXYAPATITE)	Ca5(PO4)3(OH)	Hexagonal
CHLORAPATITE	Ca5(PO4)3Cl	Monoclinic
FLUORAPATITE	Ca5(PO4)3F	Hexagonal
CARBONATEHYDROXYLAPATITE	Ca5(PO4,CO3)3(OH)	Hexagonal
CARBONATEFLUORAPATITE	Ca5(PO4,CO3)3F	Hexagonal
PARASPURRITE	Ca5(SiO4)2(CO3)	Monoclinic
SPURRITE	Ca5(SiO4)2(CO3)	Monoclinic
REINHARDBRAUNSITE	Ca5(SiO4)2(OH,F)2	Monoclinic
CHLORELLESTADITE	Ca5(SiO4,PO4,SO4)3(Cl,F)	Orthorhombic
FLUORELLESTADITE	Ca5(SiO4,PO4,SO4)3(F,OH,Cl)	Hexagonal
FLUORELLESTADITE (WILKEITE)	Ca5(SiO4,PO4,SO4)3(F,OH,Cl)	Hexagonal
TURNEAUREITE	Ca5[(As,P)O4]3Cl	Hexagonal
VLADIMIRITE	Ca5H2(AsO4)4.5H2O	Monoclinic
FERRARISITE	Ca5H2(AsO4)4.9H2O	Triclinic
GUERINITE	Ca5H2(AsO4)4.9H2O	Monoclinic
PLOMBIERITE	Ca5H2Si6O18.6H2O	Orthorhombic
WARDSMITHITE	Ca5MgB24O42.30H2O	Hexagonal
TILLEYITE	Ca5Si2O7(CO3)2	Monoclinic
CLINOTOBERMORITE	Ca5Si5(O,OH)18.5H2O	Monoclinic
RIVERSIDEITE	Ca5Si6O16(OH)2.2H2O	Orthorhombic
TOBERMORITE	Ca5Si6O16(OH)2.4H2O	Orthorhombic
OKENITE	Ca5Si9O23.9H2O	Triclinic
CHARLESITE	Ca6(Al,Si)2(SO4)2B(OH)4(OH,O)12.26H2O	Hexagonal
DEFERNITE	Ca6(CO3)2-x(SiO4)x(OH)7(Cl,OH)1-2x(x=0,5)	Orthorhombic
BENTORITE	Ca6(Cr,Al)2(SO4)3(OH)12.26H2O	Hexagonal
STURMANITE	Ca6(Fe+++,Al,Mn++)2(SO4)2[B(OH)4](OH)12.25 H2O	Hexagonal
ETTRINGITE	Ca6Al2(SO4)3(OH)12.26H2O	Hexagonal
TATARSKITE	Ca6Mg2(SO4)2(CO3)2Cl4(OH)4.7H2O	Orthorhombic
HARSTIGITE	Ca6MnBe4(SiO4)2(Si2O7)2(OH)2	Orthorhombic
DELLAITE	Ca6Si3O11(OH)2	Monoclinic
HILLEBRANDITE	Ca6Si3O9(OH)6	Monoclinic ps Orthorhombic
XONOTLITE	Ca6Si6O17(OH)2	Monoclinic/Triclinic
BREDIGITE	Ca7Mg(SiO4)4	Orthorhombic ps Hexagonal
SCAWTITE	Ca7Si6(CO3)O18.2H2O	Monoclinic
OKHOTSKITE	Ca8(Mn++,Mg)4(Mn+++,Al,Fe+++)8Si12O46(OH)12	Monoclinic
CAFARSITE	Ca8(Ti,Fe++,Fe+++,Mn)6-7(As+++O3)12.4H2O	Isometric
WHITLOCKITE	Ca9(Mg,Fe++)(PO4)6[PO3(OH)]	Trigonal
SHERWOODITE	Ca9Al2V++++4V+++++24O80.56H2O	Tetragonal
PRINGLEITE	Ca9B26O34(OH)24Cl4.13H2O	Triclinic
RUITENBERGITE	Ca9B26O34(OH)24Cl4.13H2O	Monoclinic
JENNITE	Ca9H2Si6O18(OH)8.6H2O	Triclinic

Mineral Name	Formula	Crystal System
GEARKSUTITE	CaAl(OH)F4.H2O	Monoclinic
FOGGITE	CaAL(PO4)(OH)2.H2O	Orthorhombic
AMSTALLITE	CaAl(Si,Al)4O8(OH)4.(H2O,Cl)	Monoclinic
MATULAITE	CaAl18(PO4)12(OH)20.28H2O	Monoclinic
MARGARITE	CaAl2(Al2Si2)O10(OH)2	Monoclinic
ALUMOHYDROCALCITE	CaAl2(CO3)2(OH)4.3H2O	Triclinic
PARAALUMOHYDROCALCITE	CaAl2(CO3)2(OH)4.6H2O	
PROSOPITE	CaAL2(F,OH)8	Monoclinic
GATUMBAITE	CaAl2(PO4)2(OH)2.H2O	Monoclinic
LAWSONITE	CaAl2Si2O7(OH)2.H2O	Orthorhombic
DMISTEINBERGITE	CaAl2Si2O8	Hexagonal
SVYATOSLAVITE	CaAl2Si2O8	Orthorhombic
SCOLECITE	CaAl2Si3O10.3H2O	Monoclinic
COWLESITE	CaAl2Si3O10.5-6H2O	Orthorhombic
WAIRAKITE	CaAl2Si4O12.2H2O	Monoclinic
LAUMONTITE	CaAl2Si4O12.4H2O	Monoclinic
CHABAZITE	CaAl2Si4O12.6H2O	Trigonal
YUGAWARALITE	CaAl2Si6O16.4H2O	Monoclinic
EPISTILBITE	CaAl2Si6O16.5H2O	Monoclinic
GOOSECREEKITE	CaAl2Si6O16.5H2O	Monoclinic
STELLERITE	CaAl2Si7O18.7H2O	Orthorhombic
CHANTALITE	CaAl2SiO4(OH)4	Tetragonal
WOODHOUSEITE	CaAl3(PO4)(SO4)(OH)6	Trigonal
CRANDALLITE	CaAl3(PO4)2(OH)5.H2O	Trigonal
GROSSITE	CaAl4O7	Monoclinic
JOHACHIDOLITE	CaAlB3O7	Orthorhombic
VUAGNATITE	CaAlSiO4(OH)	Orthorhombic
FROLOVITE	CaB2(OH)8	Triclinic
DANBURITE	CaB2(SiO4)2	Orthorhombic
PENTAHYDROBORITE	CaB2O(OH)6.2H2O	Triclinic
URALBORITE	CaB2O2(OH)4	Monoclinic
VIMSITE	CaB2O2(OH)4	Monoclinic
CALCIBORITE	CaB2O4	Orthorhombic
KORZHINSKITE	CaB2O4.H2O	Monoclinic
FABIANITE	CaB3O5(OH)	Monoclinic
GOWERITE	CaB6O10.5H2O	Monoclinic
NOBLEITE	CaB6O9(OH)2.3H2O	Monoclinic
BERGSLAGITE	CaBe(AsO4)(OH)	Monoclinic
HYDROXYLHERDERITE	CaBe(PO4)(OH)	Monoclinic
HERDERITE	CaBe(PO4)F	Monoclinic
HURLBUTITE	CaBe2(PO4)2	Monoclinic
WEINEBENEITE	CaBe3(PO4)2(OH)2.5H2O	Monoclinic
GLUCINE	CaBe4(PO4)2(OH)4.0,5H2O	
KETTNERITE	CaBi(CO3)OF	Tetragonal
DATOLITE	CaBSiO4(OH)	Monoclinic

Mineral Name	Formula	Crystal System
DOLLASEITE-(Ce)	CaCeMg24AlSi3O11(OH,F)2	Monoclinic
SINJARITE	CaCl2.2H2O	Tetragonal
ANTARCTICITE	CaCl2.6H2O	Trigonal
COBALTAUSTINITE	CaCo(AsO4)(OH)	Monoclinic
ARAGONITE	CaCO3	Orthorhombic
CALCITE	CaCO3	Trigonal
VATERITE	CaCO3	Hexagonal
IKAITE	CaCO3.6H2O	Monoclinic
MONOHYDROCALCITE	CaCO3.H2O	Hexagonal
CHROMATITE	CaCrO4	Tetragonal
CONICHALCITE	CaCu(AsO4)(OH)	Orthorhombic
ULRICHITE	CaCu(UO2)(PO4)2.4H2O	Monoclinic
TANGEITE	CaCu(VO4)(OH)	Orthorhombic
DEVILLINE (DEVILLITE)	CaCu4(SO4)2(OH)6.3H2O	Monoclinic
TYROLITE	CaCu5(AsO4)2(CO3)(OH)4.6H2O	Orthorhombic
AGARDITE-(Ca)	CaCu6(AsO4)3(OH)6.3H2O	Hexagonal
PAPAGOITE	CaCuAlSi2O6(OH)3	Monoclinic
CUPRORIVAITE	CaCuSi4O10	Tetragonal
STRINGHAMITE	CaCuSiO4.2H2O	Monoclinic
FLUORITE	CaF2	Isometric
STENHUGGARITE	CaFe+++(As+++O2)(As+++Sb+++O5)	Tetragonal
PHYLLOTUNGSTITE	CaFe+++3H(WO4)6.10H2O	Orthorhombic
DELVAUXITE	CaFe+++4(PO4,SO4)2(OH)8.4-6H2O	Amorphous
CALCIOCOPIAPITE	CaFe+++4(SO4)6(OH)2.19H2O	Triclinic
ESSENEITE	CaFe+++AlSiO6	Monoclinic
MELKOVITE	CaFe+++H6(MoO4)4(PO4).6H2O	Monoclinic
ILVAITE (LIEVRITE)	CaFe++2Fe+++Si2O7O(OH)	Orthorhombic/Monoclinic
MELONJOSEPHITE	CaFe++Fe+++(PO4)2(OH)	Orthorhombic
HEDENBERGITE	CaFe++Si2O6	Monoclinic
KIRSCHSTEINITE	CaFe++SiO4	Orthorhombic
RANKACHITE	CaFe++V+++++4W++++++8O36.12H2O	Orthorhombic
WEILITE	CaHAsO4	Triclinic
PHARMACOLITE	CaHAsO4.2H2O	Monoclinic
HAIDINGERITE	CaHAsO4.H2O	Orthorhombic
SIBIRSKITE	CaHBO3	Monoclinic
MONETITE	CaHPO4	Triclinic
BRUSHITE	CaHPO4.2H2O	Monoclinic
BITYITE	CaLiAl2(AlBeSi2)O10(OH)2	Monoclinic
COLQUIRIITE	CaLiAlF6	Trigonal
ADELITE	CaMg(AsO4)(OH)	Orthorhombic
CAMGASITE	CaMg(AsO4)(OH).5H2O	Monoclinic
TILASITE	CaMg(AsO4)F	Monoclinic
DOLOMITE	CaMg(CO3)2	Trigonal
PANASQUEIRAITE	CaMg(PO4)(OH,F)	Monoclinic
ISOKITE	CaMg(PO4)F	Monoclinic
SWARTZITE	CaMg(UO2)(CO3)3.12H2O	Monoclinic

Mineral Data

Mineral Name	Formula	Crystal System
INDERBORITE	CaMg[B3O3(OH)5]2.6H2O	Monoclinic
TACHYHYDRITE	CaMg2Cl6.12H2O	Trigonal
HUNTITE	CaMg3(CO3)4	Trigonal
NANLINGITE	CaMg4(AsO3)2F4	Trigonal
HENEUITE	CaMg5(PO4)3(CO3)(OH)	Triclinic
OVERITE	CaMgAl(PO4)2(OH).4H2O	Orthorhombic
ALDZHANITE	CaMgB2O4Cl.7H2O	Orthorhombic
CHELKARITE	CaMgB2O4Cl2.7H2O	Orthorhombic
HYDROBORACITE	CaMgB6O8(OH)6.3H2O	Monoclinic
SEGELERITE	CaMgFe+++(PO4)2(OH).4H2O	Orthorhombic
DIOPSIDE	CaMgSi2O6	Monoclinic
MONTICELLITE	CaMgSiO4	Orthorhombic
JAHNSITE-(CaMnMg)	CaMn(Mg,Fe++)2Fe+++2(PO4)4(OH)2.8H2O	Monoclinic
MAROKITE	CaMn+++2O4	Orthorhombic
BOSTWICKITE	CaMn+++6Si3O16.7H2O	Orthorhombic
NELTNERITE	CaMn+++6SiO12	Tetragonal
RUIZITE	CaMn+++Si2O6(OH).2H2O	Monoclinic
MOZARTITE	CaMn+++SiO4(OH)	Orthorhombic
SANTACLARAITE	CaMn++4Si5O14(OH)2.H2O	Triclinic
WILHELMVIERLINGITE	CaMn++Fe+++(PO4)2(OH).2H2O	Orthorhombic
JAHNSITE-(CaMnFe)	CaMn++Fe++2Fe+++2(PO4)4(OH)2.8H2O	Monoclinic
CHIAVENNITE	CaMnBe2Si5O13(OH)2.2H2O	Orthorhombic
FLUCKITE	CaMnH2(AsO4)2.2H2O	Triclinic
WHITEITE-(CaMnMg)	CaMnMg2Al2(PO4)4(OH)2.8H2O	Monoclinic
JAHNSITE-(CaMnMn)	CaMnMn2Fe+++2(PO4)4(OH)2.8H2O	Monoclinic
JOHANNSENITE	CaMnSi2O6	Monoclinic
GLAUCOCHROITE	CaMnSiO4	Orthorhombic
POWELLITE	CaMoO4	Tetragonal
CANAPHITE	CaNa2P2O7.4H2O	Monoclinic
LIME (CALCIUMOXIDE,CHAUX)	CaO	Isometric
LEPERSONNITE-(Gd)	CaO.(Gd,Dy)2O3.24UO3.8CO2.4SiO2.60H2O	Orthorhombic
BAZHENOVITE	CaS5.CaS2O3.6Ca(OH)2.20H2O	Monoclinic
PERETAITE	CaSb+++4O4(OH)2(SO4)2.2H2O	Monoclinic
SARABAUITE	CaSb10O10S6	Monoclinic
WOLLASTONITE-1T	CaSiO3	Triclinic
WOLLASTONITE-2M	CaSiO3	Monoclinic
WOLLASTONITE-7T	CaSiO3	Triclinic
BURTITE	CaSn(OH)6	Isometric
NORDENSKIOLDINE (NORDENSKIOELDINE)	CaSnB2O6	Trigonal
STOKESITE	CaSnSi3O9.2H2O	Orthorhombic
MALAYAITE	CaSnSiO5=CaSnOSiO4	Monoclinic
ANHYDRITE	CaSO4	Orthorhombic
GYPSUM (GYPSE)	CaSO4.2H2O	Monoclinic
METADELRIOITE	CaSrV2O6(OH)2	Triclinic

Mineral Data

Mineral Name	Formula	Crystal System
DELRIOITE	CaSrV2O6(OH)2.3H2O	Monoclinic
CALCIOTANTITE	CaTa4O11	Hexagonal
MROSEITE	CaTe++++(CO3)O2	Orthorhombic
CARLFRIESITE	CaTe++++2Te++++++O8	Monoclinic
BRABANTITE	CaTh(PO4)2	Monoclinic
KASSITE	CaTi2O4(OH)2	Orthorhombic
PEROVSKITE	CaTiO3	Orthorhombic ps Cubic
TITANITE (SPHENE)	CaTiSiO5=CaTiOSiO4	Monoclinic
WYARTITE	CaU++++(UO2)6(CO3)2(OH)18.3-5H2O	Orthorhombic
TENGCHONGITE	CaU++++++6Mo++++++2O25.12H2O	Orthorhombic
SINCOSITE	CaV++++2(PO4)2(OH)4.3H2O	Tetragonal
MELANOVANADITE	CaV++++2V+++++2O10.5H2O	Triclinic
SIMPLOTITE	CaV++++4O9.5H2O	Monoclinic
METAROSSITE	CaV2O6.2H2O	Triclinic
ROSSITE	CaV2O6.4H2O	Triclinic
METAHEWETTITE	CaV6O16.3H2O	Monoclinic
HEWETTITE	CaV6O16.9H2O	Monoclinic
SCHEELITE	CaWO4	Tetragonal
KAMPHAUGITE-	CaY(CO3)6(OH).1-1,5H2O	Tetragonal
KIMURAITE-	CaY2(CO3)4.6H2O	Orthorhombic
MINASGERAISITE-	CaY2Be2Si2O10	Monoclinic
LOKKAITE-	CaY4(CO3)7.9H2O	Orthorhombic
AUSTINITE	CaZn(AsO4)(OH)	Orthorhombic
MINRECORDITE	CaZn(CO3)2	Trigonal
ZNUCALITE	CaZn11(UO2)(CO3)3(OH)20.4H2O	Triclinic
PROSPERITE	CaZn2(AsO4)2.H2O	Monoclinic
PARASCHOLZITE	CaZn2(PO4)2.2H2O	Monoclinic
SCHOLZITE	CaZn2(PO4)2.2H2O	Orthorhombic
JUNITOITE	CaZn2Si2O7.H2O	Orthorhombic
LOTHARMEYERITE	CaZnMn+++(As+++++O3OH)2(OH)3	Monoclinic
CLINOHEDRITE	CaZnSiO4.H2O	Monoclinic
CALZIRTITE	CaZr3TiO9	Tetragonal
PAINITE	CaZrBAl9O18	Hexagonal
GITTINSITE	CaZrSi2O7	Monoclinic
CALCIUMCATAPLEIITE	CaZrSi3O9.2H2O	Hexagonal
CALCIOHILAIRITE	CaZrSi3O9.3H2O	Trigonal
ARMSTRONGITE	CaZrSi6O15.2.5H2O	Monoclinic
ZIRCONOLITE-2M	CaZrTi2O7	Monoclinic
ZIRCONOLITE-3T	CaZrTi2O7	Tetragonal
CADMIUM	Cd	Hexagonal
KEYITE	Cd++2Cu++3(Zn,Cu)4(AsO4).2H2O	Monoclinic
OTAVITE	CdCO3	Trigonal
MONTEPONITE	CdO	Isometric
GREENOCKITE	CdS	Hexagonal
HAWLEYITE	CdS	Isometric
CADMOSELITE	CdSe	Hexagonal

Mineral Data

Mineral Name	Formula	Crystal System
CERITE-(Ce)	Ce+++9Fe+++(SiO4)6[(SiO3)(OH)](OH)3	Trigonal
FLORENCITE-(Ce)	CeAl3(PO4)2(OH)6	Trigonal
GASPARITE-(Ce)	CeAsO4	Monoclinic
CEROTUNGSTITE-(Ce)	CeW2O6(OH)3	Monoclinic
ACETAMIDE	CO(CH3)(NH2)	Trigonal
UREA	CO(NH2)2	Tetragonal
METAKIRCHHEIMERITE	Co(UO2)2(AsO4)2.8H2O	Tetragonal
HETEROGENITE-2H	Co+++O(OH)	Hexagonal
HETEROGENITE-3R	Co+++O(OH)	Trigonal
COBALTZIPPEITE	Co++0,5(UO2)2(SO4)(OH)3.H2O	Monoclinic
LINNAEITE (LINNEITE)	Co++Co+++2S4	Isometric
BORNHARDTITE	Co++Co+++2Se4	Isometric
ERYTHRITE	Co3(AsO4)2.8H2O	Monoclinic
COBALTPENTLANDITE	Co9S8	Isometric
SAFFLORITE	CoAs2	Orthorhombic
COBALTITE	CoAsS	Isometric
SPHAEROCOBALTITE (COBALTOCALCITE)	CoCO3	Trigonal
WAIRAUITE	CoFe	Isometric
JAIPURITE	CoS	Hexagonal
CATTIERITE	CoS2	Isometric
KIEFTITE	CoSb3	Isometric
COSTIBITE	CoSbS	Orthorhombic
PARACOSTIBITE	CoSbS	Orthorhombic
FREBOLDITE	CoSe	Hexagonal
HASTITE	CoSe2	Orthorhombic
TROGTALITE	CoSe2	Isometric
COBALTOMENITE	CoSeO3.2H2O	Monoclinic
BIEBERITE	CoSO4.7H2O	Monoclinic
MATTAGAMITE	CoTe2	Orthorhombic
CHROMIUM	Cr	Isometric
BRACEWELLITE	Cr+++O(OH)	Orthorhombic
GRIMALDIITE	Cr+++O(OH)	Trigonal
ESKOLAITE	Cr2O3	Trigonal
TONGBAITE	Cr3C2	Orthorhombic
FERCHROMIDE	Cr3Fe1-x(x=0,6)	Isometric
BREZINAITE	Cr3S4	Monoclinic
CARLSBERGITE	CrN	Isometric
GUYANAITE	CrO(OH)	Orthorhombic
NANPINGITE	CsAl2(Si,Al)4O10(OH,F)2	Monoclinic
GAINESITE-(NaCs)	CsNa(Be,Li)Zr2(PO4)4.2H2O	Tetragonal
COPPER (CUIVRE,KUPFER)	Cu	Isometric
YVONITE	Cu(AsO3.OH).2H2O	Triclinic
GEMINITE	Cu(AsO3.OH).H2O	Triclinic
CARROLLITE	Cu(Co,Ni)2S4	Isometric
FLORENSOVITE	Cu(Cr1,5Sb0,5)S4	Isometric

Mineral Data

Mineral Name	Formula	Crystal System
SALESITE	Cu(IO3)(OH)	Orthorhombic
FLETCHERITE	Cu(Ni,Co)2S4	Isometric
SPERTINIITE	Cu(OH)2	Orthorhombic
CALUMETITE	Cu(OH,Cl)2.2H2O	Orthorhombic
ANTHONYITE	Cu(OH,Cl)2.3H2O	Monoclinic
MALANITE	Cu(Pt,Ir)2S4	Isometric
BAMBOLLAITE	Cu(Se,Te)2	Tetragonal
VANDENBRANDEITE	Cu(UO2)(OH)4	Triclinic
ZEUNERITE	Cu(UO2)2(AsO4)2.10-16H2O	Tetragonal
METAZEUNERITE	Cu(UO2)2(AsO4)2.8H2O	Tetragonal
TORBERNITE	Cu(UO2)2(PO4)2.8-12H2O	Tetragonal
METATORBERNITE	Cu(UO2)2(PO4)2.8H2O	Tetragonal
JOHANNITE	Cu(UO2)2(SO4)2(OH)2.8H2O	Triclinic
MARTHOZITE	Cu(UO2)3(SeO3)3(OH)2.7H2O	Orthorhombic
NAMIBITE	Cu++(BiO)2VO4OH	Monoclinic
MOOLOOITE	Cu++(C2O4).0,5H2O	Orthorhombic
CHALCOPHYLLITE	Cu++18Al2(AsO4)3(SO4)3(OH)27.33H2O	Trigonal
BELLINGERITE	Cu++3(IO3)6.2H2O	Triclinic
CUPROTUNGSTITE	Cu++3(WO4)2(OH)2	Tetragonal
LYONSITE	Cu++3Fe+++4(VO4)6	Orthorhombic
JENSENITE	Cu++3Te++++++O6.2H2O	Monoclinic
MCALPINEITE	Cu++3Te++++++O6.H2O	Isometric
VOLBORTHITE	Cu++3V+++++2O7(OH)2.2H2O	Monoclinic
CLARINGBULLITE	Cu++4(OH)6(Cl,OH)2	Hexagonal
DELORYITE	Cu++4(UO2)(MoO4)2(OH)6	Monoclinic
REICHENBACHITE	Cu++5(PO4)2(OH)4	Monoclinic
NABOKOITE	Cu++7Te++++O4(SO4)5.KCl	Tetragonal
HENTSCHELITE	Cu++Fe+++2(PO4)2(OH)2	Monoclinic
HENTSCHELITE (ANDREWSITE)	Cu++Fe+++2(PO4)2(OH)2	Monoclinic
VOLFSONITE	Cu+10Cu++Fe++Fe+++2Sn++++3S16	Tetragonal
ASTROCYANITE-(Ce)	Cu+2(Ce,Nd,La,Pr,Sm,Ca,Y)2(UO2)(CO3)5(OH)2.3H2O	Hexagonal
PARAMELACONITE	Cu+2Cu++2O3	Tetragonal
MAWSONITE	Cu+6Fe+++2Sn++++S8	Tetragonal
DELAFOSSITE	Cu+Fe+++O2	Trigonal
DIXENITE	Cu+Mn++14Fe+++(As+++O3)5(SiO4)2(As+++++O4)(OH)6	Trigonal
ALGODONITE	Cu1-xAsx	Hexagonal
BETEKHTINITE	Cu10(Fe,Pb)S6	Orthorhombic
CUPROBISMUTITE	Cu10Bi12S23	Monoclinic
FINGERITE	Cu11(VO4)6O2	Triclinic
VINCIENNITE	Cu11Fe4Sn(As,Sb)S16	Tetragonal ps Cubic
GOLDFIELDITE	Cu12(Te,Sb,As)4S13	Isometric
COLUSITE	Cu12-13V(As,Sb,Sn,Ge)3S16	Isometric
GERMANOCOLUSITE	Cu13V(Ge,As)3S16	Isometric
STIBIOCOLUSITE	Cu13V(Sb,As,Sn)3S16	Isometric

Mineral Data

Mineral Name	Formula	Crystal System
NEKRASOVITE	Cu13V(Sn,As,Sb)3S16	Isometric
KUTINAITE	Cu14Ag6As7	Isometric
PUTORANITE	Cu16-18(Fe,Ni)18-19S32	Isometric
ISOCHALCOPYRITE	Cu16Fe17S32or(Fe,Cu)SorCu8Fe9S16	Isometric
BUTTGENBACHITE	Cu19Cl4(NO3)2(OH)32.2H2O	Hexagonal
CONNELLITE	Cu19Cl4(SO4)(OH)32.3H2O	Hexagonal
EUCHROITE	Cu2(AsO4)(OH).3H2O	Orthorhombic
ARHBARITE	Cu2(AsO4)(OH).6H2O	Monoclinic
MALACHITE	Cu2(CO3)(OH)2	Monoclinic
FERROKESTERITE	Cu2(Fe,Zn)SnS4	Tetragonal
GERHARDTITE	Cu2(NO3)(OH)3	Orthorhombic
CLINOATACAMITE	Cu2(OH)3Cl	Monoclinic
LIBETHENITE	Cu2(PO4)(OH)	Orthorhombic
CUPROSTIBITE	Cu2(Sb,Tl)	Tetragonal
DOLEROPHANITE	Cu2(SO4)O	Monoclinic
SENGIERITE	Cu2(UO2)2V2O8.6H2O	Monoclinic
ROUBAULTITE	Cu2(UO2)3(CO3)2O2(OH)2.4H2O	Triclinic
BRIARTITE	Cu2(Zn,Fe)GeS4	Tetragonal
KESTERITE	Cu2(Zn,Fe)SnS4	Tetragonal
GERMANITE	Cu26Fe4Ge4S32	Isometric
LIROCONITE	Cu2Al(AsO4)(OH)4.4H2O	Monoclinic
LUETHEITE	Cu2Al2(AsO4)2(OH)4.H2O	Monoclinic
CERULEITE	Cu2Al7(AsO4)4(OH)13.12H2O	Triclinic
PAXITE	Cu2As3	Monoclinic ps Orthorhombic
OLIVENITE	Cu2AsO4(OH)	Orthorhombic
CERNYITE	Cu2CdSnS4	Tetragonal
ATACAMITE	Cu2Cl(OH)3	Orthorhombic
BOTALLACKITE	Cu2Cl(OH)3	Monoclinic
GEORGEITE	Cu2CO3(OH)2	Amorphous
CHENEVIXITE	Cu2Fe+++2(AsO4)2(OH)4.H2O	Monoclinic
RHODOSTANNITE	Cu2FeSn3S8	HexagonalTetragonal
STANNITE	Cu2FeSnS4	Tetragonal
VELIKITE	Cu2HgSnS4	Tetragonal
CALLAGHANITE	Cu2Mg2(CO3)(OH)6.2H2O	Monoclinic
NISSONITE	Cu2Mg2(PO4)2(OH)2.5H2O	Monoclinic
CUPRITE	Cu2O	Isometric
MELANOTHALLITE	Cu2OCl2	Orthorhombic
CHALCOCITE	Cu2S	Monoclinic
PARTZITE	Cu2Sb2(O,OH)7	Cubic
BELLIDOITE	Cu2Se	Tetragonal
BERZELIANITE	Cu2Se	Isometric
MOHITE	Cu2SnS3	Triclinic
FRANKHAWTHORNEITE	Cu2Te++++++O4(OH)2	Monoclinic
BLOSSITE	Cu2V++++2O7	Orthorhombic
ZIESITE	Cu2V++++2O7	Monoclinic
ARSENOSULVANITE	Cu3(As,V)S4	Isometric

Mineral Data

Mineral Name	Formula	Crystal System
CLINOCLASE	Cu3(AsO4)(OH)3	Monoclinic
GILMARITE	Cu3(AsO4)(OH)3	Triclinic
AZURITE	Cu3(CO3)2(OH)2	Monoclinic
LINDGRENITE	Cu3(MoO4)2(OH)2	Monoclinic
LIKASITE	Cu3(NO3)(OH)5.2H2O	Orthorhombic
CORNETITE	Cu3(PO4)(OH)3	Orthorhombic
STIBIOENARGITE	Cu3(Sb,As)S4	Orthorhombic
ANTLERITE	Cu3(SO4)(OH)4	Orthorhombic
MCBIRNEYITE	Cu3(VO4)2	Triclinic
LAMMERITE	Cu3[(As,P)O4]2	Monoclinic
DJURLEITE	Cu31S16	Monoclinic
SPIONKOPITE	Cu39S28	Hexagonal
SIELECKIITE	Cu3Al4(PO4)2(OH)12.2H2O	Triclinic
ZAPATALITE	Cu3Al4(PO4)3(OH)9.4H2O	Tetragonal
DOMEYKITE	Cu3As	Isometric
MGRIITE	Cu3AsS3	Isometric
ENARGITE	Cu3AsS4	Orthorhombic
LUZONITE	Cu3AsS4	Tetragonal
AURICUPRIDE	Cu3Au	Orthorhombic
VESIGNIEITE	Cu3Ba(VO4)2(OH)2	Monoclinic
FRANCISITE	Cu3Bi(SeO3)2O2Cl	Orthorhombic
WITTICHENITE	Cu3BiS3	Orthorhombic
IDAITE	Cu3FeS4	Hexagonal
FUKUCHILITE	Cu3FeS8	Isometric
SZENICSITE	Cu3MoO4(OH)4	Orthorhombic
SKINNERITE	Cu3SbS3	Monoclinic
FAMATINITE	Cu3SbS4	Tetragonal
PERMINGEATITE	Cu3SbSe4	Tetragonal
UMANGITE	Cu3Se2	Tetragonal
KURAMITE	Cu3SnS4	Tetragonal
XOCOMECATLITE	Cu3Te++++++O4(OH)4	Orthorhombic
SULVANITE	Cu3VS4	Isometric
BRASS (LAITON,MESSING)	Cu3Zn2	Cubic
WATANABEITE	Cu4(As,Sb)2S5	Orthorhombic ps Cubic
BROCHANTITE	Cu4(SO4)(OH)6	Monoclinic
LANGITE	Cu4(SO4)(OH)6.2H2O	Monoclinic
WROEWOLFEITE	Cu4(SO4)(OH)6.2H2O	Monoclinic
POSNJAKITE	Cu4(SO4)(OH)6.H2O	Monoclinic
DERRIKSITE	Cu4(UO2)(SeO3)2(OH)6	Orthorhombic
HENRYITE	Cu4Ag3Te4	Isometric
CARBONATECYANOTRICHITE	Cu4Al2(CO3,SO4)(OH)12.2H2O	Orthorhombic
WOODWARDITE	Cu4Al2(SO4)(OH)12.2-4H2O	Hexagonal
CYANOTRICHITE	Cu4Al2(SO4)(OH)12.2H2O	Orthorhombic
CAMEROLAITE	Cu4Al2[HSbO4,SO4]OH10(CO3).2H2O	Monoclinic
HAYCOCKITE	Cu4Fe5S8	Orthorhombic

Mineral Name	Formula	Crystal System
CAMPIGLIAITE	Cu4Mn(SO4)2(OH)6.4H2O	Monoclinic
ARZRUNITE	Cu4Pb2SO4O2Cl6.4H2O	Orthorhombic
CORNUBITE	Cu5(AsO4)2(OH)4	Triclinic
CORNWALLITE	Cu5(AsO4)2(OH)4	Monoclinic
LUDJIBAITE	Cu5(PO4)2(OH)4	Triclinic
PSEUDOMALACHITE	Cu5(PO4)2(OH)4	Monoclinic
SHATTUCKITE	Cu5(SiO3)4(OH)2	Orthorhombic
CESBRONITE	Cu5(TeO3)2(OH)6.2H2O	Orthorhombic
TURANITE	Cu5(VO4)2(OH)4	Orthorhombic
CYANOPHILLITE	Cu5Al2Sb+++3O12(OH).12H2O	Orthorhombic
KOUTEKITE	Cu5As2	Hexagonal
BORNITE	Cu5FeS4	Orthorhombic ps Cubic
HORSFORDITE	Cu5Sb	Cubic
ATHABASCAITE	Cu5Se4	Orthorhombic
GILALITE	Cu5Si6O17.7H2O[Ino.]	Monoclinic
STOIBERITE	Cu5V+++++2O10	Monoclinic
THEISITE	Cu5Zn5(As+++++,Sb+++++)2O8(OH)14	Orthorhombic
SPANGOLITE	Cu6Al(SO4)(OH)12Cl.3H2O	Trigonal
CUALSTIBITE	Cu6Al3Sb+++++3O18.16H2O	Trigonal
SINNERITE	Cu6As4S9	Triclinic
ATLASOVITE	Cu6Fe+++Bi+++O4(SO4)5.KCl	Tetragonal
CHATKALITE	Cu6Fe++Sn2S8	Tetragonal
AKTASHITE	Cu6Hg3As4S12	Triclinic
GRUZDEVITE	Cu6Hg3Sb4S12	Trigonal
BAYANKHANITE	Cu6HgS4	Hexagonal
HEMUSITE	Cu6SnMoS8	Isometric
KIDDCREEKITE	Cu6SnWS8	Isometric
SABATIERITE	Cu6TlSe4	Orthorhombic
NOWACKIITE	Cu6Zn3As4S12	Trigonal
CROOKESITE	Cu7(Tl,Ag)Se4	Tetragonal
BELENDORFFITE	Cu7Hg6	Trigonal ps Cubic
KOLYMITE	Cu7Hg6	Isometric
ANILITE	Cu7S4	Orthorhombic
RICKARDITE	Cu7Te5	Orthorhombic ps Tetragonal
STRASHIMIRITE	Cu8(AsO4)4(OH)4.5H2O	Monoclinic
STANNOIDITE	Cu8(Fe,Zn)3Sn2S12	Orthorhombic
NAKAURIITE	Cu8(SO4)4(CO3)(OH)6.48H2O	Orthorhombic
HODRUSHITE	Cu8Bi12S22	Monoclinic
GEERITE	Cu8S5	ps Cubic
PLANCHEITE	Cu8Si8O22(OH)4.H2O	Orthorhombic
PARNAUITE	Cu9(AsO4)2(SO4)(OH)10.7H2O	Orthorhombic
TALNAKHITE	Cu9(Fe,Ni)8S16	Isometric
BALKANITE	Cu9Ag5HgS8	Orthorhombic
MOOIHOEKITE	Cu9Fe9S16	Tetragonal
DIGENITE	Cu9S5	Isometric
ROXBYITE	Cu9S5	Monoclinic

Mineral Name	Formula	Crystal System
YARROWITE	Cu9S8	Hexagonal
APACHITE	Cu9Si10O29.11H2O[Ino.]	Monoclinic
EUCAIRITE	CuAgSe	Orthorhombic
AUBERTITE	CuAl(SO4)2Cl.14H2O	Triclinic
CHALCOALUMITE	CuAl4(SO4)(OH)12.3H2O	Monoclinic
TURQUOISE	CuAl6(PO4)4(OH)8.4H2O	Triclinic
TRIPPKEITE	CuAs+++2O4	Tetragonal
LAUTITE	CuAsS	Orthorhombic
KOSTOVITE	CuAuTe4	Orthorhombic
BANDYLITE	CuB(OH)4Cl	Tetragonal
KUSACHIITE	CuBi2O4	Tetragonal
EMPLECTITE	CuBiS2	Orthorhombic
NANTOKITE	CuCl	Isometric
TOLBACHITE	CuCl2	Monoclinic
ERIOCHALCITE	CuCl2.2H2O	Orthorhombic
DAYINGITE	CuCoPtS4	Isometric
MCCONNELLITE	CuCrO2	Trigonal
CHAIDAMUNITE	CuFe(SO4)2(OH).4H2O	Triclinic ps Monoclinic
GUILDITE	CuFe+++(SO4)2(OH).4H2O	Monoclinic
ARTHURITE	CuFe+++2(AsO4,PO4,SO4)2(O,OH)2.4H2O	Monoclinic
RANSOMITE	CuFe+++2(SO4)4.6H2O	Monoclinic
CUPROCOPIAPITE	CuFe+++4(SO4)6(OH)2.20H2O	Triclinic
CHALCOSIDERITE	CuFe+++6(PO4)4(OH)8.4H2O	Triclinic
CUBANITE	CuFe2S3	Orthorhombic
ISOCUBANITE	CuFe2S3	Isometric
CHALCOPYRITE	CuFeS2	Tetragonal
ESKEBORNITE	CuFeSe2	Tetragonal
GALLITE	CuGaS2	Tetragonal
MARSHITE	CuI	Isometric
ROQUESITE	CuInS2	Tetragonal
CUPROIRIDSITE	CuIr2S4	Isometric
CREDNERITE	CuMnO2	Monoclinic
CASTAINGITE	CuMo2S5	Hexagonal
MUCKEITE (MUECKEITE)	CuNiBiS3	Orthorhombic
LAPIEITE	CuNiSbS3	Orthorhombic
TENORITE	CuO	Monoclinic
DAOMANITE	CuPtAsS2	Orthorhombic
CUPRORHODSITE	CuRh2S4	Isometric
COVELLITE	CuS	Hexagonal
CHALCOSTIBITE	CuSbS2	Orthorhombic
KLOCKMANNITE	CuSe	Hexagonal
KRUTAITE	CuSe2	Isometric
CHALCOMENITE	CuSeO3.2H2O	Orthorhombic
CLINOCHALCOMENITE	CuSeO3.2H2O	Monoclinic
DIOPTASE	CuSiO2(OH)2	Trigonal
CHALCOCYANITE	CuSO4	Orthorhombic

Mineral Data

Mineral Name	Formula	Crystal System
BONATTITE	$CuSO_4.3H_2O$	Monoclinic
CHALCANTHITE	$CuSO_4.5H_2O$	Triclinic
BOOTHITE	$CuSO_4.7H_2O$	Monoclinic
VULCANITE	$CuTe$	Orthorhombic
WEISSITE	$CuTe$	Hexagonal Cubic
RAJITE	$CuTe{++++}2O5$	Monoclinic
BALYAKINITE	$CuTeO_3$	Orthorhombic
TEINEITE	$CuTeO_3.2H_2O$	Orthorhombic
GRAEMITE	$CuTeO_3.H_2O$	Orthorhombic
DANBAITE	$CuZn_2$	Isometric
HYPERSTHENE	$(Fe,Mg)SiO_3$	Orthorhombic
IRON (FER,EISEN)	Fe	Isometric
BERNALITE	$Fe(OH)_3$	Orthorhombic ps Cubic
HUMBOLDTINE	$Fe{++}(C_2O_4).2H_2O$	Monoclinic
KAHLERITE	$Fe{++}(UO_2)_2(AsO_4)_2.10-12H_2O$	Tetragonal
METAKAHLERITE	$Fe{++}(UO_2)_2(AsO_4)_2.8H_2O$	Tetragonal
BASSETITE	$Fe{++}(UO_2)_2(PO_4)_2.8H_2O$	Monoclinic
STURTITE =HYDROUSMnSILICATE	$Fe{+++}(Mn{++},Ca,Mg)Si_4O_{10}(OH)_3.10H_2O$	Amorphous
AKAGANEITE	$Fe{+++}(O,OH,Cl)$	Monoclinic ps Tetragonal
BUTLERITE	$Fe{+++}(SO_4)(OH).2H_2O$	Monoclinic
PARABUTLERITE	$Fe{+++}(SO_4)(OH).2H_2O$	Orthorhombic
AMARANTITE	$Fe{+++}(SO_4)(OH).3H_2O$	Triclinic
FIBROFERRITE	$Fe{+++}(SO_4)(OH).5H_2O$	Monoclinic
XITIESHANITE	$Fe{+++}(SO_4)(OH).7H_2O$	Monoclinic
DERBYLITE	$Fe{+++},Fe{++},Ti)_7Sb{+++}O_{13}(OH)$	Monoclinic
SCHWERTMANNITE	$Fe{+++}16O16(OH)12(SO_4)2$	Tetragonal
SARMIENTITE	$Fe{+++}2(AsO_4)(SO_4)(OH).5H_2O$	Monoclinic
BUKOVSKYITE	$Fe{+++}2(AsO_4)(SO_4)(OH).7H_2O$	Monoclinic
FERRIMOLYBDITE	$Fe{+++}2(MoO_4)3.8H_2O$	Orthorhombic
DIADOCHITE (DESTINEZITE)	$Fe{+++}2(PO_4)(SO_4)(OH).5H_2O$	Triclinic
METAHOHMANNITE	$Fe{+++}2(SO_4)2(OH)2.3H_2O$	
HOHMANNITE	$Fe{+++}2(SO_4)2(OH)2.7H_2O$	Triclinic
QUENSTEDTITE	$Fe{+++}2(SO_4)3.10H_2O$	Triclinic
LAUSENITE	$Fe{+++}2(SO_4)3.6H_2O$	Monoclinic
KORNELITE	$Fe{+++}2(SO_4)3.7H_2O$	Monoclinic
COQUIMBITE	$Fe{+++}2(SO_4)3.9H_2O$	Trigonal
PARACOQUIMBITE	$Fe{+++}2(SO_4)3.9H_2O$	Trigonal
POUGHITE	$Fe{+++}2(TeO_3)2(SO_4).3H_2O$	Orthorhombic
SCHUBNELITE	$Fe{+++}2-x(V{+++++},V{++++})2O4(OH)4$	Triclinic
FERRICOPIAPITE	$Fe{+++}2/3Fe{+++}4(SO_4)6(OH)2.20H_2O$	Triclinic
COCONINOITE	$Fe{+++}2Al2(UO_2)2(PO_4)4(SO_4)(OH)2.20H_2O$	Monoclinic
KARIBIBITE	$Fe{+++}2As{+++}4(O,OH)9$	Orthorhombic
MANDARINOITE	$Fe{+++}2Se3O9.6H_2O$	Monoclinic
HISINGERITE	$Fe{+++}2Si_2O_5(OH)4.2H_2O$	Monoclinic
FERRIPYROPHYLLITE	$Fe{+++}2Si_4O_{10}(OH)2$	Monoclinic

Mineral Name	Formula	Crystal System
CUZTICITE	Fe+++2Te++++++O6.3H2O	Hexagonal
EMMONSITE	Fe+++2Te++++3O9.2H2O	Triclinic
PSEUDORUTILE	Fe+++2Ti3O9	Hexagonal
FERRISYMPLESITE	Fe+++3(AsO4)2(OH)3.5H2O	Amorphous
ANGELELLITE	Fe+++4(AsO4)2O3	Triclinic
ZYKAITE	Fe+++4(AsO4)3(SO4)(OH).15H2O	Orthorhombic
TINTICITE	Fe+++4(PO4)3(OH)3.5H2O	Monoclinic
FERVANITE	Fe+++4(VO4)4.5H2O	Monoclinic
KAZAKHSTANITE	Fe+++5V++++3V+++++12O39(OH)9.9H2O	Monoclinic
TOOELEITE	Fe+++7,5-8(AsO4,SO4)6(OH)6.5H2O	Orthorhombic
SIGLOITE	Fe+++Al2(PO4)2(OH)3.5H2O	Triclinic
KAATIALAITE	Fe+++As+++++3O9.6-8H2O	Monoclinic
SCORODITE	Fe+++AsO4.2H2O	Orthorhombic
KANKITE	Fe+++AsO4.3.5H2O	Monoclinic
MOLYSITE	Fe+++Cl3	Hexagonal
FERRISTRUNZITE	Fe+++Fe+++2(PO4)2(OH)3.5H2O	Triclinic
FEROXYHYTE	Fe+++O(OH)	Hexagonal
GOETHITE	Fe+++O(OH)	Orthorhombic
HETEROSITE	Fe+++PO4	Orthorhombic
PHOSPHOSIDERITE	Fe+++PO4.2H2O	Monoclinic
STRENGITE	Fe+++PO4.2H2O	Orthorhombic
KONINCKITE	Fe+++PO4.3H2O	Tetragonal
SONORAITE	Fe+++Te++++O3(OH).H2O	Monoclinic
MACKAYITE	Fe+++Te2O5(OH)	Tetragonal
ZWIESELITE	Fe++,Mn)2(PO4)F	Monoclinic
CRONSTEDTITE	Fe++2Fe+++(SiFe+++)O5(OH)4	Monoclinic
VONSENITE	Fe++2Fe+++BO5	Orthorhombic
FAYALITE	Fe++2SiO4	Orthorhombic
SCHOONERITE	Fe++2ZnMnFe+++(PO4)3(OH)2.9H2O	Orthorhombic
PARASYMPLESITE	Fe++3(AsO4)2.8H2O	Monoclinic
SYMPLESITE	Fe++3(AsO4)2.8H2O	Triclinic
VIVIANITE	Fe++3(PO4)2.8H2O	Monoclinic
METAVIVIANITE	Fe++3-xFe+++x(PO4)2(OH)x.(8-x)H2O	Triclinic
ALMANDINE (ALMANDITE)	Fe++3Al2(SiO4)3	Isometric
GORMANITE	Fe++3Al4(PO4)4(OH)6.2H2O	Triclinic
LAUBMANNITE	Fe++3Fe+++6(PO4)4(OH)12	Orthorhombic
DANALITE	Fe++4Be3(SiO4)3S	Isometric
VERSILIAITE	Fe+++4Fe+++8Sb+++12S2	Orthorhombic
ALMBOSITE	Fe++5Fe+++4V+++++4Si3O27	
TRANQUILLITYITE	Fe++8(Zr,Y)2Ti3Si3O24	Hexagonal
CHILDRENITE	Fe++Al(PO4)(OH)2.H2O	Monoclinic
VAUXITE	Fe++Al2(PO4)2(OH)2.6H2O	Triclinic
METAVAUXITE	Fe++Al2(PO4)2(OH)2.8H2O	Monoclinic
PARAVAUXITE	Fe++Al2(PO4)2(OH)2.8H2O	Triclinic
HALOTRICHITE	Fe++Al2(SO4)4.22H2O	Monoclinic
CARBOIRITE	Fe++Al2GeO5(OH)2	Triclinic

Mineral Data

Mineral Name	Formula	Crystal System
HERCYNITE	Fe++Al2O4	Isometric
ROKUHNITE (ROKUEHNITE)	Fe++Cl2.H2O	Monoclinic
SIDERITE	Fe++CO3	Trigonal
CHROMITE	Fe++Cr2O4	Isometric
DAUBREELITE	Fe++Cr2S4	Isometric
BARBOSALITE	Fe++Fe+++2(PO4)2(OH)2	Monoclinic
WHITMOREITE	Fe++Fe+++2(PO4)2(OH)2.4H2O	Monoclinic
FERROSTRUNZITE	Fe++Fe+++2(PO4)2(OH)2.6H2O	Triclinic
FERROLAUEITE	Fe++Fe+++2(PO4)2(OH)2.8H2O	Triclinic
LAIHUNITE	Fe++Fe+++2(SiO4)2	Monoclinic
ROMERITE (ROEMERITE)	Fe++Fe+++2(SO4)4.14H2O	Triclinic
BILINITE	Fe++Fe+++2(SO4)4.22H2O	Monoclinic
MAGNETITE	Fe++Fe+++2O4	Isometric
GREIGITE (MELNIKOVITE)	Fe++Fe+++2S4	Isometric
SCHNEIDERHOHNITE (SCHNEIDERHOEHNITE)	Fe++Fe+++3As+++5O13	Triclinic
DUFRENITE	Fe++Fe+++4(PO4)3(OH)5.2H2O	Monoclinic
GINIITE	Fe++Fe+++4(PO4)4(OH)2.2H2O	Monoclinic
COPIAPITE (FERROCOPIAPITE)	Fe++Fe+++4(SO4)6(OH)2.20H2O	Triclinic
APUANITE	Fe++Fe+++4Sb+++4O12S	Tetragonal
BERAUNITE	Fe++Fe+++5(PO4)4(OH)5.4H2O	Monoclinic
STOTTITE	Fe++Ge(OH)6	Tetragonal
INDITE	Fe++In2S4	Isometric
FERROCOLUMBITE	Fe++Nb2O6	Orthorhombic
VIOLARITE	Fe++Ni+++2S4	Isometric
TRIPUHYITE	Fe++Sb++++2O6	Tetragonal
SCHAFARZIKITE	Fe++Sb+++2O4	Tetragonal
NATANITE	Fe++Sn++++(OH)6	Isometric
ROZENITE	Fe++SO4.4H2O	Monoclinic
SIDEROTIL	Fe++SO4.5H2O	Triclinic
FERROHEXAHYDRITE	Fe++SO4.6H2O	Monoclinic
MELANTERITE	Fe++SO4.7H2O	Monoclinic
SZOMOLNOKITE	Fe++SO4.H2O	Monoclinic
FERROTANTALITE	Fe++Ta2O6	Orthorhombic
ILMENITE	Fe++TiO3	Trigonal
COULSONITE	Fe++V+++2O4	Isometric
FERBERITE	Fe++WO4	Monoclinic
PITTICITE	Fe,AsO4,SO4,H2O	Amorphous
PYRRHOTITE	Fe1-xS(x=0-0,17)	Monoclinic/Hexagonal
MAGHEMITE	Fe2.67O4	
KAMIOKITE	Fe2Mo3O8	Hexagonal
HEMATITE (OLIGISTE)	Fe2O3	Trigonal
CHROMFERIDE	Fe3Cr1-x(x=0,6)	Isometric
GUPEIITE	Fe3Si	Isometric
ROALDITE	Fe4N	Isometric

Mineral Data

Mineral Name	Formula	Crystal System
SIDERAZOT	Fe5N2	Hexagonal
XIFENGITE	Fe5Si3	Hexagonal
LOLLINGITE (LOELLINGITE)	FeAs2	Orthorhombic
ARSENOPYRITE (MISPICKEL)	FeAsS	Monoclinic ps Orthorhombic
HYDROMOLYSITE	FeCl3.6H2O	
TETRATAENITE	FeNi	Tetragonal
WUSTITE (WUESTITE)	FeO	Isometric
TROILITE	FeS	Hexagonal
VYALSOVITE	FeS.Ca(OH)2.Al(OH)3	Orthorhombic
MARCASITE	FeS2	Orthorhombic
PYRITE	FeS2	Isometric
BERTHIERITE	FeSb2S4	Orthorhombic
GARAVELLITE	FeSbBiS4	Orthorhombic
GUDMUNDITE	FeSbS	Monoclinic
ACHAVALITE	FeSe	Trigonal
DZHARKENITE	FeSe2	Isometric
FERROSELITE	FeSe2	Orthorhombic
FERSILICITE	FeSi	Isometric
FERDISILICITE	FeSi2	Isometric
FROHBERGITE	FeTe2	Orthorhombic
MAHLMOODITE	FeZr(PO4)2.4H2O	Monoclinic
SOHNGEITE (SOEHNGEITE)	Ga(OH)3	Isometric
ARGUTITE	GeO2	Tetragonal
WALENTAITE	H(Ca,Mn,Fe++)Fe+++3[(AsO4,PO4)]4.7H2O	Orthorhombic
STERCORITE	H(NH4)Na(PO4).4H2O	Triclinic
FRANSOLETITE	H2Ca3Be2(PO4)4.4H2O	Monoclinic
PICROPHARMACOLITE	H2Ca4Mg(AsO4)4.11H2O	Triclinic
SINKANKASITE	H2MnAl(PO4)2(OH).6H2O	Triclinic
ICE (GLACE,EIS)	H2O	Hexagonal
HYDROTUNGSTITE	H2WO4.H2O	Monoclinic
SASSOLITE (BORICACID)	H3BO3	Triclinic
RODALQUILARITE	H3Fe+++2(Te++++O3)4Cl	Triclinic
OBRADOVICITE	H4(K,Na)Cu++Fe+++2(AsO4)(MoO4)5.12H2O	Orthorhombic
MOLURANITE	H4U++++(UO2)3(MoO4)7.18H2O	Amorphous
TLAPALLITE	H6(Ca,Pb)2(Cu,Zn)3(SO4)(Te++++O3)4(Te++++++O6)	Monoclinic
FRANCOANELLITE	H6(K,Na)3(Al,Fe+++)5(PO4)8.13H2O	Trigonal
TARANAKITE	H7K2(Al,Fe+++)5(PO4)8.20H2O	Trigonal
DOLORESITE	H8V6O16	Monoclinic
ARSENURANOSPATHITE	HAl(UO2)4(AsO4)4.40H2O	Tetragonal
SABUGALITE	HAl(UO2)4(PO4)4.16H2O	Monoclinic ps Tetragonal
URANOSPATHITE	HAl(UO2)4(PO4)4.40H2O	Tetragonal
RANUNCULITE	HAl(UO2)PO4(OH)3.4H2O	Monoclinic ps Orthorhombic
METABORITE	HBO2	Isometric
LANNONITE	HCa4Mg2Al4(SO4)8F9.3H2O	Tetragonal

Mineral Data

Mineral Name	Formula	Crystal System
RHOMBOCLASE	HFe+++(SO4)2.4H2O	Orthorhombic
MCAUSLANITE	HFe++3Al2(PO4)4F.18H2O	Triclinic
HAFNON	HfSiO4	Tetragonal
MERCURY (MERCURE,QUECKSILBER)	Hg	Trigonal
EDOYLERITE	Hg++3Cr++++++O4S2	Monoclinic
PINCHITE	Hg++5O4Cl2	Orthorhombic
COCCINITE	Hg++I2	Tetragonal
SZYMANSKIITE	Hg+16(Ni,Mg)6(CO3)12(OH)12(H3O8).3H2O	Hexagonal
DEANESMITHITE	Hg+2Hg++3Cr++++++O5S2	Triclinic
MAGNOLITE	Hg+2Te++++O3	Orthorhombic
PETERBAYLISSITE	Hg+3(CO3)(OH).2H2O	Orthorhombic
WATTERSITE	Hg+4Hg++Cr++++++O6	Monoclinic
SHAKHOVITE (SHAHOVITE)	Hg+4Sb+++++O3(OH)3	Monoclinic
EDGARBAILEYITE	Hg+6Si2O7	Monoclinic
CHURSINITE	Hg+Hg++(AsO4)	Monoclinic
TERLINGUAITE	Hg+Hg++ClO	Monoclinic
TVALCHRELIDZEITE	Hg12(Sb,As)8S15	Monoclinic
COMANCHEITE	Hg13(Cl,Br)8O9	Orthorhombic
KUZMINITE	Hg2(Br,Cl)2	Tetragonal
CALOMEL	Hg2Cl2	Tetragonal
MOSCHELITE	Hg2I2	Tetragonal
KLEINITE	Hg2N(Cl,SO4).nH2O	Hexagonal
MOSESITE	Hg2N(Cl,SO4,MoO4,CO3).H2O	Isometric
SCHUETTEITE	Hg3(SO4)O2	Hexagonal
KELYANITE	Hg36Sb3(Cl,Br)9O28	Monoclinic
KUZNETSOVITE	Hg3Cl(AsO4)	Isometric
POYARKOVITE	Hg3ClO	Monoclinic
ARZAKITE	Hg3S2(Br,Cl)2	Monoclinic/Triclinic
GRECHISHCHEVITE	Hg3S2(Br,Cl,I)2	Tetragonal
LAVRENTIEVITE	Hg3S2(Cl,Br)2	Monoclinic/Triclinic
CORDEROITE	Hg3S2Cl2	Isometric
RADTKEITE	Hg3S2ClI	Orthorhombic
KADYRELITE	Hg4(Br,Cl)2O	Isometric
GIANELLAITE	Hg4(SO4)N2	Isometric
PERROUDITE	Hg5Ag4S5(Cl,I,Br)4	Orthorhombic
EGLESTONITE	Hg6Cl3O(OH)	Isometric
CAPGARONNITE	HgAg(Cl,Br,I)S	Orthorhombic
MONTROYDITE	HgO	Orthorhombic
LEADAMALGAM	HgPb2	Tetragonal
CINNABAR (CINABRE,ZINNOBER)	HgS	Trigonal
HYPERCINNABAR	HgS	Hexagonal
METACINNABAR	HgS	Isometric
LIVINGSTONITE	HgSb4S8	Monoclinic
TIEMANNITE	HgSe	Isometric

Mineral Name	Formula	Crystal System
COLORADOITE	HgTe	Isometric
BOLTWOODITE	HK(UO2)(SiO4).1,5H2O	Monoclinic
TANCOITE	HNa2LiAl(PO4)2(OH)	Orthorhombic
ASHBURTONITE	HPb4Cu++4Si4O12(HCO3)4(OH)4Cl	Tetragonal
INDIUM	In	Tetragonal
YANOMAMITE	In(AsO4).2H2O	Orthorhombic
DZHALINDITE	In(OH)3	Isometric
CHENGDEITE	Ir3Fe	Isometric
GAOTAIITE	Ir3Te8	Isometric
MAYINGITE	IrBiTe	Isometric
TOLOVKITE	IrSbS	Isometric
SHUANGFENGITE	IrTe2	Trigonal
MISERITE	K(Ca,Ce)6Si8O22(OH,F)2	Triclinic
CHAROITE	K(Ca,Na)2Si4O10(OH,F).H2O	Monoclinic
TUSCANITE	K(Ca,Na)6(Si,Al)10O22(SO4,CO3,(OH)2).H2O	Monoclinic
YIMENGITE	K(Cr,Ti,Fe,Mg)12O19	Hexagonal
FERRIANNITE (FERRI-ANNITE)	K(Fe++,Mg)3(Fe+++,Al)Si3O10(OH)2	Monoclinic
STILPNOMELANE	K(Fe++,Mg,Fe+++,Al)8(Si,Al)12(O,OH)27.2H2O	Monoclinic/Triclinic
ZUSSMANITE	K(Fe++,Mg,Mn)13(Si,Al)18O42(OH)14	Trigonal
IRAQITE-(La)	K(La,Ce,Th)2(Ca,Na)4(Si,Al)16O40	Tetragonal
LEPIDOLITE	K(Li,Al)3(Si,Al)4O10(F,OH)2	Monoclinic
MASUTOMILITE	K(Li,Al,Mn++)3(Si,Al)4O10(F,OH)2	Monoclinic
HYDROBIOTITE (BIOTITE-VERMICULITE)	K(Mg,Fe)6(Si,Al)8O20(OH)4.xH2O	Monoclinic
CELADONITE	K(Mg,Fe++)(Fe+++,Al)Si4O10(OH)2	Monoclinic
BIOTITE	K(Mg,Fe++)3(Al,Fe+++)Si3O10(OH,F)2	Monoclinic
CHAYESITE	K(Mg,Fe++)4Fe+++(Si12O30)	Hexagonal
PAULKERRITE	K(Mg,Mn)2(Fe+++,Al)2Ti(PO4)4(OH)3.15H2O	Orthorhombic
CRYPTOMELANE	K(Mn++++,Mn++)8O16	Monoclinic ps Tetragonal
NORRISHITE	K(Mn+++2Li)Si4O12	Monoclinic
JOHNWALKITE	K(Mn++,Fe+++,Fe++)2(Nb,Ta)(PO4)2O2(H2O,OH)2	Orthorhombic
COOMBSITE	K(Mn++,Fe++,Mg)13(Si,Al)18O42(OH)14	Trigonal
ABERNATHYITE	K(UO2)(AsO4).4H2O	Tetragonal
ZIPPEITE	K(UO2)2(SO4)(OH)3.H2O	Monoclinic
ROSCOELITE	K(V,Al,Mg)2AlSi3O10(OH)2	Monoclinic
HENDRICKSITE	K(Zn,Mn)3Si3AlO10(OH)2	Monoclinic
TIPTOPITE	K2(Na,Ca)2Li3Be6(PO4)6(OH)2.H2O	Hexagonal
ODINTSOVITE	K2(Na,Li)4Ca3Be4Ti++++2Si6O20	Orthorhombic
METAANKOLEITE	K2(UO2)2(PO4)2.6H2O	Tetragonal
WEEKSITE	K2(UO2)2Si6O15.4H2O	Orthorhombic
CARNOTITE	K2(UO2)2V2O8.3H2O	Monoclinic
COMPREIGNACITE	K2(UO2)6O4(OH)6.8H2O	Orthorhombic
BUTSCHLIITE (BUETSCHLIITE)	K2Ca(CO3)2	Trigonal
FAIRCHILDITE	K2Ca(CO3)2	Hexagonal

Mineral Name	Formula	Crystal System
SYNGENITE	K2Ca(SO4)2.H2O	Monoclinic
LEIGHTONITE	K2Ca2Cu(SO4)4.2H2O	Triclinic ps Orthorhombic
POLYHALITE	K2Ca2Mg(SO4)4.2H2O	Triclinic
TOKKOITE	K2Ca4[Si7O18(OH)](F,OH)	Triclinic
MILARITE	K2Ca4Al2Be4Si24O60.H2O	Hexagonal
GORGEYITE (GOERGEYITE)	K2Ca5(SO4)6.H2O	Monoclinic
MAZZITE	K2CaMg2(Al,Si)36O72.28H2O	Hexagonal
RAMEAUITE	K2CaU++++++6O20.9H2O	Monoclinic
LOPEZITE	K2Cr2O7	Triclinic
TARAPACAITE	K2CrO4	Orthorhombic
CYANOCHROITE	K2Cu(SO4).6H2O	Monoclinic
CHLOROTHIONITE	K2Cu(SO4)Cl2	Orthorhombic
FEDOTOVITE	K2Cu++3O(SO4)3or(K,Na)2(Cu,Zn,Pb)3S3O13	Monoclinic
PIYPITE	K2Cu2(SO4)2O	Tetragonal
MURUNSKITE	K2Cu3FeS4	Tetragonal
MITSCHERLICHITE	K2CuCl4.2H2O	Tetragonal
MEREITERITE	K2Fe++(SO4)2.4H2O	Monoclinic
ERYTHROSIDERITE	K2Fe+++Cl5.H2O	Orthorhombic
VOLTAITE	K2Fe++5Fe+++4(SO4)12.18H2O	Isometric
DOUGLASITE	K2Fe++Cl4.2H2O	Monoclinic
BAYLISSITE	K2Mg(CO3)2.4H2O	Monoclinic
LEONITE	K2Mg(SO4)2.4H2O	Monoclinic
PICROMERITE	K2Mg(SO4)2.6H2O	Monoclinic
LANGBEINITE	K2Mg2(SO4)3	Isometric
MANGANOLANGBEINITE	K2Mn2(SO4)3	Isometric
TINAKSITE	K2Na(Ca,Fe++,Mn++,Mg)2(Ti,Fe)Si7O19(OH)	Triclinic
MEGACYCLITE	K2Na16Si18O36(OH)18.32H2O	Monoclinic
AMICITE	K2Na2Al4Si4O16.5H2O	Monoclinic
CLINOUNGEMACHITE	K2Na3Fe(SO4)6(OH)3.9H2O	Monoclinic psTrigonal
LOVDARITE	K2Na6Be4Si14O36.9H2O	Orthorhombic
METAVOLTINE	K2Na6Fe++Fe+++6(SO4)12O2.18H2O	Hexagonal
ELPASOLITE	K2NaAlF6	Isometric
PSEUDOCOTUNNITE	K2PbCl4	Orthorhombic
HIERATITE	K2SiF6	Isometric
ARCANITE	K2SO4	Orthorhombic
ARCANITE (TAYLORITE)	K2SO4	Orthorhombic
MISENITE	K2SO4.6KHSO4	Monoclinic
KALISTRONTITE	K2Sr(SO4)2	Trigonal
DAVANITE	K2TiSi6O15	Triclinic
ZINCVOLTAITE	K2Zn5Fe+++3Al(SO4)12.18H2O	Isometric
KHIBINSKITE	K2ZrSi2O7	Monoclinic psTrigonal
KOSTYLEVITE	K2ZrSi3O9.H2O	Monoclinic
UMBITE	K2ZrSi3O9.H2O	Orthorhombic
DALYITE	K2ZrSi6O15	Triclinic
ALUMOKLYUCHEVSKITE	K3Cu3(Al,Fe+++)O2(SO4)4	Monoclinic
KLYUCHEVSKITE	K3Cu3(Fe+++,Al)O2(SO4)4	Monoclinic

Mineral Data

Mineral Name	Formula	Crystal System
MINGUZZITE	K3Fe+++(C2O4)3.3H2O	Monoclinic
BARTONITE	K3Fe10S14	Tetragonal
GRIMSELITE	K3Na(UO2)(CO3)3.H2O	Hexagonal
ENGLISHITE	K3Na2Ca10Al15(PO4)21(OH)7.26H2O	Orthorhombic
IQUIQUEITE	K3Na4Mg(Cr++++++O4)B24O39(OH).12H2O	Hexagonal
HUMBERSTONITE	K3Na7Mg2(SO4)6(NO3)2.6H2O	Trigonal
UNGEMACHITE	K3Na8Fe+++(SO4)6(NO3)2.6H2O	Trigonal
RINNEITE	K3NaFe++Cl6	Trigonal
PARAUMBITE	K3Zr2HSi6O18.nH2O	Orthorhombic
PONOMAREVITE	K4Cu++4OCl10	Monoclinic
KAFEHYDROCYANITE	K4Fe++(CN)6.3H2O	Tetragonal
CHLOROMANGANOKALITE	K4MnCl6	Trigonal
ASHCROFTINE-	K5Na5(Y,Ca)12Si28O70(OH)2(CO3)8.8H2O	Tetragonal
CARLOSRUIZITE	K6(Na,K)4Na6Mg10(Se++++++O4)12(IO3)12.12H2O	Trigonal
FUENZALIDAITE	K6(Na,K)4Na6Mg10(SO4)12(IO3)12.12H2O	Trigonal
LENNILENAPEITE	K6-7(Mg,Mn,Fe++,Fe+++,Zn)48(Si,Al)72(O,OH)216.16H2O	Triclinic
KALBORSITE	K6Al4Si6BO20(OH)4Cl	Tetragonal
LITHOSITE	K6Al4Si8O25.2H2O	Monoclinic ps Orthorhombic
DJERFISHERITE	K6Na(Fe,Cu)24S26Cl	Isometric
PERLIALITE	K8Tl4Al12Si24O72.20H2O	Hexagonal
KALINITE	KAl(SO4)2.11H2O	Monoclinic
POTASSIUMALUM	KAl(SO4)2.12H2O	Isometric
TINSLEYITE	KAl2(PO4)2(OH).2H2O	Monoclinic
MINYULITE	KAl2(PO4)2(OH,F).4H2O	Orthorhombic
MUSCOVITE	KAl2(Si3Al)O10(OH,F)2	Monoclinic ps Hexagonal
BOROMUSCOVITE	KAl2(Si3B)O10(OH,F)2	Monoclinic
ALUNITE	KAl3(SO4)2(OH)6	Trigonal
BOKITE	KAl3Fe+++6V++++6V+++++20O76.30H2O	
ALUMOPHARMACOSIDERITE	KAl4(AsO4)3(OH)4.6.5H2O	Isometric
SVEITE	KAl7(NO3)4Cl2(OH)16.8H2O	Monoclinic
LEUCITE	KAlSi2O6	Tetragonal
MICROCLINE	KAlSi3O8	Triclinic
ORTHOCLASE (ORTHOSE)	KAlSi3O8	Monoclinic
KALIOPHILITE	KAlSiO4	Hexagonal
KALSILITE	KAlSiO4	Hexagonal
SANTITE	KB5O6(OH)4.2H2O	Orthorhombic
MAGBASITE	KBa(Al,Sc)(Mg,Fe++)6Si6O20F2	
PHOSPHURANYLITE	KCa(H3O)3(UO2)7(PO4)4O4.8H2O	Orthorhombic
BANNISTERITE	KCa(Mn,Fe++,Zn)21(Si,Al)32O76(OH)16.12H2O	Monoclinic
HYDRODELHAYELITE	KCa2AlSi7O17(OH)2.6H2O	Orthorhombic
VOLKOVSKITE	KCa4[B5O8(OH)4][B(OH)3]Cl.4H2O	Triclinic
HYDROXYAPOPHYLLITE	KCa4Si8O20(OH,F).8H2O	Tetragonal
BARATOVITE	KCa7Li3(Ti,Zr)2(Si6O18)2(OH,F)2	Monoclinic ps Hexagonal
WILLHENDERSONITE	KCaAl3Si3O12.5H2O	Triclinic

Mineral Name	Formula	Crystal System
CHLOROCALCITE	KCaCl3	Orthorhombic ps Cubic
SYLVITE	KCl	Isometric
KAMCHATKITE	KCu++3OCl(SO4)2	Orthorhombic
PHOSPHOFIBRITE	KCuFe+++15(PO4)12(OH)12.12H2O	Orthorhombic
CAROBBIITE	KF	Isometric
YAVAPAIITE	KFe+++(SO4)2	Monoclinic
GOLDICHITE	KFe+++(SO4)2.4H2O	Monoclinic
KRAUSITE	KFe+++(SO4)2.H2O	Monoclinic
LEUCOPHOSPHITE	KFe+++2(PO4)2(OH).2H2O	Monoclinic
JAROSITE	KFe+++3(SO4)2(OH)6	Trigonal
PHARMACOSIDERITE	KFe+++4(AsO4)3(OH)4.6-7H2O	Cubic/Tetragonal
OLMSTEADITE	KFe++2(Nb,Ta)(PO4)2O2.2H2O	Orthorhombic
SIDEROPHYLLITE	KFe++2Al(Al2Si2)O10(F,OH)2	Monoclinic
ANNITE	KFe++3AlSi3O10(OH,F)2	Monoclinic
RASVUMITE	KFe2S3	Orthorhombic
KALICINITE	KHCO3	Monoclinic
KALIBORITE	KHMg2B12O16(OH)10.4H2O	Monoclinic
MERCALLITE	KHSO4	Orthorhombic
POLYLITHIONITE	KLi2AlSi4O10(F,OH)2	Monoclinic
ZINNWALDITE	KLiFe++Al(AlSi3)O10(F,OH)2	Monoclinic
TAENIOLITE	KLiMg2Si4O10F2	Monoclinic
MANTIENNEITE	KMg2Al2Ti(PO4)4(OH)3.15H2O	Orthorhombic
PHLOGOPITE	KMg3(Si3Al)O10(F,OH)2	Monoclinic
CARNALLITE	KMgCl3.6H2O	Orthorhombic
HUMMERITE	KMgV+++++5O14.8H2O	Triclinic
SUGILITE	KNa2(Fe++,Mn++,Al)2Li3Si12O30	Hexagonal
HANKSITE	KNa22(SO4)9(CO3)2Cl	Hexagonal
SATIMOLITE	KNa2Al4B6O15Cl3.13H2O	Orthorhombic
POUDRETTEITE	KNa2B3Si12O30	Hexagonal
NEPTUNITE	KNa2Li(Fe++,Mn)2Ti2Si8O24	Monoclinic
MANGANNEPTUNITE	KNa2Li(Mn,Fe++)2Ti2Si8O24	Monoclinic
DARAPIOSITE	KNa2Li(Mn,Zn)2ZrSi12O30	Hexagonal
EIFELITE	KNa3Mg4Si12O30	Hexagonal
CARLETONITE	KNa4Ca4Si8O18(CO3)4(OH,F).H2O	Tetragonal
ARROJADITE	KNa4CaMn++4Fe++10Al(PO4)12(OH,F)2	Monoclinic
EUCHLORINE	KNaCu++3(SO4)3O	Monoclinic
LITIDIONITE	KNaCuSi4O10	Triclinic
MANAKSITE	KNaMn++Si4O10	Triclinic
GEORGECHAOITE	KNaZrSi3O9.2H2O	Orthorhombic
NITER (SALPETER,NITRE)	KNO3	Orthorhombic
BRANNOCKITE	KSn2Li3Si12O30	Hexagonal
KOSNARITE	KZr++++2(PO4)3	Trigonal ps Cubic
DZHEZKAZGANITE	Lead or copper rhenium sulfide	Amorphous
FERRISICKLERITE	Li(Fe+++,Mn++)PO4	Orthorhombic
SICKLERITE	Li(Mn++,Fe+++)PO4	Orthorhombic
LITHIOTANTITE	Li(Ta,Nb)3O8	Monoclinic

Mineral Data

Mineral Name	Formula	Crystal System
SIMFERITE	Li0,5(Mg,Mn+++)5(PO4)3	Orthorhombic
LUNIJIANLAITE	Li0,72Al6(Si7AlO20)(OH,O)10	Monoclinic
FERROCLINOHOLMQUISTITE	Li2(Fe++,Mg)3Al2Si8O22(OH)2	Monoclinic
FERROHOLMQUISTITE	Li2(Fe++,Mg)3Al2Si8O22(OH)2	Orthorhombic
CLINOHOLMQUISTITE	Li2(Mg,Fe++)3Al2Si8O22(OH)2	Monoclinic
HOLMQUISTITE	Li2(Mg,Fe++)3Al2Si8O22(OH)2	Orthorhombic
MAGNESIOCLINOHOLMQUISTITE	Li2(Mg,Fe++)3Al2Si8O22(OH)2	Monoclinic
MAGNESIOHOLMQUISTITE	Li2(Mg,Fe++)3Al2Si8O22(OH)2	Orthorhombic
MANANDONITE	Li2Al4[(Si2AlB)O10](OH)8	Orthorhombic
DIOMIGNITE	Li2B4O7	Tetragonal
LIBERITE	Li2BeSiO4	Monoclinic
ZABUYELITE	Li2CO3	Monoclinic
LITHIOPHOSPHATE	Li3PO4	Orthorhombic
LITHIOPHORITE	Li6Al14Mn++3Mn++++18O42(OH)42	Hexagonalrthorhombic Monoclinic
MONTEBRASITE	LiAl(PO4)(OH,F)	Triclinic
COOKEITE	LiAl4(Si3Al)O10(OH)8	Monoclinic
SPODUMENE	LiAlSi2O6	Monoclinic
BIKITAITE	LiAlSi2O6.H2O	Monoclinic
PETALITE	LiAlSi4O10	Monoclinic
EUCRYPTITE	LiAlSiO4	Trigonal
LITHIOMARSTURITE	LiCa2Mn2HSi5O15	Triclinic
GRICEITE	LiF	Isometric
TAVORITE	LiFe+++(PO4)(OH)	Triclinic
TRIPHYLITE	LiFe++PO4	Orthorhombic
LITHIOPHILITE	LiMnPO4	Orthorhombic
VIRGILITE	LixAlxSi3-xO6	Hexagonal
GLUSHINSKITE	Mg(C2O4).2H2O	Monoclinic
YUANFULIITE	Mg(Fe+++,Fe++,Al,Ti,Mg)(BO3)O	Orthorhombic
NESQUEHONITE	Mg(HCO3)(OH).2H2O	Monoclinic
NITROMAGNESITE	Mg(NO3)2.6H2O	Monoclinic
BRUCITE	Mg(OH)2	Trigonal
NOVACEKITE	Mg(UO2)2(AsO4)2.12H2O	Tetragonal
METANOVACEKITE	Mg(UO2)2(AsO4)2.4-8H2O	Tetragonal
SEELITE	Mg(UO2)2(AsO4,AsO3)2.4-7H2O	Monoclinic
SALEEITE	Mg(UO2)2(PO4)2.10H2O	Monoclinic ps Tetragonal
MAGNESIUMZIPPEITE	Mg++0,5(UO2)2(SO4)(OH)3.H2O	Monoclinic
MAGNESIOCOULSONITE	Mg++V+++2O4	Isometric
COALINGITE	Mg10Fe+++2(CO3)(OH)24.2H2O	Trigonal
HOLTEDAHLITE	Mg12(PO3OH,CO3)(PO4)5(OH,O)6	Hexagonal
SUDOITE	Mg2(Al,Fe+++)3Si3AlO10(OH)8	Monoclinic
CANAVESITE	Mg2(CO3)(HBO3).5H2O	Monoclinic
POKROVSKITE	Mg2(CO3)(OH)2.0,5H2O	Monoclinic
ARTINITE	Mg2(CO3)(OH)2.3H2O	Monoclinic
CHESTERMANITE	Mg2(Fe+++,Mg,Al,Sb+++++)BO3O2	Orthorhombic

Mineral Name	Formula	Crystal System
FREDRIKSSONITE	Mg2(Mn+++,Fe+++)BO5	Orthorhombic
KOVDORSKITE	Mg2(PO4)(OH).3H2O	Monoclinic
ALTHAUSITE	Mg2(PO4)(OH,F,O)	Orthorhombic
BAYLEYITE	Mg2(UO2)(CO3)3.18H2O	Monoclinic
HALURGITE	Mg2[B4O5(OH)4]2.H2O	Monoclinic
AMESITE	Mg2Al(SiAl)O5(OH)4	Triclinic
INDIGIRITE	Mg2Al2(CO3)4(OH)2.15H2O	Monoclinic
CORDIERITE	Mg2Al4Si5O18	Orthorhombic
INDIALITE	Mg2Al4Si5O18	Hexagonal
MCALLISTERITE	Mg2B12O14(OH)12.9H2O	Trigonal
SUANITE	Mg2B2O5	Monoclinic
KORSHUNOVSKITE	Mg2Cl(OH)3.3,5-4H2O	Triclinic
LUDWIGITE	Mg2Fe+++BO5	Orthorhombic
CLINOENSTATITE	Mg2Si2O6	Monoclinic
ENSTATITE	Mg2Si2O6	Orthorhombic
FORSTERITE	Mg2SiO4	Orthorhombic
HORNESITE (HOERNESITE)	Mg3(AsO4)2.8H2O	Monoclinic
FLUOBORITE	Mg3(BO3)(F,OH)3	Hexagonal
MAJORITE	Mg3(Fe,Al,Si)2(SiO4)3	Isometric
FARRINGTONITE	Mg3(PO4)2	Monoclinic
BOBIERRITE	Mg3(PO4)2.8H2O	Monoclinic
NORBERGITE	Mg3(SiO4)(F,OH)2	Orthorhombic
KORNERUPINE	Mg3-4(Al,Fe+++)5,5-6(SiO4,BO4)5(O,OH)2-3	Orthorhombic
PYROPE	Mg3Al2(SiO4)3	Isometric
TAAFFEITE	Mg3Al8BeO16	Hexagonal
PREOBRAZHENSKITE	Mg3B11O15(OH)9	Orthorhombic
LUNEBURGITE (LUENEBURGITE)	Mg3B2(PO4)2(OH)6.5H2O	Monoclinic
SULFOBORITE	Mg3B2(SO4)(OH)8(OH,F)2	Orthorhombic
KOTOITE	Mg3B2O6	Orthorhombic
BORACITE	Mg3B7O13Cl	Orthorhombic ps Cubic
KNORRINGITE	Mg3Cr2(SiO4)3	Isometric
CLINOCHRYSOTILE	Mg3Si2O5(OH)4	Monoclinic
LIZARDITE	Mg3Si2O5(OH)4	Trigonal/H
ORTHOCHRYSOTILE	Mg3Si2O5(OH)4	Orthorhombic
PARACHRYSOTILE	Mg3Si2O5(OH)4	Orthorhombic
TALC (STEATITE)	Mg3Si4O10(OH)2	Monoclinic/Triclinic
IOWAITE	Mg4Fe+++(OH)8OCl.2-4H2O	Hexagonal
SEPIOLITE	Mg4Si6O15(OH)2.6H2O	Orthorhombic
SHABYNITE	Mg5(BO3)(Cl,OH)2(OH)5.4H2O	Monoclinic
WIGHTMANITE	Mg5(BO3)O(OH)5.2H2O	Monoclinic
HYDROMAGNESITE	Mg5(CO3)4(OH)2.4H2O	Monoclinic
DYPINGITE	Mg5(CO3)4(OH)2.5H2O	Monoclinic
GIORGIOSITE	Mg5(CO3)4(OH)2.5H2O	
ALDERMANITE	Mg5Al12(PO4)8(OH)22.32H2O	Orthorhombic
HYDROTALCITE	Mg6Al2(CO3)(OH)16.4H2O	Trigonal

Mineral Data

Mineral Name	Formula	Crystal System
MANASSEITE	Mg6Al2(CO3)(OH)16.4H2O	Hexagonal
MEIXNERITE	Mg6Al2(OH)18.4H2O	Trigonal
BARBERTONITE	Mg6Cr2(CO3)(OH)16.4H2O	Hexagonal
STICHTITE	Mg6Cr2(CO3)(OH)16.4H2O	Trigonal
BRUGNATELLITE	Mg6Fe+++(CO3)(OH)13.4H2O	Hexagonal
PYROAURITE	Mg6Fe+++2(CO3)(OH)16.4H2O	Trigonal
SJOGRENITE (SJOEGRENITE)	Mg6Fe++2(CO3)(OH)14.5H2O	Hexagonal
DESAUTELSITE	Mg6Mn+++2(CO3)(OH)16.4H2O	Trigonal
ELLENBERGERITE	Mg6TiAl6Si8O28(OH)10	Hexagonal
DOZYITE	Mg7(Al,Fe+++,Cr)2Al2Si4O15(OH)12	Monoclinic
CAMINITE	Mg7(SO4)5(OH)4.H2O	Tetragonal
MUSKOXITE	Mg7Fe+++4O13.10H2O	Trigonal
WILCOXITE	MgAl(SO4)2F.18H2O	Triclinic
LAZULITE	MgAl2(PO4)2(OH)2	Monoclinic
GORDONITE	MgAl2(PO4)2(OH)2.8H2O	Triclinic
PICKERINGITE	MgAl2(SO4)4.22H2O	Monoclinic
SPINEL	MgAl2O4	Isometric
MAGNESIOCARPHOLITE	MgAl2Si2O6(OH)4	Orthorhombic
MAGNESIOCHLORITOID	MgAl2SiO5(OH)2	Triclinic
SINHALITE	MgAlBO4	Orthorhombic
PINNOITE	MgB2O4.3H2O	Tetragonal
INDERITE	MgB3O3(OH)5.5H2O	Monoclinic
KURNAKOVITE	MgB3O3(OH)5.5H2O	Triclinic
HUNGCHAOITE	MgB4O5(OH)4.7H2O	Triclinic ps Hexagonal
ADMONTITE	MgB6O10.7H2O	Monoclinic
AKSAITE	MgB6O7(OH)6.2H2O	Orthorhombic
SZAIBELYITE	MgBO2(OH)	Monoclinic
CHLOROMAGNESITE	MgCl2	Tetragonal
BISCHOFITE	MgCl2.6H2O	Monoclinic
MAGNESITE (GIOBERTITE)	MgCO3	Trigonal
BARRINGTONITE	MgCO3.2H2O	Triclinic
LANSFORDITE	MgCO3.5H2O	Monoclinic
MAGNESIOCHROMITE	MgCr2O4	Isometric
SELLAITE	MgF2	Tetragonal
BOTRYOGEN	MgFe+++(SO4)2(OH).7H2O	Monoclinic
USHKOVITE	MgFe+++2(PO4)2(OH)2.8H2O	Triclinic
MAGNESIOFERRITE	MgFe+++2O4	Isometric
MAGNESIOCOPIAPITE	MgFe+++4(SO4)6(OH)2.20H2O	Triclinic
BRASSITE	MgHAsO4.4H2O	Orthorhombic
ROSSLERITE (ROESSLERITE)	MgHAsO4.7H2O	Monoclinic
NEWBERYITE	MgHPO4.3H2O	Orthorhombic
PHOSPHORROSSLERITE (PHOSPHORROESSLERITE)	MgHPO4.7H2O	Monoclinic
PERICLASE	MgO	Isometric
BYSTROMITE	MgSb2O6	Tetragonal

Mineral Name	Formula	Crystal System
(BYSTROEMITE)		
SPADAITE	MgSiO2(OH)2.H2O	
SCHOENFLIESITE	MgSn++++(OH)6	Isometric
SANDERITE	MgSO4.2H2O	
STARKEYITE	MgSO4.4H2O	Monoclinic
PENTAHYDRITE	MgSO4.5H2O	Triclinic
HEXAHYDRITE	MgSO4.6H2O	Monoclinic
EPSOMITE	MgSO4.7H2O	Orthorhombic
KIESERITE	MgSO4.H2O	Monoclinic
KAINITE	MgSO4.KCl.3H2O	Monoclinic
GEIKIELITE	MgTiO3	Trigonal
COUSINITE	MgU2Mo2O13.6H2O	
PYROCHROITE	Mn(OH)2	Trigonal
MELANOSTIBITE	Mn(Sb+++++,Fe+++)O3	Trigonal
FRITZSCHEITE	Mn(UO2)2[(P,V)O4]2.10H2O	Tetragonal
IWAKIITE	Mn++(Fe+++,Mn+++)2O4	Tetragonal
GRAVEGLIAITE	Mn++(SO3).3H2O	Orthorhombic
CIANCIULLIITE	Mn++++(Mg,Mn++)2Zn+2(OH)10.2-4H2O	Monoclinic
JANGGUNITE	Mn++++5-x(Mn++,Fe+++)1+xO8(OH)6,x=0,2	Orthorhombic
AKHTENSKITE	Mn++++O2	Hexagonal
FEITKNECHTITE	Mn+++O(OH)	Hexagonal
GROUTITE	Mn+++O(OH)	Orthorhombic
PURPURITE	Mn+++PO4	Orthorhombic
REPPIAITE	Mn++[(V,As)O4(OH)2]2	Monoclinic
SARKINITE	Mn++2(AsO4)(OH)	Monoclinic
TIRODITE	Mn++2(Mg,Fe++)5Si8O22(OH)2	Monoclinic
ARMANGITE	Mn++26As+++18O50(OH)4(CO3)=Mn26(AsO2(OH))4(AsO3)14	Trigonal
SURSASSITE	Mn++2Al3(SiO4)(Si2O7)(OH)3	Monoclinic
FLINKITE	Mn++2Mn+++(AsO4)(OH)4	Orthorhombic
STERLINGHILLITE	Mn++3(AsO4)2.4H2O	
REDDINGITE	Mn++3(PO4)2.3H2O	Orthorhombic
OREBROITE (OEREBROITE)	Mn++3(Sb+++++,Fe+++)Si(O,OH)7	Hexagonal
WELINITE	Mn++3(W++++++,Mg)0,7SiO4(O,OH)3	Hexagonal
SPESSARTINE (SPESSARTITE)	Mn++3Al2(SiO4)3	Isometric
MANGANARSITE	Mn++3As+++2O4(OH)4	Trigonal
JAROSEWICHITE	Mn++3Mn+++(AsO4)(OH)6	Orthorhombic
FRANCISCANITE	Mn++3V+++++(SiO4)(O,OH)7	Hexagonal
TIRAGALLOITE	Mn++4As+++++Si3O12(OH)	Monoclinic
GATEHOUSEITE	Mn++5(PO4)2(OH)4	Orthorhombic
MAGNUSSONITE	Mn++5As+++3O9(OH,Cl)=Mn10(AsO3)6(OH,Cl)2	Cubic/Tetragonal
VISTEPITE	Mn++5Sn++++B2(SiO4)5	Orthorhombic
HOLDAWAYITE	Mn++6(CO3)2(OH)7(Cl,OH)	Monoclinic
YEATMANITE	Mn++9Zn6Sb+++++2Si4O28	Triclinic
STRUNZITE	Mn++Fe+++2(PO4)2(OH)2.6H2O	Triclinic ps Monoclinic

Mineral Data

Mineral Name	Formula	Crystal System
PSEUDOLAUEITE	Mn++Fe+++2(PO4)2(OH)2.7-8H2O	Monoclinic
LAUEITE	Mn++Fe+++2(PO4)2(OH)2.8H2O	Triclinic
STEWARTITE	Mn++Fe+++2(PO4)2(OH)2.8H2O	Triclinic
FRONDELITE	Mn++Fe+++4(PO4)3(OH)5	Orthorhombic
BERMANITE	Mn++Mn+++2(PO4)2(OH)2.4H2O	Monoclinic
HAUSMANNITE	Mn++Mn+++2O4	Tetragonal
BRAUNITE	Mn++Mn+++6SiO12	Tetragonal
RITTMANNITE	Mn++Mn++Fe++Al2(OH)2(PO4)4.8H2O	Monoclinic
TETRAWICKMANITE	Mn++Sn++++(OH)6	Tetragonal
WICKMANITE	Mn++Sn++++(OH)6	Isometric
MALLARDITE	Mn++SO4.7H2O	Monoclinic
LEHNERITE	Mn[UO2/PO4]2.8H2O	Monoclinic
EVEITE	Mn2(AsO4)(OH)	Orthorhombic
DANNEMORITE	Mn2(Fe++,Mg)5Si8O22(OH)2	Monoclinic
RETZIAN-(Nd)	Mn2(Nd,Ce,La)(AsO4)(OH)4	Orthorhombic
RETZIAN-(Ce)	Mn2Ce(AsO4)(OH)4	Orthorhombic
KEMPITE	Mn2Cl(OH)3	Orthorhombic
TEPHROITE	Mn2SiO4	Orthorhombic
METASWITZERITE	Mn3(PO4)2.4H2O	Monoclinic
SEAMANITE	Mn3(PO4)B(OH)6	Orthorhombic
JIMBOITE	Mn3B2O6	Orthorhombic
CHAMBERSITE	Mn3B7O13Cl	Orthorhombic
WISERITE	Mn4B2O5(OH,Cl)4	Tetragonal
HELVITE	Mn4Be3(SiO4)3S	Isometric
AKROCHORDITE	Mn4Mg(AsO4)2(OH)4.4H2O	Monoclinic
ARSENOCLASITE	Mn5(AsO4)2(OH)4	Orthorhombic
GEIGERITE	Mn5(H2O)8(AsO3OH)2(AsO4)2.2H2O	Triclinic
HUREAULITE	Mn5(PO4)2[PO3(OH)]2.4H2O	Monoclinic
ALLEGHANYITE	Mn5(SiO4)2(OH)2	Monoclinic
PENNANTITE	Mn5Al(Si3Al)O10(OH)8	Monoclinic
ALLACTITE	Mn7(AsO4)2(OH)8	Monoclinic
LEUCOPHOENICITE	Mn7(SiO4)3(OH)2	Monoclinic
KOLICITE	Mn7Zn4(AsO4)2(SiO4)2(OH)8	Orthorhombic
BEMENTITE	Mn8Si6O15(OH)10	Monoclinic
FRIEDELITE	Mn8Si6O15(OH,Cl)10	Monoclinic psTrigonal
JERRYGIBBSITE	Mn9(SiO4)4(OH)2	Orthorhombic
SONOLITE	Mn9(SiO4)4(OH,F)2	Monoclinic
EOSPHORITE	MnAl(PO4)(OH)2.H2O	Monoclinic
APJOHNITE	MnAl2(SO4)4.22H2O	Monoclinic
CARPHOLITE	MnAl2Si2O6(OH)4	Orthorhombic
DAVREUXITE	MnAl6Si4O17(OH)2	Monoclinic
KRAUTITE	MnAs+++++O3(OH).H2O	Monoclinic
VAYRYNENITE	MnBe(PO4)(OH,F)	Monoclinic
SUSSEXITE	MnBO2(OH)	Monoclinic
SCACCHITE	MnCl2	Trigonal
RHODOCHROSITE	MnCO3	Trigonal

Mineral Name	Formula	Crystal System
(DIALOGITE)		
KRYZHANOVSKITE	MnFe+++2(PO4)2(OH)2.H2O	Orthorhombic
MANGANOSITE	MnO	Isometric
MANGANITE	MnO(OH)	Monoclinic
PYROLUSITE	MnO2	Tetragonal
RAMSDELLITE	MnO2	Orthorhombic
ALABANDITE (ALABANDINE)	MnS	Cubic
HAUERITE	MnS2	Isometric
PYROXMANGITE	MnSiO3	Triclinic
TUSIONITE	MnSn++++(BO3)2	Trigonal
JOKOKUITE	MnSO4.5H2O	Triclinic
SZMIKITE	MnSO4.H2O	Monoclinic
MANGANOTANTALITE	MnTa2O6	Orthorhombic
PYROPHANITE	MnTiO3	Trigonal
HUBNERITE (HUEBNERITE)	MnWO4	Monoclinic
HODGKINSONITE	MnZn2SiO4(OH)2	Monoclinic
DRYSDALLITE	Mo(Se,S)2	Hexagonal
ILSEMANNITE	Mo3O8.nH2O	Amorphous
TUGARINOVITE	MoO2	Monoclinic
MOLYBDITE	MoO3	Orthorhombic
SIDWILLITE	MoO3.2H2O	Monoclinic
JORDISITE	MoS2	Amorphous
MOLYBDENITE (MOLYBDENITE-2H)	MoS2	Hexagonal
MOLYBDENITE-3R	MoS2	Trigonal
TSAREGORODTSEVITE	N(CH3)4AlSi5O12	Orthorhombic ps Cubic
JADEITE	Na(Al,Fe+++)Si2O6	Monoclinic
MCKELVEYITE-(Nd)	Na(Ba,Sr)3Ca(Nd,Ce,La)(CO3)6.3H2O	Triclinic
MENDOZAVILITE	Na(Ca,Mg)2Fe+++6(PO4)2(P+++++Mo++++++11 O39)(OH,Cl)10.33H2O	Monoclinic/Triclinic
VIITANIEMIITE	Na(Ca,Mn++)Al(PO4)(F,OH)3	Monoclinic
CARYINITE	Na(Ca,Pb)(Ca,Mn)(Mn,Mg)2(AsO4)3	Monoclinic
CALCJARLITE	Na(Ca,Sr)3Al3(F,OH)16	Monoclinic
CHVILEVAITE	Na(Cu,Fe,Zn)2S2	Hexagonal
NATRODUFRENITE	Na(Fe+++,Fe++)(Fe+++,Al)5(PO4)4(OH)6.2H2O	Monoclinic
HOWIEITE	Na(Fe++,Mn)10(Fe+++,Al)2Si12O31(OH)13	Triclinic
ELBAITE	Na(Li,Al)3Al6(BO3)3Si6O18(OH)4	Trigonal
SODIUMGEDRITE	Na(Mg,Fe++)6Al(Si6Al2)O22(OH)2	Orthorhombic
SODIUMANTHOPHYLLITE	Na(Mg,Fe++)7(Si7Al)O22(OH)2	Orthorhombic
JOHILLERITE	Na(Mg,Zn)3Cu(AsO4)3	Monoclinic
SERANDITE	Na(Mn++,Ca)2Si3O8(OH)	Triclinic
TANEYAMALITE	Na(Mn++,Mg,Fe++)12Si12(O,OH)44	Triclinic
GAINESITE-(NaNa)	Na(Na,K)(Be,Li)Zr2(PO4)4.1,5-2H2O	Tetragonal
ZDENEKITE	Na(Pb,Ca)Cu5(AsO4)4Cl.5H2O	Tetragonal
NASTROPHITE	Na(Sr,Ba)(PO4).9H2O	Isometric
OLGITE	Na(Sr,Ba)PO4	Hexagonal

Mineral Name	Formula	Crystal System
SODIUMZIPPEITE	Na(UO2)2(SO4)(OH)3.H2O	Monoclinic
NATALYITE	Na(V+++,Cr+3)Si2O6	Monoclinic
ODANIELITE	Na(Zn,Mg)3H2(AsO4)3	Monoclinic
SARCOLITE	Na,Ca6Al4Si6O24F	Tetragonal
FOITITE	Na<0,5(Fe++,Al)3Al6Si6O18(BO3)3(OH)4	Trigonal
HECTORITE	Na0,3(Mg,Li)3Si4O10(F,OH)2	Monoclinic
KULKEITE	Na0,35Mg8Al(AlSi7)O20(OH)10	Monoclinic
SCHOLLHORNITE (SCHOELLHORNITE)	Na0,3CrS2.H2O	Trigonal
NONTRONITE	Na0,3Fe+++2(Si,Al)4O10(OH)2.nH2O	Monoclinic
SAUCONITE	Na0,3Zn3(Si,Al)4O10(OH)2.4H2O	Monoclinic
TOSUDITE (CHLORITE-SMECTITE)	Na0,5(Al,Mg)6(Si,Al)8O18(OH)12.5H2O	Orthorhombic
VICANITE-(Ce)	Na0,5(Ce,Ca,Th)15Fe+++As+++0,5As+++++B4Si6O40F7	Trigonal
SALIOTITE	Na0,5Li0,5Al3AlSi3O10(OH)5	Monoclinic
NICKENICHITE	Na0,8Ca0,4(Mg,Fe+++,Al)3Cu0,4(AsO4)3	Monoclinic
SOBOLEVITE	Na11(Na,Ca)4(Mg,Mn)Ti++++4(Si4O12)(PO4)4O5F3	Triclinic
VUONNEMITE	Na11Nb2Ti+++Si4O12(PO4)2O5F2	Triclinic
IMANDRITE	Na12Ca3Fe+++2Si12O36	Orthorhombic
LOWEITE (LOEWEITE)	Na12Mg7(SO4)13.15H2O	Trigonal
QUADRUPHITE	Na14Ca(Mg,Mn)(Ti,Mn,Zr,Nb)4Si4O12(PO4)4O6F2	Triclinic
STEENSTRUPINE-(Ce)	Na14Ce6Mn++Mn+++Fe++2(Zr,Th)(Si6O18)2(PO4)7.3H2O	Trigonal
GALEITE	Na15(SO4)5F4Cl	Trigonal
POLYPHITE	Na17Ca3(Mg,Mn)(Ti,Mn,Zr,Nb)4Si4O12(PO4)6O4F6	Triclinic
ALLUAIVITE	Na19(Ca,Mn++)6(Ti,Nb)3(Si3O9)2(Si10O28)2Cl.2H2O	Trigonal
FERROGLAUCOPHANE	Na2(Fe++,Mg)3Al2Si8O22(OH)2	Monoclinic
RIEBECKITE	Na2(Fe++,Mg)3Fe+++2Si8O22(OH)2	Monoclinic
CROSSITE	Na2(Mg,Fe++)3(Al,Fe+++)2Si8O22(OH)2	Monoclinic
GLAUCOPHANE	Na2(Mg,Fe++)3Al2Si8O22(OH)2	Monoclinic
MAGNESIORIEBECKITE	Na2(Mg,Fe++)3Fe+++2Si8O22(OH)2	Monoclinic
SANEROITE	Na2(Mn++,Mn+++)10Si11VO34(OH)4	Triclinic
NICKELBLODITE (NICKELBLOEDITE)	Na2(Ni,Mg)(SO4)2.4H2O	Monoclinic
GERSTLEYITE	Na2(Sb,As)8S13.2H2O	Monoclinic
LEIFITE	Na2(Si,Al,Be)7(O,OH,F)14	Trigonal
LAMPROPHYLLITE	Na2(Sr,Ba)2Ti3(SiO4)4(OH,F)2	Monoclinic
JARLITE	Na2(Sr,Na)14Al12Mg2F64(OH,H2O)4	Monoclinic
IRTYSHITE	Na2(Ta,Nb)4O11	Hexagonal
FREUDENBERGITE	Na2(Ti,Fe)8O16	Monoclinic ps Hexagonal
NARSARSUKITE	Na2(Ti,Fe+++)Si4(O,F)11	Tetragonal
NATISITE	Na2(TiO)SiO4	Tetragonal

Mineral Data

Mineral Name	Formula	Crystal System
PARANATISITE	Na2(TiO)SiO4	Orthorhombic
SODIUMAUTUNITE	Na2(UO2)2(PO4)2.8H2O	Tetragonal
STRELKINITE	Na2(UO2)2V2O8.6H2O	Orthorhombic
KELDYSHITE	Na2-xHxZrSi2O7.nH2O	Triclinic
METAMUNIRITE	Na2[V+++++2O6]orNaVO3	Orthorhombic
SCHAIRERITE	Na21(SO4)7F6Cl	Trigonal
DANSITE	Na21Mg(SO4)10Cl3	Isometric
MINEEVITE-	Na25Ba(Y,Gd,Dy)2(HCO3)4(CO3)11(SO4)2ClF2	Hexagonal
ABENAKIITE-(Ce)	Na26(Ce,Nd,La,Pr,Th,Sm)6(SiO3)6(PO4)6(CO3)6(S++++O2)O	Trigonal
NATROLITE	Na2Al2Si3O10.2H2O	Orthorhombic
TETRANATROLITE	Na2Al2Si3O10.2H2O	Tetragonal
PARANATROLITE	Na2Al2Si3O10.3H2O	Monoclinic ps Tetragonal
USSINGITE	Na2AlSi3O8(OH)	Triclinic
TEEPLEITE	Na2B(OH)4Cl	Tetragonal
TINCALCONITE	Na2B4O5(OH)4.3H2O	Trigonal
BORAX	Na2B4O5(OH)4.8H2O	Monoclinic
KERNITE	Na2B4O6(OH)2.3H2O	Monoclinic
NASINITE	Na2B5O8(OH).2H2O	Orthorhombic
BIRINGUCCITE	Na2B5O8(OH).H2O	Monoclinic
CHKALOVITE	Na2BeSi2O6	Orthorhombic
NATROFAIRCHILDITE	Na2Ca(CO3)2	Orthorhombic
NYEREREITE	Na2Ca(CO3)2	Orthorhombic ps Hexagonal
PIRSSONITE	Na2Ca(CO3)2.2H2O	Orthorhombic
GAYLUSSITE	Na2Ca(CO3)2.5H2O	Monoclinic
TARAMITE	Na2Ca(Fe++,Mg)3Al2(Si6Al2)O22(OH)2	Monoclinic
ALUMINOKATOPHORITE	Na2Ca(Fe++,Mg)4Al(Si7Al)O22(OH)2	Monoclinic
FERRIKATOPHORITE	Na2Ca(Fe++,Mg)4Fe+++(Si7Al)O22(OH)2	Monoclinic
FERRORICHTERITE	Na2Ca(Fe++,Mg)5Si8O22(OH)2	Monoclinic
MAGNESIOTARAMITE	Na2Ca(Mg,Fe++)3Al2Si6Al2O22(OH)2	Monoclinic
MAGNESIOALUMINOKATOPHORITE	Na2Ca(Mg,Fe++)4Al(Si7Al)O22(OH)2	Monoclinic
MAGNESIOFERRIKATOPHORITE	Na2Ca(Mg,Fe++)4Fe+++Si7AlO22(OH)2	Monoclinic
RICHTERITE	Na2Ca(Mg,Fe++)5Si8O22(OH)2	Monoclinic
CHLADNIITE	Na2Ca(Mg,Fe++)7(PO4)6	Trigonal
JOHNSOMERVILLEITE	Na2Ca(Mg,Fe++,Mn)7(PO4)6	Trigonal
FILLOWITE	Na2Ca(Mn,Fe++)7(PO4)6	Monoclinic
NACAPHITE	Na2Ca(PO4)F	Orthorhombic
GLAUBERITE	Na2Ca(SO4)2	Monoclinic
WATTEVILLITE	Na2Ca(SO4)2.4H2O	Orthorhombic
ANDERSONITE	Na2Ca(UO2)(CO3)3.6H2O	Trigonal
LOVOZERITE	Na2Ca(Zr,Ti)Si6(O,OH)18	Trigonal
SHORTITE	Na2Ca2(CO3)3	Orthorhombic
MESOLITE	Na2Ca2Al6Si9O30.8H2O	Monoclinic
COMBEITE	Na2Ca2Si3O9	Trigonal
HEIDORNITE	Na2Ca3B5O8(SO4)2Cl(OH)2	Monoclinic

Mineral Data

Mineral Name	Formula	Crystal System
MOSANDRITE	Na2Ca4(Ce,Y)(Ti,Zr)(Si2O7)2OF3	Monoclinic
ARCTITE	Na2Ca4(PO4)3F	Trigonal
HAINITE	Na2Ca4(Ti,Zr,Mn,Fe)2(Si2O7)F4	Triclinic
GARRONITE	Na2Ca5Al12Si20O64.27H2O	Orthorhombic ps Tetragonal
LISETITE	Na2CaAl4Si4O16	Orthorhombic
GONNARDITE	Na2CaAl4Si6O20.7H2O	Orthorhombic
BRIANITE	Na2CaMg(PO4)2	Monoclinic
BURPALITE	Na2CaZrSi2O7F2	Monoclinic
SAZHINITE-(Ce)	Na2CeSi6O14(OH).nH2O,(nca.5)	Orthorhombic
JULIENITE	Na2Co++(SCN)4.8H2O	Tetragonal
NATRITE	Na2CO3	Monoclinic
NATRON (SODA)	Na2CO3.10H2O	Monoclinic
THERMONATRITE	Na2CO3.H2O	Orthorhombic
WHEATLEYITE	Na2Cu(C2O4)2.2H2O	Triclinic
CHALCONATRONITE	Na2Cu(CO3)2.3H2O	Monoclinic
KROHNKITE (KROEHNKITE)	Na2Cu(SO4)2.2H2O	Monoclinic
SIDERONATRITE	Na2Fe+++(SO4)2(OH).3H2O	Orthorhombic
METASIDERONATRITE	Na2Fe+++(SO4)2(OH).H2O	Orthorhombic
WILKINSONITE	Na2Fe++4Fe+++2Si6O20	Triclinic
AENIGMATITE	Na2Fe++5TiSi6O20	Triclinic
NAHPOITE	Na2HPO4	Monoclinic
DORFMANITE	Na2HPO4.2H2O	Orthorhombic
SITINAKITE	Na2K(Ti,Nb)4O4(SiO4)2(O,OH).4H2O	Tetragonal
PERRAULTITE	Na2KBaMn++8(Ti,Nb)4Si8O32(OH,F,H2O)7	Monoclinic
EITELITE	Na2Mg(CO3)2	Trigonal
BLODITE (BLOEDITE)	Na2Mg(SO4)2.4H2O	Monoclinic
KONYAITE	Na2Mg(SO4)2.5H2O	Monoclinic
MOTUKOREAITE	Na2Mg38Al24(CO3)13(SO4)8(OH)108.56H2O	Hexagonal
LOUGHLINITE	Na2Mg3Si6O16.8H2O	Orthorhombic
WEBERITE	Na2MgAlF7	Orthorhombic
ARISTARAINITE	Na2MgB12O20.8H2O	Monoclinic
BOBFERGUSONITE	Na2Mn++5Fe+++Al(PO4)6	Monoclinic
JOHNINNESITE	Na2Mn++9(Mg,Mn++)7Si12O34(AsO4)2(OH)8	Triclinic
QINGHEIITE	Na2NaMn2Mg2(Al,Fe+++)2(PO4)6	Monoclinic
FRANCONITE	Na2Nb4O11.9H2O	Monoclinic
KENYAITE	Na2Si22O41(OH)8.6H2O	Monoclinic
NATROSILITE	Na2Si2O5	Monoclinic
REVDITE	Na2Si2O5.5H2O	Triclinic
MAKATITE	Na2Si4O8(OH)2.4H2O	Monoclinic
ERTIXIITE	Na2Si4O9	Isometric
MALLADRITE	Na2SiF6	Trigonal
THENARDITE	Na2SO4	Orthorhombic
MIRABILITE	Na2SO4.10H2O	Monoclinic
BOGVADITE	Na2SrBa2Al4F20	Orthorhombic
LORENZENITE (RAMSAYITE)	Na2Ti2Si2O9	Orthorhombic
BARNESITE	Na2V6O16.3H2O	Monoclinic

Mineral Name	Formula	Crystal System
VOGGITE	Na2Zr(PO4)(CO3)(OH).2H2O	Monoclinic
PARAKELDYSHITE	Na2ZrSi2O7	Triclinic
CATAPLEIITE	Na2ZrSi3O9.2H2O	Hexagonal
GAIDONNAYITE	Na2ZrSi3O9.2H2O	Orthorhombic
HILAIRITE	Na2ZrSi3O9.3H2O	Trigonal
VLASOVITE	Na2ZrSi4O11	Monoclinic/Triclinic
ELPIDITE	Na2ZrSi6O15.3H2O	Orthorhombic
PHOSINAITE	Na3(Ca,Ce)PSiO7	Orthorhombic
CALCIOBURBANKITE	Na3(Ca,Ce,Sr,La,Nd)3(CO3)5	Hexagonal
ROUVILLEITE	Na3(Ca,Mn++)2(CO3)3(F,OH)	Monoclinic
TUNDRITE-(Ce)	Na3(Ce,La)4(Ti,Nb)2(SiO4)2(CO3)3O4(OH).2H2O	Triclinic
REMONDITE-(Ce)	Na3(Ce,La,Ca,Na,Sr)3(CO3)5	Monoclinic ps Hexagonal
VITUSITE-(Ce)	Na3(Ce,La,Nd)(PO4)2	Orthorhombic
TRONA	Na3(CO3)(HCO3).2H2O	Monoclinic
FERROECKERMANNITE	Na3(Fe++,Mg)4AlSi8O22(OH)2	Monoclinic
ARFVEDSONITE	Na3(Fe++,Mg)4Fe+++Si8O22(OH)2	Monoclinic
ECKERMANNITE	Na3(Mg,Fe++)4AlSi8O22(OH)2	Monoclinic
MAGNESIOARFVEDSONITE	Na3(Mg,Fe++)4Fe+++Si8O22(OH)2	Monoclinic
TUNDRITE-(Nd)	Na3(Nd,La)4(Ti,Nb)2(SiO4)2(CO3)3O4(OH).2H2O	Triclinic
DARAPSKITE	Na3(SO4)(NO3).H2O	Monoclinic
KOGARKOITE	Na3(SO4)F	Monoclinic
MURMANITE	Na3(Ti,Nb)4(Si4O12)O4.4H2O	Triclinic
PEISLEYITE	Na3Al16(SO4)2(PO4)10(OH)17.20H2O	Monoclinic
CRYOLITE	Na3AlF6	Monoclinic
CESANITE	Na3Ca2(SO4)3(OH)	Hexagonal
CLINOPHOSINAITE	Na3CaPSiO7	Monoclinic
BONSHTEDTITE	Na3Fe++(PO4)(CO3)	Monoclinic ps Orthorhombic
FERRINATRITE	Na3Fe+++(SO4)3.3H2O	Trigonal
TISINALITE	Na3H3(Mn++,Ca,Fe)TiSi6(O,OH)18.2H2O	Trigonal
CRYOLITHIONITE	Na3Li3Al2F12	Isometric
LINTISITE	Na3LiTi2Si4O14.2H2O	Monoclinic
NORTHUPITE	Na3Mg(CO3)2Cl	Isometric
BRADLEYITE	Na3Mg(PO4)(CO3)	Monoclinic
SIDORENKITE	Na3Mn(PO4)(CO3)	Monoclinic ps Orthorhombic
KOZULITE	Na3Mn4(Fe+++,Al)Si8O22(OH,F)2	Monoclinic
CARACOLITE	Na3Pb2(SO4)3Cl	Monoclinic ps Hexagonal
CRAWFORDITE	Na3Sr(PO4)(CO3)	Monoclinic ps Orthorhombic
UMBOZERITE	Na3Sr4ThSi8(O,OH)24	Amorphous
SHOMIOKITE-	Na3Y(CO3)3.3H2O	Orthorhombic
HYDROXYCANCRINITE	Na4(AlSiO4)3(OH).H2O	Hexagonal
EUDIALYTE	Na4(Ca,Ce)2(Fe++,Mn,Y)ZrSi8O22(OH,Cl)2	Trigonal
GOBBINSITE	Na4(Ca,Mg,K2)Al6Si10O32.12H2O	Orthorhombic ps Tetragonal
KVANEFJELDITE	Na4(Ca,Mn)Si6O14(OH)2	Orthorhombic
PENKVILKSITE	Na4(Ti++++,Zr)Si8O22.4H2O	Monoclinic

Mineral Name	Formula	Crystal System
MARIALITE	Na4Al3Si9O24Cl	Tetragonal
TUGTUPITE	Na4AlBeSi4O12Cl	Tetragonal
EZCURRITE	Na4B10O17.7H2O	Triclinic
EUGSTERITE	Na4Ca(SO4)3.2H2O	Monoclinic
HYDROGLAUBERITE	Na4Ca(SO4)3.2H2O	Monoclinic
CHESSEXITE	Na4Ca2(Mg,Zn)3Al8(SiO4)2(SO4)10(OH)10.40H2O	Orthorhombic
GRIPHITE	Na4Ca6(Mn,Fe++,Mg)19Li2Al8(PO4)24(F,OH)8	Isometric
GRANTSITE	Na4CaxV++++2xV+++++12-2xO32.8H2O	Monoclinic
LAPLANDITE-(Ce)	Na4CeTiPSi7O22.5H2O	Orthorhombic
DAVYNE	Na4K2Ca2Si6Al6O24(SO4)Cl2	Hexagonal
ERSHOVITE	Na4K3(Fe++,Mn++,Ti)2Si8O20(OH)4.4H2O	Triclinic
EMELEUSITE	Na4Li2Fe+++2Si12O30	Orthorhombic ps Hexagonal
KOENENITE	Na4Mg9Al4Cl12(OH)22	Trigonal
HUEMULITE	Na4MgV10O28.24H2O	Triclinic
ZAKHAROVITE	Na4Mn++5Si10O24(OH)6.6H2O	Trigonal
BIRNESSITE	Na4Mn14O27.9H2O	Orthorhombic
RAITE	Na4Mn3Si8(O,OH)24.9H2O	Orthorhombic
SORENSENITE	Na4SnBe2Si6O18.2H2O	Monoclinic
GAULTITE	Na4Zn2Si7O18.5H2O	Orthorhombic
TERSKITE	Na4Zr(H4Si6O18)	Orthorhombic
SABINAITE	Na4Zr2TiO4(CO3)4	Monoclinic
WEGSCHEIDERITE	Na5(CO3)(HCO3)3	Triclinic
CHIOLITE	Na5Al3F14	Tetragonal
NEFEDOVITE	Na5Ca4(PO4)4F	Triclinic
OLYMPITE	Na5Li(PO4)2	Orthorhombic
EPISTOLITE	Na5Ti+++Nb2Si4O12O5F.5H2O	Triclinic
PETARASITE	Na5Zr2Si6O18(Cl,OH).2H2O	Monoclinic
KOASHVITE	Na6(Ca,Mn)(Ti,Fe)Si6O18.H2O	Orthorhombic
ZIRSINALITE	Na6(Ca,Mn,Fe++)ZrSi6O18	Trigonal
BURKEITE	Na6(CO3)(SO4)2	Orthorhombic
NAUJAKASITE	Na6(Fe++,Mn)Al4Si8O26	Monoclinic
MANGANOTYCHITE	Na6(Mn++,Fe++,Mg)2(SO4)(CO3)4	Isometric
KAZAKOVITE	Na6(Mn,H2)TiSi6O18	Trigonal
SULPHOHALITE	Na6(SO4)2FCl	Isometric
ZORITE	Na6(Ti,Nb)2(Si6O17)2(O,OH).11H2O	Orthorhombic
TULIOKITE	Na6BaTh(CO3)6.6H2O	Hexagonal
CANCRINITE	Na6Ca2Al6Si6O24(CO3)2	Hexagonal
FERROTYCHITE	Na6Fe++2(SO4)(CO3)4	Isometric
PITIGLIANOITE	Na6K2Si6Al6O24(SO4).2H2O	Hexagonal
VANTHOFFITE	Na6Mg(SO4)4	Monoclinic
TYCHITE	Na6Mg2(CO3)4(SO4)	Isometric
RIVADAVITE	Na6MgB24O40.22H2O	Monoclinic
KUKISVUMITE	Na6ZnTi4Si8O28.4H2O	Orthorhombic
NATROPHOSPHATE	Na7(PO4)2F.19H2O	Isometric
CANCRISILITE	Na7Al5Si7O24(C03).3H2O	Hexagonal

Mineral Name	Formula	Crystal System
BARENTSITE	Na7AlH2(CO3)4F4	Triclinic
VARENNESITE	Na8(Mn,Fe+++,Ti)2Si10O25(OH,Cl)2.12H2O	Orthorhombic
NOSEAN	Na8Al6Si6O24(SO4)	Isometric
SODALITE	Na8Al6Si6O24Cl2	Isometric
LOMONOSOVITE	Na8Mn++Ti++++3Si4O12(PO4)2O4	Triclinic
VINOGRADOVITE	Na8Ti8O8(Si2O6)4[(Si3Al)O10]2[(H2O),(Na,K)2]	Monoclinic
HECTORFLORESITE	Na9(IO3)(SO4)4	Monoclinic
DURANGITE	NaAl(AsO4)F	Monoclinic
DAWSONITE	NaAL(CO3)(OH)2	Orthorhombic
LACROIXITE	NaAl(PO4)F	Monoclinic
MENDOZITE	NaAl(SO4)2.11H2O	Monoclinic
SODIUMALUM	NaAl(SO4)2.12H2O	Isometric
TAMARUGITE	NaAl(SO4)2.6H2O	Monoclinic
DIAOYUDAOITE	NaAl11O17	Hexagonal
PARAGONITE	NaAl2(Si3Al)O10(OH)2	Monoclinic
BRAZILIANITE	NaAl3(PO4)2(OH)4	Monoclinic
WARDITE	NaAl3(PO4)2(OH)4.2H2O	Tetragonal
NATROALUNITE	NaAl3(SO4)2(OH)6	Trigonal
OLENITE	NaAl3Al6(BO3)3(Si6O18)(O,OH)4	Trigonal
PARAMENDOZAVILITE	NaAl4Fe+++7(PO4)5(P+++++Mo++++++12O40)(OH)16.56H2O	Monoclinic/Triclinic
VANALITE	NaAl8V10O38.30H2O	Monoclinic
ANALCIME (ANALCITE)	NaAlSi2O6.H2O	Isometric
ALBITE	NaAlSi3O8	Triclinic
WYCHEPROOFITE	NaAlZr(PO4)2(OH)2.H2O	Triclinic
AMEGHINITE	NaB3O3(OH)4	Monoclinic
SBORGITE	NaB5O6(OH)4.3H2O	Monoclinic
BYELORUSSITE-(Ce) (BELORUSSITE-(Ce))	NaBa2(Ce,La)2Mn++Ti2Si8O26(F,OH).H2O	Orthorhombic
STRAKHOVITE	NaBa3(Mn++,Mn+++)4Si6O19(OH)3	Monoclinic
NABAPHITE	NaBaPO4.9H2O	Isometric
SWEDENBORGITE	NaBe4SbO7	Hexagonal
BERYLLONITE	NaBePO4	Monoclinic
EPIDIDYMITE	NaBeSi3O7(OH)	Orthorhombic
EUDIDYMITE	NaBeSi3O7(OH)	Monoclinic
FERRUCCITE	NaBF4	Orthorhombic
SEARLESITE	NaBSi2O5(OH)2	Monoclinic
REEDMERGNERITE	NaBSi3O8	Triclinic
FERROBARROISITE	NaCa(Fe++,Mg)3Al2(Si7Al)O22(OH)2	Monoclinic
FERROWINCHITE	NaCa(Fe++,Mg)4AlSi8O22(OH)2	Monoclinic
BARROISITE	NaCa(Mg,Fe++)3Al2(Si7Al)O22(OH)2	Monoclinic
WINCHITE	NaCa(Mg,Fe++)4AlSi8O22(OH)2	Monoclinic
EDENITE	NaCa(Mg,Fe++)5(Si7Al)O22(OH)2	Monoclinic
NORMANDITE	NaCa(Mn++,Fe++)(Ti,Nb,Zr)Si2O7(O,F)2	Monoclinic
ZINCLAVENDULAN	NaCa(Zn,Cu)5(AsO4)4Cl.4-5H2O	Orthorhombic
GYROLITE	NaCa16Si23AlO60(OH)8.64H2O	Triclinic ps Hexagonal

Mineral Data

Mineral Name	Formula	Crystal System
BOGGSITE	NaCa2(Al5Si19O48).17H2O	Orthorhombic
FERROPARGASITE	NaCa2(Fe++,Mg)4Al(Si6Al2)O22(OH)2	Monoclinic
HASTINGSITE	NaCa2(Fe++,Mg)4Fe+++(Si6Al2)O22(OH)2	Monoclinic
FERROKAERSUTITE	NaCa2(Fe++,Mg)4Ti(Si6Al2)O22(OH)2	Monoclinic
FERROEDENITE	NaCa2(Fe++,Mg)5(Si7Al)O22(OH)2	Monoclinic
WICKSITE	NaCa2(Fe++,Mn++)4MgFe+++(PO4)6.2H2O	Orthorhombic
PARGASITE	NaCa2(Mg,Fe++)4Al(Si6Al2)O22(OH)2	Monoclinic
MAGNESIOHASTINGSITE	NaCa2(Mg,Fe++)4Fe+++(Si6Al2)O22(OH)2	Monoclinic
KAERSUTITE	NaCa2(Mg,Fe++)4Ti(Si6Al2)O22(OH)2	Monoclinic
WOHLERITE (WOEHLERITE)	NaCa2(Zr,Nb)Si2O7(O,OH,F)2	Monoclinic
MORINITE	NaCa2Al2(PO4)2(F,OH)5.2H2O	Monoclinic
TUNISITE	NaCa2Al4(CO3)4(OH)8Cl	Tetragonal
STILBITE (DESMINE)	NaCa2Al5Si13O36.14H2O	Monoclinic
THOMSONITE	NaCa2Al5Si5O20.6H2O	Orthorhombic
STUDENITSITE	NaCa2B9O14(OH)4.2H2O	Monoclinic
GIRVASITE	NaCa2Mg3(PO4)2[PO2(OH)2](CO3)(OH)2.4H2O	Monoclinic
GRISCHUNITE	NaCa2Mn++5Fe+++(AsO4)6.2H2O	Orthorhombic
PECTOLITE	NaCa2Si3O8(OH)	Triclinic
PECTOLITE-M2abc	NaCa2Si3O8(OH)	Monoclinic
AGRELLITE	NaCa2Si4O10F	Triclinic
SCHROCKINGERITE (SCHROECKINGERITE)	NaCa3(UO2)(CO3)3(SO4)F.10H2O	Triclinic
NATROAPOPHYLLITE	NaCa4Si8O20F.8H2O	Orthorhombic
UNGURSAITE =SODIANCALCIOTANTITE	NaCa5(Ta,Nb)24O65(OH)	Hexagonal
MCNEARITE	NaCa5H4(AsO4)5.4H2O	Triclinic
LOUDOUNITE	NaCa5Zr4Si16O40(OH)11.8H2O	
PACHNOLITE	NaCaAlF6.H2O	Monoclinic
THOMSENOLITE	NaCaAlF6.H2O	Monoclinic
TUZLAITE	NaCaB508(0H)2.3H2O	Monoclinic
ULEXITE	NaCaB5O6(OH)6.5H2O	Triclinic
PROBERTITE	NaCaB5O7(OH)4.3H2O	Monoclinic
LAVENDULAN	NaCaCu5(AsO4)4Cl.5H2O	Orthorhombic
SAMPLEITE	NaCaCu5(PO4)4Cl.5H2O	Orthorhombic
FERROALLUAUDITE	NaCaFe++(Fe++,Mn,Fe+++,Mg)2(PO4)3	Monoclinic
ALLUAUDITE	NaCaFe++(Mn,Fe++,Fe+++,Mg)2(PO4)3	Monoclinic
FERRIWINCHITE	NaCaMg4Fe+++Si8O22(OH)2	Monoclinic
BOLDYREVITE	NaCaMgAl3F14.4H2O	Amorphous
HAGENDORFITE	NaCaMn(Fe++,Fe+++,Mg)2(PO4)3	Monoclinic
VARULITE	NaCaMn(Mn,Fe++,Fe+++)2(PO4)3	Monoclinic
ARSENIOPLEITE	NaCaMn(Mn,Mg)2(AsO4)3	Monoclinic
MARSTURITE	NaCaMn3Si5O14(OH)	Triclinic
BUCHWALDITE	NaCaPO4	Orthorhombic
GAGARINITE-	NaCaY(F,Cl)6	Hexagonal
HALITE (ROCKSALT,SELGEMME)	NaCl	Isometric
HYDROHALITE	NaCl.2H2O	Monoclinic

Mineral Name	Formula	Crystal System
KOSMOCHLOR	NaCr+++Si2O6	Monoclinic
CASWELLSILVERITE	NaCrS2	Trigonal
MCCRILLISITE	NaCs(Be,Li)Zr2(PO4)4.1-2H2O	Tetragonal
HOWARDEVANSITE	NaCu++Fe+++2(VO4)3	Triclinic
NATROCHALCITE	NaCu2(SO4)2(OH).H2O	Monoclinic
VILLIAUMITE	NaF	Isometric
AMARILLITE	NaFe++(SO4)2.6H2O	Monoclinic
CYRILOVITE	NaFe+++3(PO4)2(OH)4.2H2O	Tetragonal
NATROJAROSITE	NaFe+++3(SO4)2(OH)6	Trigonal
BUERGERITE	NaFe+++3Al6(BO3)3Si6O21F	Trigonal
TUPERSSUATSIAITE	NaFe+++3Si8O20.5H2O	Monoclinic
KIDWELLITE	NaFe+++9(PO4)6(OH)10.5H2O	Monoclinic
AEGIRINE (ACMITE)	NaFe+++Si2O6	Monoclinic
SCHORL	NaFe++3Al6(BO3)3Si6O18(OH)4	Trigonal
MARICITE	NaFe++PO4	Orthorhombic
COYOTEITE	NaFe3S5.2H2O	Triclinic
ERDITE	NaFeS2.2H2O	Monoclinic
QILIANSHANITE	NaH4(CO3)(BO3).2H2O	Monoclinic
NAHCOLITE	NaHCO3	Monoclinic
KANEMITE	NaHSi2O4(OH)2.2H2O	Orthorhombic
GRUMANTITE	NaHSi2O5.H2O	Orthorhombic
MATTEUCCITE	NaHSO4.H2O	Monoclinic
SELWYNITE	NaK(Be,Al)Zr2(PO4)4.2H2O	Tetragonal
NALIPOITE	NaLi2PO4	Orthorhombic
EPHESITE	NaLiAl2(Al2Si2)O10(OH)2	Monoclinic/Triclinic
SILINAITE	NaLiSi2O5.2H2O	Monoclinic
ZEKTZERITE	NaLiZrSi6O15	Orthorhombic
ZHEMCHUZHNIKOVITE	NaMg(Al,Fe+++)(C2O4)3.8H2O	Trigonal
UKLONSKOVITE	NaMg(SO4)F.2H2O	Monoclinic
PREISWERKITE	NaMg2Al3Si2O10(OH)2	Monoclinic
KRINOVITE	NaMg2CrSi3O10	Triclinic
SLAVIKITE	NaMg2Fe+++5(SO4)7(OH)6.33H2O	Trigonal
CHROMDRAVITE	NaMg3(Cr,Fe+++)6(BO3)3Si6O18(OH)4	Trigonal
DRAVITE	NaMg3Al6(BO3)3Si6O18(OH)4	Trigonal
SODIUMPHLOGOPITE	NaMg3Si3AlO10(OH)2	Monoclinic
NEIGHBORITE	NaMgF3	Orthorhombic
STEPANOVITE	NaMgFe+++(C2O4)3.8-9H2O	Trigonal
MAGHAGENDORFITE	NaMgMn(Fe++,Fe+++)2(PO4)3	Monoclinic
NAMANSILITE	NaMn+++(Si2O6)	Monoclinic
SHIGAITE	NaMn++6Al3(SO4)2(OH)18.12H20	Trigonal
SVERIGEITE	NaMnMgSn++++Be2Si3O12(OH)	Orthorhombic
NATROPHILITE	NaMnPO4	Orthorhombic
LANDAUITE	NaMnZn2(Ti,Fe+++)6Ti12O38	Monoclinic psTrigonal
LEAKEITE	NaNa2(Mg2Fe+++2Li)Si8O22(OH)2	Monoclinic
NYBOITE	NaNa2Mg3Al2(Si7Al)O22(OH)2	Monoclinic
LUESHITE	NaNbO3	Monoclinic ps Cubic

Mineral Name	Formula	Crystal System
NATRONIOBITE	NaNbO3	Monoclinic
KAMBALDAITE	NaNi4(CO3)3(OH)3.3H2O	Hexagonal
NITRATINE (SODANITER,NITRONATRITE)	NaNO3	Trigonal
MOPUNGITE	NaSb(OH)6	Tetragonal
BRIZZIITE	NaSb+++++O3	Trigonal
MAGADIITE	NaSi7O13(OH)3.4H2O	Monoclinic
NATROTANTITE	NaTa3O8	Monoclinic
MUNIRITE	NaVO3.(2-x)H2O	Monoclinic
RALSTONITE	NaxMgxAl2-x(F,OH)6.H2O	Isometric
NACARENIOBSITE-(Ce)	NbNa3Ca3(Ce,La,Pr,Nd)(Si2O7)2OF3	Monoclinic
CHURCHITE-(Nd)	Nd(PO4).2H2O	Monoclinic
SALAMMONIAC (SALMIAC)	NH4Cl	Isometric
NICKEL	Ni	Isometric
THEOPHRASTITE	Ni(OH)2	Trigonal
PARAOTWAYITE	Ni(OH)2-x(SO4,CO3)0,5x	Monoclinic
ABELSONITE	Ni++[C32H36N4]	Triclinic
NICKELZIPPEITE	Ni++0,5(UO2)2(SO4)(OH)3.H2O	Monoclinic
MAUCHERITE	Ni11As8	Tetragonal
AERUGITE	Ni17As6O32	Trigonal
ARSENOHAUCHECORNITE	Ni18Bi3AsS16	Tetragonal
NULLAGINITE	Ni2(CO3)(OH)2	Monoclinic
OTWAYITE	Ni2(CO3)(OH)2.H2O	Orthorhombic
BONACCORDITE	Ni2Fe+++BO5	Orthorhombic
OREGONITE	Ni2FeAs2	Hexagonal
AWARUITE	Ni2FetoNi3Fe	Hexagonal
XANTHIOSITE	Ni3(AsO4)2	Monoclinic
ANNABERGITE	Ni3(AsO4)2.8H2O	Monoclinic
PARKERITE	Ni3(Bi,Pb)2S2	Monoclinic
ZARATITE	Ni3(CO3)(OH)4.4H2O	Isometric
DIENERITE	Ni3As	Isometric
HEAZLEWOODITE	Ni3S2	Trigonal
TRUSTEDTITE (TRUESTEDTITE)	Ni3Se4	Isometric
WILKMANITE	Ni3Se4	Monoclinic
NEPOUITE	Ni3Si2O5(OH)4	Monoclinic
PECORAITE	Ni3Si2O5(OH)4	Monoclinic
ORCELITE	Ni5As2	Hexagonal
TAKOVITE	Ni6Al2(OH)16(CO3,OH).4H2O	Trigonal
REEVESITE	Ni6Fe+++2(CO3)(OH)16.4H2O	Trigonal
HONESSITE	Ni6Fe+++2(SO4)(OH)16.4H2O	Trigonal
HYDROHONESSITE	Ni6Fe+++2(SO4)(OH)16.7H2O	Hexagonal
DONHARRISITE	Ni8Hg3S9	Monoclinic
HAUCHECORNITE	Ni9Bi(Sb,Bi)S8	Tetragonal
BISMUTOHAUCHECORNITE	Ni9Bi2S8	Tetragonal
TELLUROHAUCHECORNITE	Ni9BiTeS8	Tetragonal

Mineral Name	Formula	Crystal System
TUCEKITE	Ni9Sb2S8	Tetragonal
NICKELINE (NICCOLITE)	NiAs	Hexagonal
KRUTOVITE	NiAs2	Isometric
PARARAMMELSBERGITE	NiAs2	Orthorhombic
RAMMELSBERGITE	NiAs2	Orthorhombic
GERSDORFFITE	NiAsS	Isometric
NICKELBISCHOFITE	NiCl2.6H2O	Monoclinic
HELLYERITE	NiCO3.6H2O	Triclinic
TREVORITE	NiFe+++2O4	Isometric
ERNIENICKELITE	NiMn++++3O7.3H2O	Trigonal
POLYDYMITE	NiNi2S4	Isometric
BUNSENITE	NiO	Isometric
MILLERITE	NiS	Trigonal
VAESITE	NiS2	Isometric
BREITHAUPTITE	NiSb	Hexagonal
BOTTINOITE	NiSb+++++2(OH)12.6H2O	Hexagonal
NISBITE	NiSb2	Orthorhombic
ULLMANNITE	NiSbS	Triclinic ps Cubic
MAKINENITE (MAEKINENITE)	NiSe	Trigonal
SEDERHOLMITE	NiSe	Hexagonal
KULLERUDITE	NiSe2	Orthorhombic
RETGERSITE	NiSO4.6H2O	Tetragonal
MORENOSITE	NiSO4.7H2O	Orthorhombic
MELONITE	NiTe2	Trigonal
KITKAITE	NiTeSe	Trigonal
ERLICHMANITE	OsS2	Isometric
LEAD (PLOMB,BLEI)	Pb	Isometric
USTARASITE	Pb(Bi,Sb)6S10	Orthorhombic
MARGAROSANITE	Pb(Ca,Mn++)2Si3O9	Triclinic
GARTRELLITE	Pb(Cu++,Fe++)2(AsO4,SO4)2(CO3,H2O)0,7	Triclinic
THOMETZEKITE	Pb(Cu,Zn)2(AsO4)2.2H2O	Monoclinic/Triclinic
LAURELITE	Pb(F,Cl,OH)2	Hexagonal
MAGNETOPLUMBITE	Pb(Fe+++,Mn+++)12O19	Hexagonal
MAWBYITE	Pb(Fe+++Zn)2(AsO4)2(OH,H2O)2	Monoclinic
CECHITE	Pb(Fe++,Mn)(VO4)(OH)	Orthorhombic
CORONADITE	Pb(Mn++++,Mn++)8O16	Monoclinic ps Tetragonal
GYSINITE-(Nd)	Pb(Nd,La)(CO3)2(OH).H2O	Orthorhombic
SURITE	Pb(Pb,Ca)(Al,Fe+++,Mg)2(Si,Al)4O10(OH)2(CO3)2	Monoclinic
GUETTARDITE	Pb(Sb,As)2S4	Monoclinic
TWINNITE	Pb(Sb,As)2S4	Orthorhombic
SCHIEFFELINITE	Pb(Te++++++,S)O4.H2O	Orthorhombic
SENAITE	Pb(Ti,Fe,Mn)21O38	Trigonal
MOCTEZUMITE	Pb(UO2)(TeO3)2	Monoclinic
PRZHEVALSKITE	Pb(UO2)2(PO4)2.4H2O	Orthorhombic
CURIENITE	Pb(UO2)2V2O8.5H2O	Orthorhombic

Mineral Data

Mineral Name	Formula	Crystal System
FOURMARIERITE	Pb(UO2)4O3(OH)4.4H2O	Orthorhombic
KASOLITE	Pb(UO2)SiO4.H2O	Monoclinic
JIXIANITE	Pb(W,Fe+++)2(O,OH)7	Isometric
MINIUM	Pb++2Pb++++O4	Tetragonal
ROSIERESITE	Pb,Cu,Al,PO4,H2O	Amorphous
PLUMBONACRITE	Pb10(CO3)6O(OH)6	Hexagonal
SUNDIUSITE	Pb10(SO4)Cl2O8	Monoclinic
DADSONITE	Pb10+xSb14-xS31-xClx	Triclinic
HEYROVSKYITE	Pb10AgBi5S18	Orthorhombic
KIRKIITE	Pb10Bi3As3S19	Orthorhombic/Monoclinic
IRANITE	Pb10Cu(CrO4)6(SiO4)2(F,OH)2	Triclinic
HEMIHEDRITE	Pb10Zn(CrO4)6(SiO4)2F2	Triclinic
BAUMHAUERITE-2a	Pb11Ag0,7As17.2Sb0,4S36or(Pb,Ag)3As4S9	Monoclinic
MENEGHINITE	Pb13CuSb7S24	Orthorhombic
JORDANITE	Pb14(As,Sb)6S23	Monoclinic
SAHLINITE	Pb14(AsO4)2O9Cl4	Monoclinic
GEOCRONITE	Pb14(Sb,As)6S23	Monoclinic
KOMBATITE	Pb14(VO4)2O9Cl4	Monoclinic
OURAYITE-P	Pb14Ag18Bi28S65	Orthorhombic
FIZELYITE	Pb14Ag5Sb21S48	Monoclinic
GEORGIADESITE	Pb16(AsO4)4Cl14O2(OH)2orPb16(AsO4)4Cl14(OH)6	Monoclinic
PLAYFAIRITE	Pb16Sb18S43	Monoclinic
MADOCITE	Pb17(Sb,As)16S41	Orthorhombic
HYTTSJOITE	Pb18Ba2Ca5Mn++2Fe++2(Si15O30)2Cl.6H2O	Trigonal
SORBYITE	Pb19(Sb,As)20S49	Monoclinic
ARDAITE	Pb19Sb13S35Cl7	Monoclinic
PHOSGENITE	Pb2(CO3)Cl2	Tetragonal
PHOENICOCHROITE	Pb2(CrO4)O	Monoclinic
DRUGMANITE	Pb2(Fe+++,Al)H(PO4)2(OH)2	Monoclinic
BURCKHARDTITE	Pb2(Fe+++,Mn+++)Te++++(AlSi3)O12(OH)2.H2O	Monoclinic ps Hexagonal
ARSENBRACKEBUSCHITE	Pb2(Fe++,Zn)(AsO4)2.H2O	Monoclinic
PLUMBOFERRITE	Pb2(Fe,Mn++,Mg)11O19	Hexagonal
LINDQVISTITE	Pb2(Mn++,Mg)2Fe+++15-16O27	Hexagonal
BRACKEBUSCHITE	Pb2(Mn,Fe++)(VO4)2.H2O	Monoclinic
VEENITE	Pb2(Sb,As)2S5	Orthorhombic
OLSACHERITE	Pb2(SeO4)(SO4)	Orthorhombic
LANARKITE	Pb2(SO4)O	Monoclinic
HALLIMONDITE	Pb2(UO2)(AsO4)2	Triclinic
WIDENMANNITE	Pb2(UO2)(CO3)3	Orthorhombic
PARSONSITE	Pb2(UO2)(PO4)2.2H2O	Triclinic
HUGELITE (HUEGELITE)	Pb2(UO2)3(AsO4)2(OH)4.3H2O	Monoclinic
DUMONTITE	Pb2(UO2)3O2(PO4)2.5H2O	Monoclinic
SAYRITE	Pb2(UO2)5O6(OH)2.4H2O	Monoclinic
JASKOLSKIITE	Pb2+xCux(Sb,Bi)2-xS5,x=0,2	Orthorhombic

Mineral Data

Mineral Name	Formula	Crystal System
MATTHEDDLEITE	Pb20(SiO4)7(SO4)4Cl4	Hexagonal
CUMENGITE (CUMENGEITE)	Pb21Cu20Cl42(OH)40	Tetragonal
TINTINAITE	Pb22Cu+4(Sb,Bi)30S69	Orthorhombic
KOBELLITE	Pb22Cu4(Bi,Sb)30S69	Orthorhombic
LAUNAYITE	Pb22Sb26S61	Monoclinic
BOLEITE	Pb26Ag10Cu24Cl62(OH)48.3H2O	Isometric
IZOKLAKEITE	Pb27(Cu,Fe)2(Sb,Bi)19S57	Orthorhombic
DIAPHORITE	Pb2Ag3Sb3S8	Monoclinic
BIDEAUXITE	Pb2AgCl3(F,OH)2	Isometric
PAULMOOREITE	Pb2As+++2O5	Monoclinic
DUFRENOYSITE	Pb2As2S5	Monoclinic
COSALITE	Pb2Bi2S5	Orthorhombic
ROEBLINGITE	Pb2Ca6(Si6O18)(SO4)2(OH)2.4H2O	Monoclinic
BLIXITE	Pb2Cl(O,OH)2	Orthorhombic
PENFIELDITE	Pb2Cl3(OH)	Hexagonal
SHANNONITE	Pb2CO3O	Orthorhombic
ARSENTSUMEBITE	Pb2Cu(AsO4)(SO4)(OH)	Monoclinic
VAUQUELINITE	Pb2Cu(CrO4)(PO4)(OH)	Monoclinic
NUFFIELDITE	Pb2Cu(Pb,Bi)Bi2S7	Orthorhombic
TSUMEBITE	Pb2Cu(PO4)(SO4)(OH)	Monoclinic
SCHMIEDERITE (SCHMEIDERITE)	Pb2Cu++2(Se++++O3)(Se++++++O4)(OH)4	Monoclinic
MOLYBDOFORNACITE	Pb2Cu[(As,P)O4][(Mo,Cr)O4](OH)	Monoclinic
HAMMARITE	Pb2Cu2Bi4S9	Orthorhombic
CREASEYITE	Pb2Cu2Fe+++2Si5O17.6H2O	Orthorhombic
LUDDENITE	Pb2Cu2Si5O14.14H2O	Monoclinic
DUHAMELITE	Pb2Cu4Bi(VO4)4(OH)3.8H2O	Orthorhombic
DEMESMAEKERITE	Pb2Cu5(UO2)2(SeO3)6(OH)6.2H2O	Triclinic
DIABOLEITE	Pb2CuCl2(OH)4	Tetragonal
MELANOTEKITE	Pb2Fe+++2Si2O9	Orthorhombic
EZTLITE	Pb2Fe+++6(Te++++O3)3(Te++++++O6)(OH)10.8 H2O	Monoclinic
MOLYBDOPHYLLITE	Pb2Mg2Si2O7(OH)2	Trigonal
ROUSEITE	Pb2Mn++(As+++O3)2.2H2O	Triclinic
KENTROLITE	Pb2Mn+++2Si2O9	Orthorhombic
SHANDITE	Pb2Ni3S2	Trigonal ps Cubic
RHODPLUMSITE	Pb2Rh3S2	Trigonal
BINDHEIMITE	Pb2Sb2O6(O,OH)	Isometric
PLUMOSITE	Pb2Sb2S5	Monoclinic
GRANDREEFITE	Pb2SO4F2	Monoclinic
CURITE	Pb2U5O17.4H2O	Orthorhombic
CHERVETITE	Pb2V2O7	Monoclinic
JAMESITE	Pb2Zn2Fe+++5(AsO4)5O4	Triclinic
BERRYITE	Pb3(Ag,Cu)5Bi7S16	Monoclinic
HYDROCERUSSITE	Pb3(CO3)2(OH)2	Trigonal
ZENZENITE	Pb3(Fe+++,Mn+++)4Mn++++3O15	Hexagonal

Mineral Data

Mineral Name	Formula	Crystal System
THORIKOSITE	Pb3(Sb+++,As+++)O3(OH)Cl2	Tetragonal
DEWINDTITE	Pb3[H(UO2)3O2(PO4)2]2.12H2O	Orthorhombic
ARAVAIPAITE	Pb3AlF9.H2O	Triclinic
BAUMHAUERITE	Pb3As4S9	Triclinic
LILLIANITE	Pb3Bi2S6	Orthorhombic
XILINGOLITE	Pb3Bi2S6	Monoclinic
WITTITE	Pb3Bi4(S,Se)9	Monoclinic
HEDYPHANE	Pb3Ca2(AsO4)3Cl	Hexagonal
WICKENBURGITE	Pb3CaAl2Si10O27.3H2O	Trigonal
MENDIPITE	Pb3Cl2O2	Orthorhombic
SEELIGERITE	Pb3Cl3(IO3)O	Orthorhombic
FIEDLERITE	Pb3Cl4(OH)2	Monoclinic
JUNOITE	Pb3Cu2Bi8(S,Se)16	Monoclinic
LINDSTROMITE (LINDSTROEMITE)	Pb3Cu3Bi7S15	Orthorhombic
NORDSTROMITE (NORDSTROEMITE)	Pb3CuBi7(S10Se4)	Monoclinic
CHLOROXIPHITE	Pb3CuCl2(OH)2O2	Monoclinic
JAGOITE	Pb3Fe+++Si4O12(Cl,OH)	Hexagonal
FLEISCHERITE	Pb3Ge(SO4)2(OH)6.3H2O	Hexagonal
ITOITE	Pb3Ge(SO4)2O2(OH)2	Orthorhombic
GIRDITE	Pb3H2(Te++++O3)(Te++++++O6)	Monoclinic
TRIGONITE	Pb3Mn(As+++O3)2(As+++O2OH)	Monoclinic
FULOPPITE	Pb3Sb8S15	Monoclinic
CYLINDRITE	Pb3Sn4FeSb2S14	Triclinic
MASUYITE	Pb3U++++++8O27.10H2O	Orthorhombic
PINALITE	Pb3WO5Cl2	Orthorhombic
DUGGANITE	Pb3Zn3Te(As,V,Si)2(O,OH)14	Hexagonal
BENAVIDESITE	Pb4(Mn,Fe)Sb6S14	Monoclinic
LEADHILLITE	Pb4(SO4)(CO3)2(OH)2	
MACPHERSONITE	Pb4(SO4)(CO3)2(OH)2	Orthorhombic
SUSANNITE	Pb4(SO4)(CO3)2(OH)2	Trigonal
POTOSIITE	Pb48Sn++++18Fe++7Sb+++16S115	Triclinic
PLUMALSITE	Pb4Al2(SiO3)7	Orthorhombic
CANNIZZARITE	Pb4Bi5(S,Se)11,5	Monoclinic
ELYITE	Pb4Cu(SO4)(OH)8	Monoclinic
CHENITE	Pb4Cu(SO4)2(OH)6	Triclinic
NEALITE	Pb4Fe++(As+++++O3)2Cl4.2H2O	Triclinic
HEMATOPHANITE	Pb4Fe+++3O8(OH,Cl)	Tetragonal
JAMESONITE	Pb4FeSb6S14	Monoclinic
PARAJAMESONITE	Pb4FeSb6S14	Orthorhombic
DAMARAITE	Pb4O3Cl2	Orthorhombic
ROBINSONITE	Pb4Sb6S13	Triclinic
QUEITITE	Pb4Zn2(SiO4)(Si2O7)(SO4)	Monoclinic
FINNEMANITE	Pb5(As+++O3)3Cl	Hexagonal
CLINOMIMETITE	Pb5(AsO4)3Cl	Monoclinic ps Hexagonal

Mineral Name	Formula	Crystal System
MIMETITE	Pb5(AsO4)3Cl	Hexagonal
EMBREYITE	Pb5(CrO4)2(PO4)2.H2O	Monoclinic
PYROMORPHITE	Pb5(PO4)3Cl	Hexagonal
CASSEDANNEITE	Pb5(VO4)2(CrO4)2.H2O	Monoclinic
VANADINITE	Pb5(VO4)3Cl	Hexagonal
BURSAITE	Pb5Bi4S11	Monoclinic
CALEDONITE	Pb5Cu2(CO3)(SO4)3(OH)6	Orthorhombic
PSEUDOBOLEITE	Pb5Cu4Cl10(OH)8.2H2O	Tetragonal
FRIEDRICHITE	Pb5Cu5Bi7S18	Orthorhombic
HEYITE	Pb5Fe++2(VO4)2O4	Monoclinic
BOULANGERITE	Pb5Sb4S11	Monoclinic
FALKMANITE	Pb5Sb4S11	Monoclinic
PLAGIONITE	Pb5Sb8S17	Monoclinic
PLUMBOTSUMITE	Pb5Si4O8(OH)10	Orthorhombic
FRANCKEITE	Pb5Sn3Sb2S14	Triclinic
LENGENBACHITE	Pb6(Ag,Cu)2As4S13	Triclinic
SCHWARTZEMBERGITE	Pb6(IO3)2Cl4O2(OH)2	Orthorhombic ps Tetragonal
ECDEMITE	Pb6As+++2O7Cl4	Tetragonal
HELIOPHYLLITE	Pb6As2O7Cl4	Orthorhombic
ASCHAMALMITE	Pb6Bi2S9	Monoclinic
NASONITE	Pb6Ca4Si6O21Cl2	Hexagonal
YEDLINITE	Pb6CrCl6(O,OH)8	Trigonal
MAMMOTHITE	Pb6Cu4AlSb+++++O2(SO4)2Cl4(OH)16	Monoclinic
OBOYERITE	Pb6H6(Te++++O3)3(Te++++++O6)2.2H2O	Triclinic
PSEUDOGRANDREEFITE	Pb6SO4F1O	Orthorhombic ps Tetragonal
NEYITE	Pb7(Cu,Ag)2Bi6S17	Monoclinic
MATHEWROGERSITE	Pb7(Fe,Cu)Al3GeSi12O36.(OH,H2O)6	Trigonal
OWYHEEITE	Pb7Ag2(Sb,Bi)8S20	Orthorhombic
MARICOPAITE	Pb7Ca2(Si,Al)48O100.32H2O	Orthorhombic
MACQUARTITE	Pb7Cu2(CrO4)4(SiO4)2(OH)2.H2O	Monoclinic
WHERRYITE	Pb7Cu2(SO4)4(SiO4)2(OH)2	Monoclinic
CHUBUTITE	Pb7O6Cl2	Orthorhombic ps Tetragonal
HETEROMORPHITE	Pb7Sb8S19	Monoclinic
ASISITE	Pb7SiO8Cl2	Tetragonal
GEBHARDITE	Pb8(As+++2O5)2OCl6	Monoclinic
KEGELITE	Pb8Al4Si8(SO4)2(CO3)4(OH)8O20	Monoclinic ps Hexagonal
FREEDITE	Pb8Cu+(As+++O3)2O3Cl5	Monoclinic
BARYSILITE	Pb8Mn(Si2O7)3	Trigonal
LEVYCLAUDITE	Pb8Sn7Cu3(Bi,Sb)3S28	
ECLARITE	Pb9(Cu,Fe)Bi12S28	Orthorhombic
LIVEINGITE	Pb9As13S28	Monoclinic
GRATONITE	Pb9As4S15	Trigonal
GANOMALITE	Pb9Ca5Mn++Si9O33	Hexagonal
ZINKENITE	Pb9Sb22S42	Hexagonal
SEMSEYITE	Pb9Sb8S21	Monoclinic
MARRITE	PbAgAsS3	Monoclinic

Mineral Name	Formula	Crystal System
GUSTAVITE	PbAgBi3S6	Orthorhombic
ANDORITE	PbAgSb3S6	Orthorhombic
KAMITUGAITE	PbAl(UO2)5[(P,As)O4]2(OH)9.9.5H2O	Triclinic
DUNDASITE	PbAl2(CO3)2(OH)4.H2O	Orthorhombic
HIDALGOITE	PbAl3(AsO4)(SO4)(OH)6	Trigonal
PHILIPSBORNITE	PbAl3(AsO4)2(OH)5.H2O	Trigonal
PLUMBOGUMMITE	PbAl3(PO4)2(OH)5.H2O	Trigonal
ORPHEITE	PbAl3(PO4,SO4)2(OH)6	Trigonal
ARTROEITE	PbAlF3(OH)2	Triclinic
SARTORITE	PbAs2S4	Monoclinic
GALENOBISMUTITE	PbBi2S4	Orthorhombic
POUBAITE	PbBi2Se2(Te,S)2	Trigonal
ALEKSITE	PbBi2Te2S2	Trigonal
KOCHKARITE	PbBi4Te7	Trigonal
POTTSITE	PbBiH(VO4)2.2H2O	Tetragonal
PERITE	PbBiO2Cl	Orthorhombic
JOESMITHITE	PbCa2(Mg,Fe++,Fe+++)5Si6Be2O22(OH)2	Monoclinic
ESPERITE	PbCa3Zn4(SiO4)4	Monoclinic
LAURIONITE	PbCl(OH)	Orthorhombic
PARALAURIONITE	PbCl(OH)	Monoclinic
COTUNNITE	PbCl2	Orthorhombic
CERUSSITE	PbCO3	Orthorhombic
CROCOITE	PbCrO4	Monoclinic
DUFTITE	PbCu(AsO4)(OH)	Orthorhombic
SCHUILINGITE-(Nd)	PbCu(Nd,Gd,Sm,Y)(CO3)3(OH).1,5H2O	Orthorhombic
LINARITE	PbCu(SO4)(OH)2	Monoclinic
CHOLOALITE	PbCu(Te++++O3)2	Isometric
MOTTRAMITE	PbCu(VO4)(OH)	Orthorhombic
BEAVERITE	PbCu++(Fe+++,Al)2(SO4)2(OH)6	Trigonal
LENINGRADITE	PbCu++3(VO4)Cl2	Orthorhombic
KHINITE	PbCu++3Te++++++O6(OH)2	Orthorhombic
PARAKHINITE	PbCu++3Te++++++O6(OH)2	Trigonal ps Hexagonal
WATKINSONITE	PbCu2Bi4(Se,S)8	Monoclinic
INAGLYITE	PbCu3(Ir,Pt)8S16	Hexagonal
KONDERITE (KONDORITE)	PbCu3(Rh,Pt,Ir)8S16	Hexagonal
LAUTENTHALITE	PbCu4(SO4)2(OH)6.3H2O	Monoclinic
MIHARAITE	PbCu4FeBiS6	Orthorhombic
MURDOCHITE	PbCu6O8-x(Cl,Br)2x	Isometric
OSARIZAWAITE	PbCuAl2(SO4)2(OH)6	Trigonal
SELIGMANNITE	PbCuAsS3	Orthorhombic
SOUCEKITE	PbCuBi(S,Se)3	Orthorhombic
PEKOITE	PbCuBi11(S,Se)18	Orthorhombic
KRUPKAITE	PbCuBi3S6	Orthorhombic
GLADITE	PbCuBi5S9	Orthorhombic
AIKINITE	PbCuBiS3	Orthorhombic
BOURNONITE	PbCuSbS3	Orthorhombic

Mineral Name	Formula	Crystal System
MATLOCKITE	PbFCl	Tetragonal
GABRIELSONITE	PbFe++(AsO4)(OH)	Orthorhombic
CARMINITE	PbFe+++2(AsO4)2(OH)2	Orthorhombic
MOUNANAITE	PbFe+++2(VO4)2(OH)2	Triclinic
BEUDANTITE	PbFe+++3(AsO4)(SO4)(OH)6	Trigonal
CORKITE	PbFe+++3(PO4)(SO4)(OH)6	Trigonal
KINTOREITE	PbFe+++3(PO4,AsO4,SO4)2(OH,H2O)6	Trigonal
SEGNITITE	PbFe+++3H(AsO4)2(OH)6	Trigonal
LUDLOCKITE	PbFe+++4(As+++5O11)2	Triclinic
PLUMBOJAROSITE	PbFe+++6(SO4)4(OH)12	Trigonal
BARTELKEITE	PbFe++Ge3O8	Monoclinic
OTJISUMEITE	PbGe4O9	Triclinic ps Hexagonal
CESAROLITE	PbH2Mn++++3O8	
SCHULTENITE	PbHAsO4	Monoclinic
PETROVICITE	PbHgCu3BiSe5	Orthorhombic
PYROBELONITE	PbMn(VO4)(OH)	Orthorhombic
KURANAKHITE	PbMn++++Te++++++O6	Orthorhombic
QUENSELITE	PbMn+++O2(OH)	Monoclinic
NASLEDOVITE	PbMn3Al4(CO3)4(SO4)O5.5H2O	
WULFENITE	PbMoO4	Tetragonal
CHANGBAIITE	PbNb2O6	Trigonal
LITHARGE	PbO	Tetragonal
MASSICOT	PbO	Orthorhombic
PLATTNERITE	PbO2	Tetragonal
GALENA	PbS	Isometric
NADORITE	PbSbO2Cl	Orthorhombic
CLAUSTHALITE	PbSe	Isometric
MOLYBDOMENITE	PbSeO3	Monoclinic
KERSTENITE	PbSeO4	Orthorhombic
ALAMOSITE	PbSiO3	Monoclinic
TEALLITE	PbSnS2	Orthorhombic
SCOTLANDITE	PbSO3	Monoclinic
ANGLESITE	PbSO4	Orthorhombic
ALTAITE	PbTe	Isometric
FAIRBANKITE	PbTe++++O3	Triclinic
PLUMBOTELLURITE	PbTe++++O3	Orthorhombic
RADHAKRISHNAITE	PbTe3(Cl,S)2	Tetragonal
KOLARITE	PbTeCl2	Orthorhombic
MACEDONITE	PbTiO3	Tetragonal
WALLISITE	PbTl(Cu,Ag)As2S5	Triclinic
RICHETITE	PbU++++++4O13.4H2O	Triclinic
VANDENDRIESSCHEITE	PbU++++++7O22.12H2O	Orthorhombic
METAVANDENDRIESSCHEITE	PbU7O22.nH2O(n<12)	Orthorhombic
RASPITE	PbWO4	Monoclinic
STOLZITE	PbWO4	Tetragonal

Mineral Data

Mineral Name	Formula	Crystal System
ARSENDESCLOIZITE	PbZn(AsO4)(OH)	Orthorhombic
DESCLOIZITE	PbZn(VO4)(OH)	Orthorhombic
HELMUTWINKLERITE	PbZn2(AsO4)2.2H2O	Triclinic
TSUMCORITE	PbZnFe++(AsO4)2.H2O	Monoclinic
LARSENITE	PbZnSiO4	Orthorhombic
URVANTSEVITE	Pd(Bi,Pb)2	Tetragonal
KOTULSKITE	Pd(Te,Bi)	Hexagonal
BORISHANSKIITE	Pd1+x(As,Pb)2,x=0-0,2	Orthorhombic
MERTIEITE-I	Pd11(Sb,As)4	ps Hexagonal
ISOMERTIEITE	Pd11Sb2As2	Isometric
PALLADSEITE	Pd17Se15	Isometric
PALLADOBISMUTHARSENIDE	Pd2(As,Bi)	Orthorhombic
KEITHCONNITE	Pd20Te7	Trigonal
PALLADOARSENIDE	Pd2As	Monoclinic
POLARITE	Pd2PbBi	Orthorhombic
PAOLOVITE	Pd2Sn	Orthorhombic
CABRIITE	Pd2SnCu	Orthorhombic
TEMAGAMITE	Pd3HgTe3	Orthorhombic
PLUMBOPALLADINITE	Pd3Pb2	Hexagonal
BOROVSKITE	Pd3SbTe4	Isometric
STIBIOPALLADINITE	Pd5Sb2	Hexagonal
ARSENOPALLADINITE	Pd8(As,Sb)3	Triclinic
MERTIEITE-II	Pd8(Sb,As)3	Trigonal
PALARSTANIDE	Pd8(Sn,As)3	Hexagonal
STILLWATERITE	Pd8As3	Hexagonal
TELLUROPALLADINITE	Pd9Te4	Monoclinic
SOBOLEVSKITE	PdBi	Hexagonal/Monoclinic
FROODITE	PdBi2	Monoclinic
PADMAITE	PdBiSe	Isometric
POTARITE	PdHg	Tetragonal
MAJAKITE (MAYAKITE)	PdNiAs	Hexagonal
PALLADIUM	Pdor(Pd,Hg)	Isometric
TESTIBIOPALLADITE	PdTe(Sb,Te)	Isometric
PLATINUM	Pt	Isometric
INSIZWAITE	Pt(Bi,Sb)2	Isometric
TULAMEENITE	Pt(Cu,Fe)	Tetragonal
FERRONICKELPLATINUM	Pt(Ni,Fe)	Tetragonal
STUMPFLITE	Pt(Sb,Bi)	Hexagonal
GEVERSITE	Pt(Sb,Bi)2	Isometric
LUBEROITE	Pt5Se4	Monoclinic
SPERRYLITE	PtAs2	Isometric
MASLOVITE	PtBiTe	Isometric
HONGSHIITE	PtCu	Trigonal
TETRAFERROPLATINUM	PtFe	Tetragonal
YIXUNITE	PtIn	Isometric

Mineral Data

Mineral Name	Formula	Crystal System
NIGGLIITE	PtSn	Hexagonal
RHENIUM	Re	Hexagonal
RHENIITE	ReS2	Hexagonal
PRASSOITE	Rh17S15	Isometric
CHEREPANOVITE	RhAs	Orthorhombic
RUARSITE	RuAsS	Monoclinic
LAURITE	RuS2	Isometric
ROSICKYITE	S	Monoclinic
SULFUR (SOUFRE,SCHWEFEL)	S	Orthorhombic
ANTIMONY	Sb	Trigonal
STIBIVANITE-2M (STIBIVANITE)	Sb+++2V++++O5	Monoclinic
STIBIVANITE-2O	Sb+++2V++++O5	Orthorhombic
KLEBELSBERGITE	Sb+++4O4(OH)2(SO4)	Orthorhombic
CHAPMANITE	Sb+++Fe+++2(SiO4)2(OH)	Monoclinic
STIBICONITE	Sb+++Sb+++++2O6(OH)	Isometric
CERVANTITE	Sb+++Sb+++++O4	Orthorhombic
PARADOCRASITE	Sb2(Sb,As)2	Monoclinic
PAAKKONENITE	Sb2AsS2	Monoclinic
SENARMONTITE	Sb2O3	Isometric
VALENTINITE	Sb2O3	Orthorhombic
KERMESITE	Sb2S2O	Triclinic ps Monoclinic
METASTIBNITE	Sb2S3	Amorphous
STIBNITE (STIBINE,ANTIMONITE)	Sb2S3	Orthorhombic
ANTIMONSELITE	Sb2Se3	Orthorhombic
TELLURANTIMONY	Sb2Te3	Trigonal
COQUANDITE	Sb6O8(SO4).H2O	Triclinic
ONORATOITE	Sb8O11Cl2	Monoclinic
STIBARSEN	SbAs	Trigonal
STIBIOCOLUMBITE	SbNbO4	Orthorhombic
STIBIOTANTALITE	SbTaO4	Orthorhombic
KOLBECKITE	ScPO4.2H2O	Monoclinic
SELENIUM	Se	Trigonal
DOWNEYITE	SeO2	Tetragonal
SILICON (SILICIUM)	Si	Isometric
SINOITE	Si2N2O	Orthorhombic
NIERITE	Si3N4	Trigonal
MOISSANITE-15R (CARBORUNDUM)	SiC	Trigonal
MOISSANITE-33R (CARBORUNDUM)	SiC	Trigonal
MOISSANITE-5H (CARBORUNDUM)	SiC	Hexagonal
MOISSANITE-6H (CARBORUNDUM)	SiC	Hexagonal

Mineral Data

Mineral Name	Formula	Crystal System
COESITE	SiO2	Monoclinic
CRISTOBALITE	SiO2	Tetragonal
LECHATELIERITE	SiO2	
MELANOPHLOGITE	SiO2	Cubic/Tetragonal
QUARTZ	SiO2	Trigonal
STISHOVITE	SiO2	Tetragonal
TRIDYMITE	SiO2	Monoclinic ps Hexagonal
OPAL	SiO2.nH2O	Amorphous
TIN (ETAIN,ZINN)	Sn	Tetragonal
FOORDITE	Sn++(Nb,Ta)2O6	Monoclinic
THOREAULITE	Sn++Ta2O6	Monoclinic
OTTEMANNITE	Sn2S3	Orthorhombic
ABHURITE	Sn3O(OH)2Cl2	Trigonal
HYDROROMARCHITE	Sn3O2(OH)2	Tetragonal
ROMARCHITE	SnO	Tetragonal
CASSITERITE	SnO2	Tetragonal
HERZENBERGITE	SnS	Orthorhombic
BERNDTITE-C27	SnS2	Hexagonal
BERNDTITE-C6	SnS2	Trigonal
STISTAITE	SnSb	Isometric
OLEKMINSKITE	Sr(Sr,Ca,Ba)(CO3)2	Trigonal
HARADAITE	Sr2(VO2)Si4O12	Orthorhombic
VEATCHITE	Sr2B11O16(OH)5.H2O	Monoclinic
VEATCHITE-A	Sr2B11O16(OH)5.H2O	Triclinic
VEATCHITE-P (P-VEATCHITE)	Sr2B11O16(OH)5.H2O	Monoclinic
BALAVINSKITE	Sr2B6O11.4H2O	
STRONTIOJOAQUINITE	Sr2Ba2(Na,Fe++)2Ti2Si8O24(O,OH)2.H2O	Monoclinic
STRONTIOORTHOJOAQUINITE	Sr2Ba2(Na,Fe++)2Ti2Si8O24(O,OH)2.H2O	Orthorhombic
BOGGILDITE (BOEGGILDITE)	Sr2Na2Al2(PO4)F9	Monoclinic
STRONTIOPYROCHLORE	Sr2Nb2(O,OH)7	Isometric
OHMILITE	Sr3(Ti,Fe+++)(Si2O6)2(O,OH).2-3H2O	Monoclinic
WELOGANITE	Sr3Na2Zr(CO3)6.3H2O	Triclinic psTrigonal
DONNAYITE-	Sr3NaCaY(CO3)6.3H2O	Triclinic psTrigonal
MONTROYALITE	Sr4Al8(CO3)3(OH,F)26.10-11H2O	Triclinic
STRONTIOWHITLOCKITE	Sr7(Mg,Ca)3(PO4)6[PO3(OH)]	Trigonal
SVANBERGITE	SrAl3(PO4)(SO4)(OH)6	Trigonal
GOYAZITE	SrAl3(PO4)2(OH)5.H2O	Trigonal
ACUMINITE	SrAlF4(OH).H2O	Monoclinic
TIKHONENKOVITE	SrAlF4(OH).H2O	Monoclinic
TUNELLITE	SrB6O9(OH)2.3H2O	Monoclinic
STRONTIOBORITE	SrB8O11(OH)4	Monoclinic
KARASUGITE	SrCaAl(F,OH)7	Monoclinic
ANCYLITE-(Ce)	SrCe(CO3)2(OH).H2O	Orthorhombic
STRONTIANITE	SrCO3	Orthorhombic

Mineral Name	Formula	Crystal System
HENNOMARTINITE	SrMn+++2Si2O7(OH)2.H2O	Orthorhombic
STRONALSITE	SrNa2Al4Si4O16	Orthorhombic
CELESTINE (CELESTITE)	SrSO4	Orthorhombic
TAUSONITE	SrTiO3	Isometric
TANTITE	Ta2O5	Triclinic
TANTALCARBIDE	TaC	Isometric
TELLURIUM	Te	Trigonal
PARATELLURITE	TeO2	Tetragonal
TELLURITE	TeO2	Orthorhombic
THORBASTNASITE (THORBASTNAESITE)	Th(Ca,Ce)(CO3)2F2.3H2O	Hexagonal
STEACYITE	Th(Ca,Na)2K1-xSi8O20	Tetragonal
THOROGUMMITE	Th(SiO4)1-x(OH)4x	Tetragonal
ALTHUPITE	ThAl(UO2)7(PO4)4(OH)5.15H2O	Triclinic
EKANITE	ThCa2Si8O20	Tetragonal
THORIANITE	ThO2	Isometric
HUTTONITE	ThSiO4	Monoclinic
TITANIUM	Ti	Hexagonal
ULVOSPINEL (ULVITE)	TiFe++2O4	Isometric
OSBORNITE	TiN	Isometric
HONGQUIITE	TiO	Isometric
ANATASE	TiO2	Tetragonal
BROOKITE	TiO2	Orthorhombic
RUTILE	TiO2	Tetragonal
WINSTANLEYITE	TiTe++++3O8	Isometric
BERNARDITE	Tl(As1-xSbx)5S8orTl(As1-x)5S9	Monoclinic
PARAPIERROTITE	Tl(Sb,As)5S8	Monoclinic
GILLULYITE	Tl2(As,Sb)8S13	Monoclinic
BUKOVITE	Tl2Cu3FeSe4	Tetragonal
ROHAITE	Tl2Cu8,5Sb2S4	Orthorhombic
AVICENNITE	Tl2O3	Isometric
CARLINITE	Tl2S	Trigonal
PIERROTITE	Tl2Sb6As4S16	Orthorhombic
ERNIGGLIITE	Tl2SnAs2S6	Hexagonal
ELLISITE	Tl3AsS3	Trigonal
FANGITE	Tl3AsS4	Orthorhombic
VRBAITE	Tl4Hg3Sb2As8S20	Orthorhombic
REBULITE	Tl5Sb5As8S22	Monoclinic
JANKOVICITE	Tl5Sb9(As,Sb)4S22	Triclinic
IMHOFITE	Tl6CuAs16S40	Monoclinic
CRIDDLEITE	TlAg2Au3Sb10S10	Monoclinic ps Tetragonal
LORANDITE	TlAsS2	Monoclinic
THALCUSITE	TlCu3FeS4	Tetragonal
PICOTPAULITE	TlFe2S3	Orthorhombic
RAGUINITE	TlFeS2	Orthorhombic ps Hexagonal
SIMONITE	TlHgAs3S6	Monoclinic

Mineral Name	Formula	Crystal System
CHRISTITE	TlHgAsS3	Monoclinic
ROUTHIERITE	TlHgAsS3	Tetragonal
VAUGHANITE	TlHgSb4S7	Triclinic
EDENHARTERITE	TlPbAs3S6	Orthorhombic
WEISSBERGITE	TlSbS2	Triclinic
SEDOVITE	U(MoO4)2	Orthorhombic
COFFINITE	U(SiO4)1-x(OH)4x	Tetragonal
VYACHESLAVITE	U++++(PO4)(OH).2.5H2O	Orthorhombic
LERMONTOVITE	U++++(PO4)(OH).H2O	Orthorhombic
LIANDRATITE	U++++++(Nb,Ta)2O8	Hexagonal
METAVANMEERSSCHEITE	U++++++(UO2)3(PO4)2(OH)6.2H2O	Orthorhombic
VANMEERSSCHEITE	U++++++(UO2)3(PO4)2(OH)6.4H2O	Orthorhombic
UVANITE	U++++++2V+++++6O21.15H2O	Orthorhombic
SWAMBOITE	U++++++H6(UO2)6(SiO4)6.30H2O	Monoclinic
URANOSILITE	U++++++Si7O17	Orthorhombic
PETSCHECKITE	U++++Fe++(Nb,Ta)2O8	Hexagonal
MOURITE	U++++Mo++++++5O12(OH)10	Monoclinic
ORTHOBRANNERITE	U++++U++++++Ti4O12(OH)2	Orthorhombic
URANINITE	UO2	Isometric
RUTHERFORDINE	UO2(CO3)	Orthorhombic
IRIGINITE	UO2Mo2O7.3H2O	Orthorhombic
PARASCHOEPITE	UO3.2H2O	Orthorhombic
SCHOEPITE	UO3.2H2O	Orthorhombic
METASCHOEPITE	UO3.nH2O(n<2)	Orthorhombic
METASTUDTITE	UO4.2H2O	Orthorhombic
STUDTITE	UO4.4H2O	Monoclinic
CLIFFORDITE	UTe3O9	Isometric
CORVUSITE	V++++2V+++++12O34.nH2O	Orthorhombic
VANOXITE	V++++4V+++++2O13.8H2O	Trigonal
DUTTONITE	V++++O(OH)2	Monoclinic
KYZYLKUMITE	V+++2Ti3O9	Monoclinic
SCHREYERITE	V+++2Ti3O9	Monoclinic
BERDESINSKIITE	V+++2TiO5	Monoclinic
TIVANITE	V+++TiO3(OH)	Monoclinic
YUSHKINITE	V1-xS.n(Mg,Al)(OH)2	Hexagonal
HAGGITE (HAEGGITE)	V2O2(OH)3	Monoclinic
KARELIANITE	V2O3	Trigonal
LENOBLITE	V2O4.2H2O	Orthorhombic
BARIANDITE	V2O4.4V2O5.12H2O	Monoclinic
SHCHERBINAITE	V2O5	Orthorhombic
NAVAJOITE	V2O5.3H2O	Monoclinic
MINASRAGRITE	VO(SO4).5H2O	Monoclinic
PARAMONTROSEITE	VO2	Orthorhombic
STANLEYITE	VOSO4.6H2O	Orthorhombic
PATRONITE	VS4	Monoclinic
MPOROROITE	W++++++AlO3(OH)3.2H2O	Triclinic

Mineral Data

Mineral Name	Formula	Crystal System
ANTHOINITE	WAl(O,OH)3	Triclinic
MEYMACITE	WO3.2H2O	Amorphous
TUNGSTITE	WO3.H2O	Orthorhombic
TUNGSTENITE-2H	WS2	Hexagonal
TUNGSTENITE-3R	WS2	Trigonal
DAVIDITE-	Y(Ti,Fe)21O38	Trigonal
IIMORIITE-	Y2(SiO4)(CO3)	Triclinic
GADOLINITE-	Y2Fe++Be2Si2O10	Monoclinic
TENGERITE-	Y3(CO3)4.2-3H2O	Orthorhombic
THALENITE-	Y3Si3O10(F,OH)	Monoclinic
TOMBARTHITE-	Y4(Si,H4)4O12-x(OH)4+2x	Monoclinic
VYUNTSPAKHKITE-	Y4Al2AlSi5O18(OH)5	Monoclinic
ROWLANDITE-	Y4Fe++Si4O14F2	Amorphous
CHERNOVITE-	YAsO4	Tetragonal
MOYDITE-	YB(OH)4(CO3)	Orthorhombic
FERGUSONITE-	YNbO4	Tetragonal
FERGUSONITE-BETA-	YNbO4	Monoclinic
XENOTIME-	YPO4	Tetragonal
CHURCHITE- (WEINSCHENKITE)	YPO4.2H2O	Monoclinic
FORMANITE-	YTaO4	Tetragonal
WAKEFIELDITE-	YVO4	Tetragonal
YTTROTUNGSTITE-	YW2O6(OH)3	Monoclinic
ZINC	Zn	Hexagonal
KORITNIGITE	Zn(As+++++O3)(OH).H2O	Triclinic
ASHOVERITE	Zn(OH)2	Tetragonal
SWEETITE	Zn(OH)2	Tetragonal
WULFINGITE (WUELFINGITE)	Zn(OH)2	Orthorhombic
METALODEVITE	Zn(UO2)2(AsO4)2.10H2O	Tetragonal
ZINCZIPPEITE	Zn++0,5(UO2)2(SO4)(OH)3.H2O	Monoclinic
KOLOVRATITE	Zn,Ni.VO4	
ADAMITE	Zn2(AsO4)(OH)	Orthorhombic/Monoclinic
PARADAMITE	Zn2(AsO4)(OH)	Triclinic
LEGRANDITE	Zn2(AsO4)(OH).H2O	Monoclinic
PHOSPHOPHYLLITE	Zn2(Fe++,Mn)(PO4)2.4H2O	Monoclinic
TARBUTTITE	Zn2(PO4)(OH)	Triclinic
SOFIITE	Zn2(Se++++O3)Cl2	Orthorhombic ps Hexagonal
ZINALSITE	Zn2AlSi2O5(OH)4.2H2O	Monoclinic
STRANSKIITE	Zn2Cu(AsO4)2	Triclinic
MAPIMITE	Zn2Fe+++3(AsO4)3(OH)4.10H2O	Monoclinic
HYDROHETAEROLITE	Zn2Mn+++4O8.H2O	Tetragonal
WILLEMITE	Zn2SiO4	Trigonal
REINERITE	Zn3(As+++O3)2	Orthorhombic
WARIKAHNITE	Zn3(AsO4)2.2H2O	Triclinic
KOTTIGITE (KOETTIGITE)	Zn3(AsO4)2.8H2O	Monoclinic
BRIANYOUNGITE	Zn3(CO3,SO4)(OH)4	Monoclinic ps Orthorhombic

Mineral Data

Mineral Name	Formula	Crystal System
HOPEITE	Zn3(PO4)2.4H2O	Orthorhombic
PARAHOPEITE	Zn3(PO4)2.4H2O	Triclinic
CHEREMNYKHITE	Zn3Pb3Te++++O6(VO4,AsO4)2	Orthorhombic
ZINCSILITE	Zn3Si4O10(OH)2.4H2O	Monoclinic
SPENCERITE	Zn4(PO4)2(OH)2.3H2O	Monoclinic
GENTHELVITE	Zn4Be3(SiO4)3S	Isometric
HEMIMORPHITE	Zn4Si2O7(OH)2.H2O	Orthorhombic
HYDROZINCITE	Zn5(CO3)2(OH)6	Monoclinic
SIMONKOLLEITE	Zn5(OH)8Cl2.H2O	Hexagonal
FAHLEITE	Zn5CaFe+++2(AsO4)6.14H2O	Orthorhombic
ZINCALUMINITE	Zn6Al6(SO4)2(OH)26.5H2O	Hexagonal
QUETZALCOATLITE	Zn8Cu4(TeO3)3(OH)18	Hexagonal
KLEEMANITE	ZnAl2(PO4)2(OH)2.3H2O	Monoclinic
GAHNITE	ZnAl2O4	Isometric
LEITEITE	ZnAs+++2O4	Monoclinic
SMITHSONITE	ZnCO3	Trigonal
ZINCOCHROMITE	ZnCr2O4	Isometric
KALININITE	ZnCr2S4	Isometric
OJUELAITE	ZnFe+++2(AsO4)2(OH)2.4H2O	Monoclinic
LISHIZHENITE	ZnFe+++2(SO4)4.14H2O	Triclinic
ZINCOCOPIAPITE	ZnFe+++4(SO4)6(OH)2.18H2O	Triclinic
HETAEROLITE	ZnMn+++2O4	Tetragonal
MATRAITE	ZnS	Trigonal
ORDONEZITE	ZnSb2O6	Tetragonal
STILLEITE	ZnSe	Isometric
VISMIRNOVITE	ZnSn++++(OH)6	Isometric
ZINKOSITE	ZnSO4	Orthorhombic
GOSLARITE	ZnSO4.7H2O	Orthorhombic
ZIRCOSULFATE	Zr(SO4)2.4H2O	Orthorhombic
BADDELEYITE	ZrO2	Monoclinic
ZIRCON	ZrSiO4	Tetragonal

References

Mineral Databases and Nomenclature:

Bayliss, P., Berry, L.G., Mrose, M.E., Sabina, A.P., Smith, D.K. (eds), 1983, "Mineral Powder Diffraction File". JCPDS, ICDD, 1005p.

Craig, J.R., Vaughan, D,J., 1981, " Ore Microscopy and Ore Petrography", John Wiley and Sons, New York, 406p.

Dana, J.D., "Manual of Mineralogy".

Deer, W.A., Howie, R.A. and Zussman, J, 1980, "An Introduction to the Rock Forming Minerals". Longman, London. 528p.

JCPDS-International Center for Diffraction Data, 1995, " Powder diffraction file. Alphabetical index, Inorganic phases". JCPDS, International Centre for Diffraction Data.

Kerr, P.F., 1977, "Optical Mineralogy", McGraw-Hill, 492p.

Le Maitre, R.W.(ed), 1989, "A Classification of Igneous Rocks and Glossary of Terms". Blackwell Scientific Publications, Oxford, UK. - Igneous plots and systematics.

Myashiro, A.,1973. "Metamorphism and Metamorphic Belts". Allen and Unwin, London,492p.

Mutschler, F.E., Rougon, D.J., Lavin, O.P., Hughes, R.D., 1981, "Petros - A Data Bank of Major Element Chemical Analyses of Igneous Rocks for Research and Teaching (Version 6.1)". NOAA-National Geophysical and Solar-Terrestrial Data Center.

Nichols, M.C., Nickel, E.H., 1991, "Mineral Reference Manual", Chapman and Hall, New York, 250p.

Wills, B.A., 1992, "Mineral Processing Technology". Pergamon, Oxford, 855p.

Web Sites and Software:

Athena Mineralogy http://un2sg1.unige.ch/www/athena/mineral/mineral.html - List of Mineral Names, Mineralogy

Geologynet..com http://www.geologynet.com/indexa.htm - Mineralogy Software and Databases

LR Ream Publishing http://www.mineralnews.com/ - Publishers of The Mineral Database (TMD)

Mineralogy Database http://webmineral.com/ - HTML Mineral Database

MinDat.org http://www.mindat.org/ - Mineralogy Database

www.ingramcontent.com/pod-product-compliance
Lightning Source LLC
Chambersburg PA
CBHW081121170526

45165CB00008B/2519